The Hydrogen Bond and the Water Molecule

The Physics and Chemistry of Water, Aqueous and Bio Media

The Hydrogen Bond and the Water Molecule

The Physics and Chemistry of Water, Aqueous and Bio Media

Yves Maréchal
DRFMC/SI3M-CEA
Grenoble, France

ELSEVIER
Amsterdam • Boston • Heidelberg • London • New York • Oxford
Paris • San Diego • San Francisco • Singapore • Sydney • Tokyo

Elsevier
Radarweg 29, PO Box 211, 1000 AE Amsterdam, The Netherlands
The Boulevard, Langford Lane, Kidlington, Oxford OX5 1GB, UK

First edition 2007

Library of Congress Cataloging-in-Publication Data
A catalog record for this book is available from the Library of Congress

British Library Cataloguing in Publication Data
A catalogue record for this book is available from the British Library

ISBN-13: 978-0-444-51957-3
ISBN-10: 0-444-51957-2

Transferred to digital print, 2007
Printed and bound by CPI Antony Rowe, Eastbourne

Contents

Acknowledgments

I am grateful to many persons for the precious aid they provided me during the research activity I had on hydrogen bonds and water molecules and during the writing of this book. First, Andrzej Witkowski under whose direction I started, a long time ago, my thesis on the IR spectra of hydrogen bonds and whose constant questioning on topics of physics and on many other intellectual issues, always revealed original ways of thinking. Then, Hans Rainer Zelsmann, with whom I worked for several decades on hydrogen bonds. Without him and his always highly valuable and rigorous advices I would never have reached anywhere in the interpretations of experiments where they can be considered as fully exploited. I am also grateful to initiators of the hydrogen bond research community, particularly Dusan Hadzi, Camille Sandorfy, Lucjan Sobczyk, Savo Bratos and Henryk Ratajczak, with whom I had numerous and often passionate discussions on hydrogen bonds. Also John E. Bertie, with whom we exchanged views on precise measurements of intensities in IR spectra, a somewhat great number of years ago, and Ludwig Hofacker for offering a fruitful collaboration, an even greater number of years ago.

I would also thank Serge Pérez, who gave me the opportunity to deliver lectures on hydrogen bonds and water molecule in doctoral teachings (Diplome d'Etudes Approfondies) on Physical Chemistry at the Joseph Fourier University in Grenoble. These lectures and corresponding exchanges with Students gave me the idea of writing this book. Before writing it, I particularly benefited from the great experiences in scientific publications of Austin Barnes and Jean Bornarel.

Discussions on special topics evocated in this book allowed me getting a sufficiently precise view of these topics. I particularly thank the following: Philippe Pruzan, who unfortunately passed away much too early, but to whom I am indebted for having sent me diagrams on the various phases of ice and articles on the spectroscopy of ice; André Grand for discussions on DFT methods; Olivier Henri-Rousseau and Paul Blaise for their readings and comments on these central chapters of this book that concern IR spectroscopy of H-bonds; Armel Guillermo, Michel Bardet and Jacques Gaillard for having raised my attention to the advent of recent NMR methods to look at hydrogen bonds and having provided me with related references; and Yoshiharu Nishiyama for references concerning the development of algae in heavy water. Life in heavy water provides a central argument discussed in this book on the fundamental role of water molecules in the bioreactivity; many other persons who did not hesitate sparing time explaining me particular points.

I finally acknowledge the invaluable support provided by my own family, particularly my wife Marie-France, during the long time that this book was written.

Preface

In one of his "News and Views" in *Nature* (332 (1988) 677), John Maddox wrote some 20 years ago: "Is the scandal, that so little is known about the interactions of macromolecules and their aqueous environment, about to be removed?" This one sentence clearly defines the aim of this book. What is the point? The water molecule, H_2O, is one of the most familiar molecules. It is the component of a species, liquid water, which we all drink daily and use in many various ways. It is therefore no surprise that H_2O is often considered a "casual" molecule. It nevertheless remains surprising that it is considered at the same time a molecule with almost no interest and which can be consequently ignored. John Maddox called this attitude a scandal, as it is indeed untenable. It actually disregards a fundamental point: life, which started developing some 3–4 billion years ago within the oceans, requires the presence of these molecules to proceed. In other words, we know that in biology this molecule plays a central role and we nevertheless often continue ignoring it. Why is this so, and will it still remain so for long? One of the reasons for this attitude is that the water molecule is much more difficult to observe than currently thought. However, it has been the object of many research activities in various fields in recent years. The development and efficiency of experimental methods that were previously severely hindered when used to observe the water molecule have conveyed new pieces of information, giving evidence of subtle and discrete properties that make it a far more active molecule than previously thought, not only in biology but also in physics and chemistry. As time goes on our knowledge of this molecule and its role thus becomes more and more precise. The aim of this book is our present view of this molecule, in the hope that it is no longer ignored where it intervenes, often decisively and much more often than ordinarily thought, and also in order to clearly show what we still have to learn about it. On reading the conclusion at the end of this book, it should be clear that in recent years our point of view on this molecule has changed fundamentally.

Understanding the subtle properties of the water molecule, which indeed make it an exceptional molecule, requires first having a precise knowledge of the molecular interaction that is at the origin of all its properties: the hydrogen bond (H-bond in this book). An important part of this book, about half of it, is therefore devoted to the properties and implications of this crucial intermolecular bond that many scientists often use and invoke for a particular property of its own without having an overview of all of its properties and implications. The geometrical and thermodynamic properties of the H-bond are well known and have been described in several classical textbooks. They are briefly but precisely reviewed and commented in the first chapters of this book that precede chapters devoted to the experimental and theoretical methods that are particularly adapted to the observation and description of H-bonds. The dynamic properties of H-bonds, at the origin of their particularly crucial reactivity, are examined in a separate chapter. Their fundamental importance has recently emerged, and their study constitutes a field of a growing interest in physics and chemistry. The description of these dynamic properties starts with that of the exceptional features it displays in its vibrational spectra. We shall see that IR spectroscopy appears to be the most precise tool to observe both H-bonds and the water molecule, an opinion that only specialists have shared until recently. It will also hopefully make it evident that this powerful tool, IR spectroscopy, is not so hard to handle as commonly thought. It should thus help stimulate more scientists to use vibrational spectroscopy with confidence, as it is now well understood. Even if it requires some care in its interpretations, it is no longer a method to be used only by specialists. The introduction of anharmonicity, a concept that naturally explains the exceptional spectroscopic properties of H-bonds, makes it moreover easy to understand how H-bonds are the path through which protons and hydrogen atoms can be transferred between molecules. Some kinds of proton transfers, such as those that are at the origin of all acid/base chemistry, are reasonably well known. Some others, which occur in such biomechanisms as photosynthesis or vision, are the object of intense research activity and are less known. Even less known, however, are transfers of H-atoms via tautomerism, which we now suspect to be crucial mechanisms in enzymatic activity, or more generally to be the basic mechanisms of bioreactivity. In these transfers, water molecules play a crucial role, and at the end of this part devoted to H-bonds, it should clearly appear that if H-bonds are at the origin of nearly all the properties of the water molecule, they could not play the central role they have in chemistry and biology if water molecules did not exist. In other words, H-bonds and water molecules are so intricately linked that they cannot be separated.

In view of the above noted contradictions and paradoxes that the simple-looking and familiar water molecule conveys, and which have only recently been recognized, it is now timely to clarify what we know, what we ignore of this crucial and ubiquitous molecule and of the H-bond which gives it nearly all its properties, and also what questions and/or long-term implications the newly revealed aspects of this molecule raise. One of these questions, a fundamental one already outlined above, is: *how is it that life occurs within water, and within water only*? An older but somewhat vague answer is that water is important in biology to provide a medium for biosystems. In the light of recent studies this answer can be made with much more precision and constitutes a guideline for the whole book. It is: water molecules, with their unique ability to develop a particularly dense, evolutive, and flexible H-bond network, not only influence the structure of many a macromolecule, but, potentially more important, play a crucial role in the reactivity of all bio-media, at neutral pH, by enabling transfers of H-atoms that are now suspected to constitute the elementary reactions in such media. This property comes in addition to the well-known one, which is that in any aqueous system they also enable transfers of protons, the origin of all acid/base chemistry.

Such a book, which attempts to make a synthesis of what is known, what is being studied and what is at stake in a field of research of growing interest (water and aqueous media are ubiquitous; H-bonds are central in molecular biology) has the ambition of being a reference book for various scientists in many different fields of interest, which extend from physics to biology and naturally includes chemistry. It is aimed at collecting from an appreciable part of the whole scientific endeavour and presenting with some unity items of knowledge all related to the water molecule. From another point of view, many scientists in completely different fields often encounter the H-bond or the water molecule in their own domain. They may be eager for more precise knowledge of what they are dealing with in order to place their own field of research in a wider domain. This book is aimed at helping them do so. With this view an appreciable part of the book concerns various methods that can be used to observe different features of H-bonds and of the water molecule. This book might thus help in defining strategies for many studies where these two entities, the H-bond and the water molecule, are encountered. It should also interest science students who have to learn physical chemistry, biophysics or biochemistry, the physics of the atmosphere, of ice or of this special liquid: water. It might also help instructors lecturing on H-bonds, water molecules and many related domains.

This book has been written with the rigor and criticism that a physicist or a chemist requires. It has also been written in such a way that a biologist should not encounter difficulties reading it, because biology is the field where H-bonds and water molecules show their fundamental and even vital importance. Biologists also require rigor and criticism in their own domains, but the objects they study being different and particularly complex, they do not put the emphasis on the same points. With this in mind, the necessary mathematical developments to describe some particular points are often given in appendices at the end of chapters. When they cannot be avoided in the text, as for instance in the description of the H-bond network of liquid water, which is still presently the object of passionate discussions in the community of chemical physicists, or in the mechanics of H-bonds necessary to understand their IR spectra, a sentence indicates what in the following developments the uninterested reader can skip and where he or she should resume reading. Will it be enough to make this goal of having a book that is intended to be read by such a wide variety of scientists of different cultures viable? No answer can be given at present but the question itself points to the challenge encountered in writing this book.

Part I

THE HYDROGEN BOND

– 1 –

The Hydrogen Bond: Formation, Thermodynamic Properties, Classification

CHEMICAL BONDS

Electrical forces that act on positively charged nuclei of various atoms and negatively charged electronic clouds that extend around these nuclei rule chemistry. The three other fundamental forces in physics, namely strong and weak interactions that act on the protons and neutrons of the nuclei and gravity, do not play any role in chemistry. The first two are much stronger than electromagnetic forces and consequently correspond to much larger energy level separations than energies due to electromagnetic interactions. It implies that in chemistry all nuclear levels are ground state levels or, in other words, nuclei are always in their fundamental state. The third fundamental force, gravity, is orders of magnitude too weak to have any detectable influence on electromagnetic levels. The elementary constituents in chemistry are therefore atoms, made of positively charged nuclei that are always in their ground nuclear state and surrounded by negatively charged electronic clouds. The precise knowledge of the structures of these electronic clouds is the object of chemistry. Atoms are the simplest arrangement of all these electrons and nuclei. They are not the most stable ones. Two H-atoms, for instance, the simplest atoms made of single protons surrounded by single electrons, are attracted to each other in such a way that their initially separated electronic clouds mix together so as to form a single cloud occupied by both electrons with different spins, which keep the two protons separated by a well-defined distance. This configuration, the H_2 molecule, is more stable by -4.5 eV than the configuration defined by the two far-away noninteracting H-atoms. This electric rearrangement of charges with an appreciable energy gain (more precisely an enthalpy gain) is called covalent interaction and is at the origin of formation of molecules. Enthalpies of covalent bonds typically fall in the range of about -5 eV, with for example the formation of O_2 from two far-away O-atoms being at the origin of an enthalpy gain of -5.2 eV, that of two H_2O from two H_2 and one O_2 molecules of -2×2.5 eV, which gives, with the enthalpy of formation of H_2 of -4.5 eV, an enthalpy for a single O–H bond of -4.8 eV. These covalent interactions are short-range interactions. In the case of the H_2 molecule, for instance, its energy is of the order of -4.5 eV when the distance between the two protons is in the vicinity of 0.8 Å, but it rapidly approaches zero when this distance increases. In this book we shall often have to deal with these covalent interactions.

Atoms may also undergo other interactions. Charged atoms that have lost or gained one or more electrons are ruled by ionic interaction that we may occasionally encounter. The magnitudes of the enthalpies of these ionic interactions are comparable to those of covalent interactions. Contrary to covalent interactions, however, ionic interactions are long-range interactions: when two ionized atoms are separated by a distance R this interaction asymptotically tends towards a Coulomb interaction in $1/R^2$ when R increases, which is a relatively slow decrease, much slower than that of a covalent bond with distance. Furthermore, ionic interactions are barely directional, contrary to covalent interactions that are strongly directional.

The energies of covalent bonds are smaller than atomic energies and much smaller than nuclear energies. Ejecting an electron from an outer orbital of an atom thus requires about 10 eV, which corresponds to the energy $h\nu$ of a near UV photon. The inner electrons require some keV to be ejected from their atomic orbitals. It corresponds to $h\nu$ of an X-ray photon having a wavelength of the order of 1 nm. We thus see that chemical interactions, with enthalpies typically of about -5 eV, only imply outer electrons of atoms, the much greater energies of the inner electrons being hardly affected by the chemical state. Nuclear energies are still greater. Thus ejecting a neutron from an atomic nucleus requires about 10 MeV. A fission reaction requires about 100 MeV. Energies involved in chemical reactions, some eV, are thus clearly much too small to induce transitions from ground state levels of nucleons towards excited states, as such transitions require at least some MeV.

INTERMOLECULAR BONDS

Van der Waals interactions

Covalent interactions are at the origin of the stability of molecules and govern their structures. Molecules are well-defined entities that appear as stable arrangements of atoms at room temperature, typically 300 K. At this temperature the energy kT typical of thermal fluctuations is equal to 0.026 eV, with the Boltzmann constant k equal to 1.38×10^{-23} J K^{-1}, as mentioned in the appendix of this chapter. Compared to enthalpies of covalent bonds, this energy is weak and temperature has consequently almost no influence on the structure of molecules as long as it is not much higher than 300 K. When two identical molecules come in close proximity they, nevertheless, suffer residual electrostatic interactions called Van der Waals interactions. These are at the origin of the condensations of gases into liquids when temperature decreases, with the notable exception, however, of liquid water where these interactions are negligible and condensation is almost entirely due to another interaction that we shall consider throughout this book and define below: the hydrogen bond. Energies of Van der Waals interactions are typically of the order of about 0.01 eV for small molecules, which is at least two orders of magnitude smaller than the energies of covalent bonds. Their origin is electric dipole–dipole interactions, also called Keesom interaction, or induction (called Debye interaction in solids), which is at the origin of a dipole moment induced in an apolar molecule that interacts with the permanent dipole moment of a polar molecule, or dispersion interaction (also called London

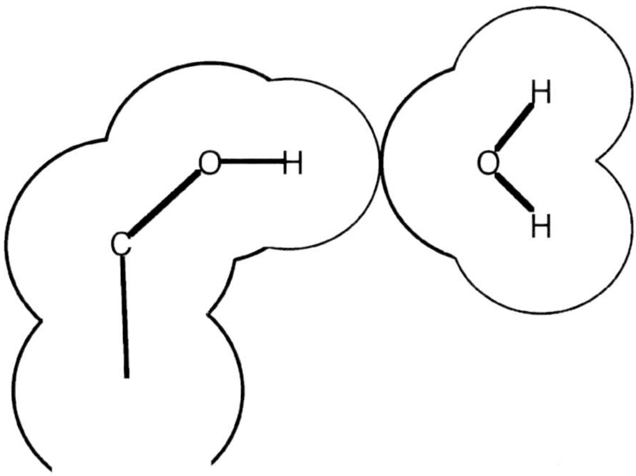

Figure 1.1 Van der Waals spheres for an alcohol and a water molecule. All atoms X are at the centres of spheres with radii R_X equal to their Van der Waals radii. With the O–H distance equal to 0.95 Å, $R_H = 1.2$ Å and $R_O = 1.5$ Å, the shortest O···O distance of this molecular Van der Waals complex is 3.65 Å.

interaction), which is at the origin of phase correlations between electronic displacements. If R labels some average distance of the two molecules, then this interaction is represented by a potential well with a minimum for some value of R. At larger R it is attractive and varies as $-R^{-n}$ with $n \geq 6$, which indicates that such an interaction most rapidly falls off with distance. At smaller values of R, on the other side of the well, it is strongly repulsive, meaning that it hinders molecules from coming into close contact. It allows all atoms to take on a "Van der Waals radius", which is the effective (approximate) size this atom occupies when it is part of any molecule. It is equivalent to approximating the various atoms of a molecule by hard spheres with radii equal to their respective Van der Waals radii. No sphere of any atom of another molecule, also characterized by its own Van der Waals radius, can penetrate this hard sphere. It thus defines the shortest distance at which atoms of various molecules can aggregate (Figure 1.1). Beyond this distance the interaction between the molecules is attractive but decays rapidly. The Van der Waals radii of most common atoms are $R_H = 1.2$ Å for H-atom, $R_O = 1.5$ Å for O-atom, $R_N = 1.55$ Å for N-atom and $R_C = 1.71$ Å for C-atom. They have been initially measured by the excluded volume method that consists of measuring the volumes of molecules in their solid state(s) and determining the greatest volume this molecule occupies if one assumes each of its atoms is a hard sphere with radius equal to its Van der Waals radius. More recent measurements are based on X-ray and neutron scattering techniques.

As already seen, thermal fluctuation of the order of kT at room temperatures corresponds to thermal energies of the order of 0.025 eV, greater than an average Van der Waals interaction of 0.01 eV. Most species made of small molecules interacting through Van der Waals interaction are consequently gases at room temperature. They become liquids when temperature is so lowered that kT becomes smaller than this average Van der Waals interaction.

Hydrogen bonds

Between these two electrical interactions—covalent between atoms and Van der Waals between molecules—exists an intermediate interaction, called the "hydrogen bond", that requires some conditions to be fulfilled. In the rest of this book we shall abbreviate "hydrogen bond" to H-bond. It occurs between a molecular group, most often O–H or N–H, which carries an H-atom and exhibits a marked electric dipole moment, and the O- or N-atom of another molecule. This latter atom is characterized by the presence of at least one nonbonding orbital that can point towards the H-atom of the polar group of the first molecule and is filled with a lone pair of electrons. This H-bond "acceptor site" may also exceptionally be an extended π electronic cloud such as is found with aromatic rings, also filled with electron pairs that point towards this H-atom. The "donating" molecular groups, O–H or N–H made of covalently bound atoms, retain their identities upon establishment of this H-bond. This property is shared with Van der Waals interactions. We represent an H-bond by a dotted line that clearly differentiates it from a covalent bond represented by a solid line. Throughout this book an H-bond will be shortly labelled in the text as X–H···Y, where X–H and Y are molecules or parts of molecules. When we have to particularly specify the atoms of X or Y that are involved in the H-bond, we shall preferably write it in the form –O–H···N– when the atom of X involved in the H-bond is an O-atom and that of Y an N-atom. Typical H-bonds are shown in Figure 1.2. The polar group that carries the H-atom is called the "donor", while the group O or N with a nonbonding orbital

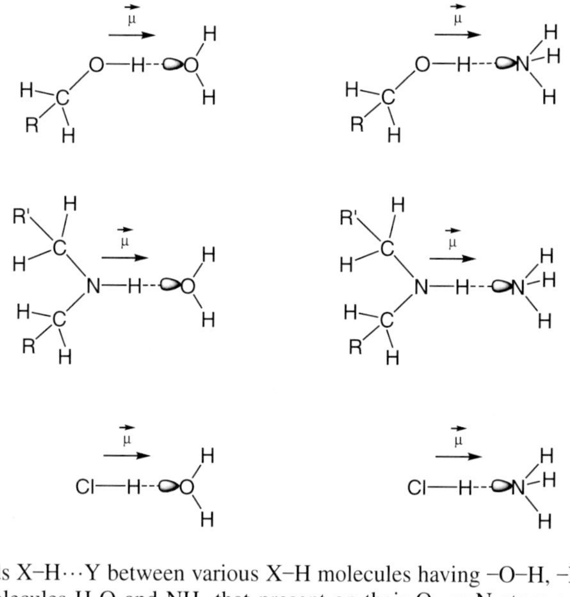

Figure 1.2 H-bonds X–H···Y between various X–H molecules having –O–H, –N–H or Cl–H polar groups and two molecules H_2O and NH_3 that present on their O- or N-atom a (greyed) nonbonding orbitals filled with lone pair electrons pointing towards the H-atom of X–H. Arrows stand for electric dipole moments. The acceptor O-atom has another masked and consequently not drawn nonbonding orbital occupied by two electrons.

is called the "acceptor". This denomination immediately calls for a caveat: the (H-bond) acceptor acts as an electron donor, and vice versa. In consequence, we shall always in this book consider *H-bond acceptors and donors* and will avoid considering electron donors or acceptors. An "acceptor" or "donor" will always implicitly be an H-bond acceptor or an H-bond donor.

As already mentioned, the establishment of an H-bond does not destroy covalent bonds. It means H-bonds are most of the time interactions between two molecules that retain their individualities. This is the reason why we classify such bonds as intermolecular interactions, even if in the following we may encounter H-bonds established inside single molecules that will then be called "intramolecular" H-bonds. These intramolecular H-bonds do not destroy the covalent bonds of the molecule they are part of. We may also note that only H-atom, with its isotopic variations D (deuterium) or T (tritium), establishes such H-bonds. It indicates that the especially small sizes of these atoms are crucial in the formation of H-bonds. These latter isotopic forms of the H-atom have identical electronic structures when they are part of a molecule, which consist of a single σ-orbital filled with two electrons. The O···O distance between the two O-atoms of an −O−H···O− bond is shorter than the distance defined by Van der Waals radii, 3.65 Å (Figure 1.1). It is typically 2.8 Å for an −O−H···O− bond, but may vary between 2.5 and 3 Å, depending on the molecules X−H and Y they belong to. The enthalpy of formation of such an H-bond is 0.1 eV for a weak H-bond and can reach 0.7 eV for a strong H-bond. These enthalpies are consequently intermediate between the enthalpies of covalent and Van der Waals bonds. As will be seen in Ch. 2, this energetic hierarchy of chemical bonds corresponds to the hierarchy of primary, secondary and tertiary structures of proteins. The enthalpy of an H-bond is thus typically somewhat less than 10 kT at room temperature, with $kT = 0.026$ eV at 300 K. We shall see in Ch. 2 and later in this book that this is one of the fundamental properties of H-bonds, at the origin of their ubiquity. In opposition to Van der Waals interactions most H-bonds are directional: the three atoms X, H and Y in X−H···Y are collinear in their equilibrium state. When they depart from linearity a force tends to restore this linearity. This force is at the origin of "bending intermolecular vibrations". We shall discuss this in Chs. 4 and 5.

THE H-BOND: HISTORICAL AND PROSPECTIVE ASPECTS, GENERAL BIBLIOGRAPHY

The concept of the H-bond slowly emerged during the 20th century and took some time to be fully accepted. H-bonds have for long been considered as anecdotic interactions. Their fundamental importance, in particular how much life rests on them, a point this book is aimed at establishing, became clear to scientists only in recent years. It suggests H-bonds might still be concealing some even more fundamental aspects that could make them even more crucial in the future. Following Lippert (1) and Jeffrey (2), who wrote precise historical accounts, Werner (3) seems to be the first who described an interaction we would now call an H-bond. In 1902, he suggested that hydrated ammonium NH_4OH should better be written $H{-}O{-}H\cdots NH_3$. He called this interaction *Nebenvalenzbindung*, a nearly covalent bond. Later, in 1910, Hantzsch (4) described the presence of such a bond in acetoacetic acid ester, while in 1912, Moore and Winmill (5) described a weak union for amines in water and in 1914, Pfeiffer (6) discovered the structure of acetic acid dimers found in acetic acid vapour. These cyclic structures, established by carboxylic acid dimers, constitute excellent models of H-bonds that we shall encounter in Ch. 4, Figure 4.4, and occasionally later. In 1920, Latimer and Rodebush (7), two students of G. N. Lewis, postulated that if an H-atom lies between two electronic octets, a weak bond appears. This was one of the first serious breaches in the then sacred rule of the octet. It was during this same period of time that the H-bond was recognized as responsible for the anomalous properties of liquid water. The concept itself and the denomination "hydrogen bond" were developed in the years after 1930 (2), and Pauling's (8) famous "*Nature of the Chemical Bond*" was the book that made H-bonds known to chemists. It followed several earlier articles by Pauling on $F{-}H\cdots F^-$, and on water and ice. Meanwhile infrared (IR) spectroscopy appeared as early as 1936 as a particularly efficient method to detect and observe H-bonds. As developed in Chs. 4 and 5 of this book, IR spectroscopy is now the most precise and sensitive tool to observe H-bonds. Knowledge of the H-bond progressed in the years following 1950, when X-rays and, somewhat later, neutron scattering established a property that will appear in the course of this book as fundamental to H-bonds: they are directional and, consequently, at the origin of organized molecular structures that are crucial in chemistry and biology (Ch. 2). These were the years of Nobel prizes rewarding Pauling for the structure of proteins and Watson and Crick for the structure of DNA, two discoveries that have been at the origin of the exploding development of biochemistry. In the 1970s, scientists became aware that the dynamical properties of H-bonds might be even more fundamental. Several chapters of this book, Chs. 4, 6, 9 and 10, deal with these specific dynamical properties of H-bonds and their importance in aqueous media. Finally, it was not before the 1990s that the ubiquity of H-bonds in our surroundings was clearly appreciated, in particular with the ubiquity of the H_2O molecule and its fundamental role in bio-reactions at the molecular level. These aqueous media have for long been considered as casual media devoid of any special property and, consequently, of any interest. In the course of this book, this perception will be challenged: they are media with subtle properties that are crucial for our knowledge of many processes, particularly life processes, but that we are still far from understanding precisely. They are basically made of assemblies of H_2O molecules that have the unique ability to develop a hyperdense "H-bond network" inside which reactivity, particularly bio-reactivity, occurs. The poor knowledge we still have of this H-bond network and of its reactivity based on transfers of protons and of H-atoms was called a scandal by Maddox in one of his "News and Views" in *Nature*, 1988 (9) (see Ch. 10). It illustrates how these most familiar water molecules are indeed still poorly known and how they are furthermore far less easily observed than the familiarity of liquid water might suggest. It points to the direction research on H-bonds and on the water molecule is likely to adopt about 100 years after the concept of the H-bond began to emerge. We may predict that it will constitute an important field of research in the near future: the precise knowledge of the dynamical properties of H-bonds is certainly a necessary achievement before we can start having a clear idea of how life proceeds at the molecular level.

The preceding paragraph shows that research on H-bonds has a history, and most likely a future, as we are far from understanding its properties, especially its dynamical properties.

H-bonds have consequently been the subject of various books. Among those that had an impact on research on H-bonds, are as follows:

- L. Pauling (8) (1939) "The Nature of the Chemical Bond". A book that marked a period of time and introduced the H-bond.
- D. Hadzi (Ed.) (1959) "Hydrogen Bonding", Pergamon Press, London. Papers presented at the first symposium on H-bonding that clearly established its basic properties.
- G. C. Pimentel and A. L. McClellan (10) (1960) "The Hydrogen Bond". The first exhaustive compilation of the basic properties of H-bonds, mainly thermodynamic, structural and spectroscopic properties. Still a reference book for these properties.
- P. Schuster, G. Zundel and C. Sandorfy (Eds.) (1976) "The Hydrogen Bond: Recent Developments in Theory and Experiments", North Holland, Amsterdam. Three volumes dealing with the state of our knowledge of H-bonds and related problems around 1975.
- H. Ratajczak and W. J. Orville-Thomas (Eds.) (1980) "Molecular interactions", John Wiley and Sons, Chichester. Our view on the nature of H-bond and understanding of its exceptional spectroscopic properties some years later.
- G. A. Jeffrey and A. Saenger (11) (1994) "Hydrogen Bonding in Biological Structures". An exhaustive modern compilation of structures of biological interest that involve H-bonds. The crystallographers' point of view on H-bonds.
- G. A. Jeffrey (2) (1997) "An Introduction to Hydrogen Bonding". A textbook on H-bonds, for a large part is devoted to structural aspect of H-bonds.

INTERMOLECULAR AND INTRAMOLECULAR H-BONDS

Most H-bonds X–H···Y are formed between two independent molecules X–H and Y, as represented in Figure 1.2. These are "intermolecular H-bonds" and when speaking of H-bonds in the following with no other specification, we always refer to this type of H-bond, which represents the large majority of them. Another category of H-bonds however exists, the "intramolecular H-bonds", where molecular groups X–H and Y are both parts of a same molecule. Even if they represent only a minority of H-bonds, these intramolecular H-bonds include quite a large variety of H-bonds. Two typical examples are shown in Figure 1.3.

These two types of H-bonds have macroscopic manifestations that are different: an intramolecular H-bond involves a single molecule, whereas an intermolecular H-bond involves two molecules that become independent upon disruption of the H-bond. As a consequence, intermolecular H-bonds, which establish relatively strong interactions between molecules in a liquid, are known to strongly influence the magnitudes of the temperature and heat of evaporation of this liquid. This is particularly marked in the case of liquid water. This is not at all so for intramolecular H-bonds that do not modify the interactions between molecules, which most often remain Van der Waals interactions. In a gas, intermolecular H-bonds are at the origin of deviations from perfect gas law, which is not so for intramolecular H-bonds. Also, in an intermolecular H-bond the relative positions of the donor and acceptor groups, X–H and Y, are only ruled by the H-bond interaction. The other groups of the molecule have almost no influence on these relative positions, as may be seen in Figure 1.2. This is not the case with an intramolecular H-bond. Thus the relative

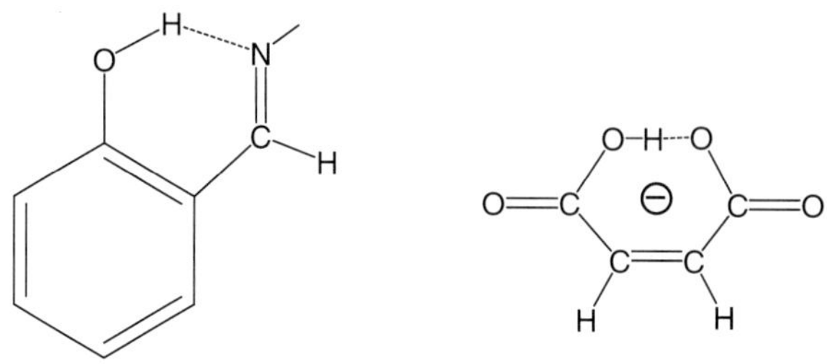

Figure 1.3 Two intramolecular H-bonds in an aromatic Schiff base (left drawing) and in maleate anion (right drawing).

positions of the three atoms that compose the –O–H···N– or –O–H···O– bonds in Figure 1.3 are first governed by the surrounding covalent bonds that are predominant and impose their own steric conditions. The –O–H···N– H-bond of the Schiff base of Figure 1.3 is, for instance, not straight, but bent. Also the symmetry of the maleate ion in this figure is C_{2v} when the H-atom of the H-bond is in its ground vibrational state. It loses this symmetry when the stretching vibration of this H-atom is in its first excited state (12), because the amplitude of vibration has then increased and has the effect of ejecting the H-atom out of the plane of symmetry of the ion that is also the plane of the drawing.

There exist many intermediate cases, particularly in polymers or macromolecules. They are intramolecular bonds that suffer only weak constraints from covalent bonds. This decrease of constraints may have several origins. One of them may be the great separation along the successive covalent bonds of the acceptor and donor groups that may occur in large molecules together with the possibility of folding that may offer a sufficiently close proximity of these groups to allow formation of the H-bond. A typical example is the β-sheet secondary structure of proteins we examine in Ch. 2 (Figures 2.5 and 2.6). Another example is given in Figure 1.4. It represents the repeat unit of chains of cellulose that we also examine in Ch. 2 (Figure 2.3). We may see that the H-bond established between $-O_3$–H···O_5– atoms is possible because of the existence within the covalent bond network of three degrees of freedom represented by rotations around axes C_1, $-O_4$, O_4–C_4 and C_3–O_3. By comparison, the intramolecular H-bonds of Figure 1.3 are more constrained because both molecules are planar, due to the conjugation of the well-developed π-orbital systems. The only degree of freedom left to establish these H-bonds are single rotations around the C–O bonds of the C–O–H groups that establish these H-bonds.

ELECTRONIC STRUCTURES OF HYDROGEN BONDS

We have seen that when two neutral atoms with their positive nuclei surrounded by spherical electronic clouds approach each other, this results in a strong distortion of the electronic clouds and finally leads to the formation of more stable molecules where the nuclei adopt fixed relative positions and are surrounded by electrons that occupy new orbitals around them. More precisely, the inner orbitals of the atoms of the molecule are but slightly modified with respect to those of the isolated original atoms. The outer orbitals are completely

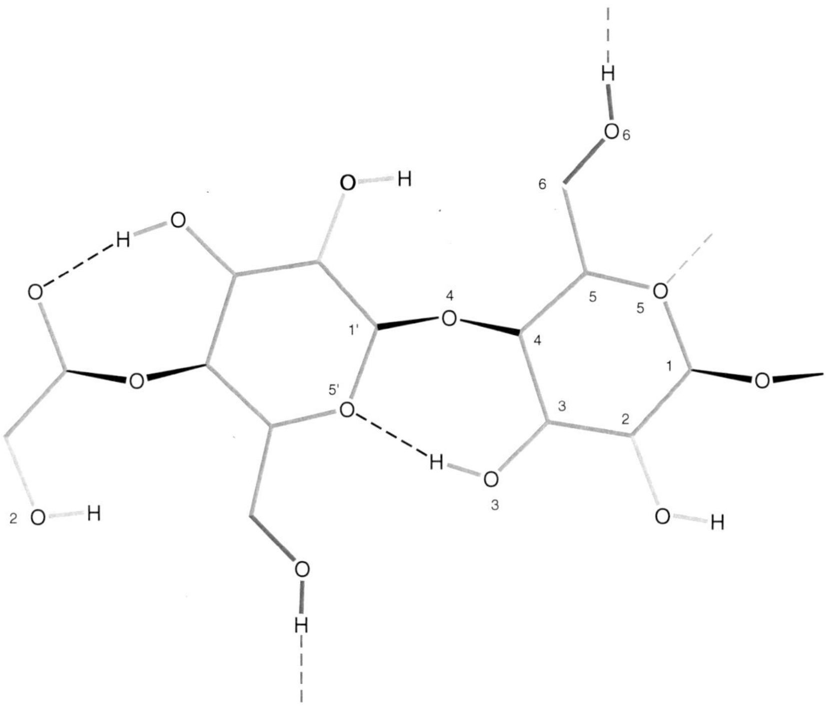

Figure 1.4 The repeat unit in cellulose Iβ. C- and H-atoms of CH or CH_2 groups are not written for clarity.

different. Instead of atomic orbitals with spherical symmetry, they are now σ- or π-type orbitals, and the whole new electronic distribution has energy lower than that of the whole initial atomic outer orbitals. This is at the origin of covalent bonds, which are nowadays fairly well described by quantum chemistry, at least for molecules having a limited number of atoms. Quantum chemistry consists of establishing an as-precise-as-possible description of the electronic structure of molecules. Incorporating H-bonds with energies roughly one order of magnitude smaller than covalent bonds requires an accuracy that is at the limits of the possibilities of these methods. H-bonds cannot be treated in the same way as Van der Waals interactions that we have seen only slightly modify the covalent orbitals by mutual polarization and can therefore be handled on the basis of perturbations of orbitals of the noninteracting molecules. H-bonds have stronger effects than covalent bonds, which means that they require treating the whole H-bonded complex as a supermolecule or molecular complex. It involves a great number of orbitals that rapidly surpasses the possibilities of any computing facility, so that a full "*ab initio*" treatment of H-bonds can only be performed on small H-bonded complexes such as $F{-}H{\cdots}OH_2$, for instance, or water dimers. *Ab initio* calculations are computations of electronic orbitals with no other hypotheses than Coulomb interactions between all electrons and nuclei with electrons obeying Fermi statistics with the Pauli exclusion principle. It leads to orbitals being occupied by pairs of electrons with opposite spin states. The precision of such a method depends on the number of basic orbitals on which all final molecular orbitals are decomposed. This number grows extremely rapidly with the number of atoms, which can therefore only be small. It means that even in the case of small dimers such as F–H···OH_2 or water dimers, a particular treatment of electronic correlations should be added to this *ab initio* treatment (13) if one wishes to keep enough accuracy. This additional treatment is responsible for a gain in energy of about 1.2 kcal mol^{-1} for an H-bond of about 5 kcal mol^{-1}, and cannot consequently be avoided.

Modern techniques of quantum chemistry, such as represented by "Density Functional Theories" often labelled DFT, give satisfactory descriptions of H-bonds. Their principle is the optimization of wavefunctions with respect to electronic densities, which is different from classical SCF (self-consistent field) approaches that optimize wavefunctions themselves. Their inconvenience, however, is that a physical interpretation is not straightforward with these methods. In order to have a simplified image of the nature of the H-bond, we have then to rely on older methods of quantum chemistry that, even if less accurate, give an image of the nature of H-bonds. Coulson (14) could thus put into evidence four mechanisms that play a role in the formation of H-bonds X–H···Y. The first one is polarization of the nonbonding orbital of Y by the dipole moment of the X–H group. It results in a deformation of this nonbonding orbital by this dipole moment. This interaction, which also exists in Van der Waals interactions, may be the only one in weak H-bonds. It is still there and may even remain predominant in strong H-bonds. It is responsible for the directionality of H-bonds (15). For stronger H-bonds, which have shorter X···Y equilibrium distances, as we shall see in Ch. 2, quantum forces appear with the overlap of electronic orbitals of X–H and Y. It results in a partial transfer of an electron from the nonbonding orbital of the acceptor atom of Y to the donor molecule X–H. In a first approximation, this transfer occurs towards the antibonding orbital of X–H (16) and is accompanied by an *s-p* rehybridization of the acceptor atom on Y, as suggested by photoelectron and X-ray spectroscopy (17). It has the effect of weakening the X–H covalent bond, which induces a weakening of the force constant that binds X and H. It also has the effect of increasing the X–H equilibrium distance and of strongly increasing the variation of the electric dipole moment $\partial\mu/\partial q$ when the distance q of X–H is varied. These are signatures of the presence of H-bonds that, as we shall see in the following chapters, appear with an exceptional intensity in the IR spectra of H-bonds (Chs. 4 and 5). Another quantum effect is due to the Pauli exchange principle of electrons of both X–H and Y that cannot be distinguished. Finally, phase correlations between electronic displacements, known as London dispersion forces, also appear. These last two interactions are typical of covalent interactions. They are especially important in strong H-bonds, such as F–H···F^-. All these quantum effects are, however, so much intermixed in this particularly strong H-bond that this decomposition becomes irrelevant. It nevertheless gives a useful image of the electronic structure of more weaker H-bonds, even if it remains an approximate one.

THERMODYNAMICS OF H-BONDS: ELECTRONIC AND VIBRATIONAL CONTRIBUTIONS TO ENTHALPIES

H-bonds may be characterized by various quantities, such as their enthalpies of formation, or their X···Y equilibrium distance, or, better, the wavenumbers of the centres of some of their characteristic spectral bands. Although it will appear at the end of this section that the enthalpies, ΔH, of H-bonds are not the best quantity to characterize them, we examine in this section a peculiarity of their own: the contribution to these enthalpies of vibrations. The main contribution to enthalpies of covalent bonds arises from rearrangements of electrons in new nuclear geometries. After such rearrangements molecular vibrations are somewhat modified, but with only relatively small changes of their energies. C–H groups, for instance, roughly keep their frequencies of vibrations, whatever be the molecule they are part of. The vibrational energies have consequently small relative contributions in the enthalpies of formation of covalent bonds. This is not so for H-bonds that see one vibration, the stretching X–H vibration ν_s, described in detail in Chs. 4 and 5, strongly modified by the establishment of an H-bond, X–H···Y. It implies a vibrational contribution to enthalpies of H-bonds that we examine in this section. We do it considering an isolated (intermolecular) H-bond and writing the minimum number of equations to make this point understandable. These equations rely on general principles of quantum description of molecules such as the Born–Oppenheimer separation of electrons and nuclei that we briefly comment upon. They are written in an intuitive form that is more rigorously established in Chs. 5 and 7. They allow establishing eq. (1.7) from which conclusions are drawn.

The enthalpy ΔH of an isolated H-bond is defined as the enthalpy of the reaction:

$$\mathrm{X-H+Y \rightleftharpoons X-H\cdots Y} \tag{1.1}$$

It is equal to

$$\Delta H = \bar{E}_{\mathrm{X-H\cdots Y}} - \bar{E}_{\mathrm{X-H}} - \bar{E}_{\mathrm{Y}} \tag{1.2}$$

where $E_{\mathrm{X-H\cdots Y}}$ is the total energy of the H-bond X–H···Y, $E_{\mathrm{X-H}}$ and E_{Y} are the total energies of the X–H and Y molecules when they are separated by a long distance and consequently do not establish H-bonds. The corresponding quantities $\bar{E}_{\mathrm{X-H\cdots Y}}$, $\bar{E}_{\mathrm{X-H}}$ and $\bar{E}_{\mathrm{Y}}$ are the thermal averages of these quantities. The calculation of $E_{\mathrm{X-H\cdots Y}}$ requires first calculating the energy $V(q,Q)$ of the ground electronic state of the complex X–H···Y that is precisely defined in eq. (7.A4) of the appendix of Ch. 7. It depends on the relative positions of the atoms of the H-bond. These positions are defined by coordinates q for the X–H distance and Q that represents the three "intermonomer coordinates" that define the relative positions of X–H and Y in X–H···Y. These intermonomer coordinates Q consist of the X···Y distance Q_s and of two angular coordinates that are defined in more detail in Ch. 2, Figure 2.1. Vibrations other than those represented by these coordinates are present in the X–H···Y system. We suppose they are little affected by the formation of the H-bond, a supposition that will indeed be revealed as realistic. Following eq. (1.2) they consequently have no contribution in ΔH, as they equally contribute to $E_{\mathrm{X-H\cdots Y}}$ and to either $E_{\mathrm{X-H}}$ or E_{Y}. In the Born–Oppenheimer approximation of separation of rapid electrons and slow nuclei of the molecules, which is examined in more detail in Ch. 7 and which is nearly always a very good approximation for molecules and molecular complexes, $V(q,Q)$, the energy of the ground electronic state acts as the potential energy for vibrations q and Q. The total average energy of the H-bonded complex X–H···Y is then

$$\bar{E}_{\mathrm{X-H\cdots Y}} = V(q_0,Q_0) + \bar{V}_{\mathrm{HB}} = V(q_0,Q_0) + \bar{E}_q^{\mathrm{HB}} + \bar{E}_Q \tag{1.3}$$

where q_0 and Q_0 are values of q and Q for which the electronic energy $V(q,Q)$ is minimum. $\bar{V}_{HB}$ is the average thermal energy of vibrations around these equilibrium values that decomposes into the sum of an average energy $\bar{E}_q^{HB}$ of the stretching vibrations of the H-atom (coordinate $q - q_0$) plus an average energy $\bar{E}_Q$ of the three intermonomer vibrations of the X–H···Y complex (coordinates $Q - Q_0$). This decomposition naturally occurs in the harmonic approximation that is valid for nearly all molecules when vibrational amplitudes are small. It is still valid in the case of H-bonds that display a strong anharmonic coupling between q and the Q_s, a characteristic feature of all H-bonds described in Chs. 4 and 5. As excited states of the stretching vibration of the H-atom ($q - q_0$) have energies above the ground state of this vibration that are much bigger than kT at room temperature, only the ground state of this vibration is populated. It implies that $\bar{E}_q^{HB}$ is equal to the energy of the ground state for the vibration in $q - q_0$.

In a similar way, the thermal averaged energy of the system composed of the two far-away X–H and Y molecules may be written as

$$\bar{E}_{X-H} + \bar{E}_Y = V(q_0,\infty) + \bar{E}_q^{free} + \tfrac{3}{2}kT \quad (1.4)$$

where $V(q_0, \infty)$ is the energy of the ground electronic state for both molecules X–H and Y when they are separated by a very great distance Q_s, which is equivalent to writing Q equal to infinity, as, in this case, the two other angular intermolecular coordinates have no influence on the energy of the system. $\bar{E}_q^{free}$ is the energy of the stretching vibration q of X–H alone, which is, for the same reason as above, equal to the energy of the ground state of this vibration. It is different from $\bar{E}_q^{HB}$, a difference that reflects the strong effect the formation of an H-bond has on this vibration. The last term of eq. (1.4) is the average thermal energy of the three relative translations or rotations of the two independent moieties X–H and Y. The position of each component X–H and Y is defined by three coordinates, which make six coordinates for the set X–H + Y. However, three of them are for the centre of gravity of this set. We do not write them in eq. (1.4) because they are exactly counterbalanced in the value of ΔH by a similar not written term in eq. (1.3), which represents the average energy of the centre of gravity of the X–H···Y complex. It is equivalent to considering centres of gravity of both X–H···Y and the set X–H + Y as fixed. The enthalpy ΔH is then equal to

$$\Delta H = V(q_0,Q_0) - V(q_0,\infty) + \bar{E}_q^{HB} - \bar{E}_q^{free} + \bar{E}_Q - \tfrac{3}{2}kT \quad (1.5)$$

The first two terms define the enthalpy gain due to the rearrangement of the electrons when the H-bond is formed. They constitute the preponderant term in ΔH. When the H-bond X–H···Y is stretched, that is the X···Y distance is increased with respect to its equilibrium value, the corresponding change in energy due to the rearrangement of electrons becomes $V(q_0,Q_s) - V(q_0,\infty)$ where Q_s is the X···Y distance that is also the coordinate of the stretching intermolecular vibration of the X–H···Y complex. It is one of the three intermolecular coordinates of the complex represented in the preceding equation by Q. In Figure 1.5 the shape of this quantity is drawn; supposing the two other angular intermolecular coordinates are fixed at their equilibrium position, we can keep X–H···Y linear. It takes on the form of a potential well. The depth of this well, which is the object of calculations of the

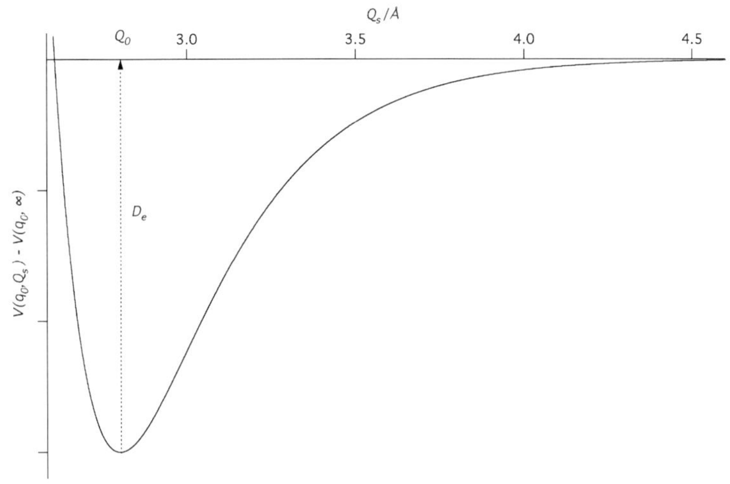

Figure 1.5 Schematic shape $V(q_0,Q_s) - V(q_0,\infty)$ of the electronic contribution to the enthalpy ΔH of an H-bond X–H···Y. The intermolecular coordinate Q_s is the distance X···Y, and q_0 is the equilibrium X–H distance. The two other intermolecular coordinates are set at their equilibrium values.

electronic structure of H-bonds at fixed atomic distances, is often labelled D_e. It is also called the binding energy of the H-bond. It is equal to $-\{V(q_0,Q_0) - V(q_0,\infty)\}$.

As evident from eq. (1.5) this binding energy is not equal to ΔH, even if it is the preponderant term. The energy involved in the third and fourth terms of eq. (1.5) is equal to the difference of the energies of the ground vibrational states of the stretching vibrations of the H-atom, defined by coordinate q, in X–H···Y and X–H. It is, in a good approximation, equal to

$$\bar{E}_q^{\mathrm{HB}} - \bar{E}_q^{\mathrm{free}} = \frac{\hbar[\omega_{\mathrm{c}}^{\mathrm{HB}} - \omega_{\mathrm{c}}^{\mathrm{free}}]}{2} = \frac{h[\nu_{\mathrm{c}}^{\mathrm{HB}} - \nu_{\mathrm{c}}^{\mathrm{free}}]}{2} \tag{1.6}$$

where ω's are equal to vibrational frequencies ν's multiplied by 2π and $\hbar = h/2\pi$ with h being the Planck's constant (see the appendix). Subscript c with ω's and ν's in eq. (1.6) implies values at centres of the bands and as before superscripts "HB" and "free" are for X–H···Y and X–H, respectively. We shall see in Ch. 4 that these stretching vibrations in X–H···Y are at the origin of exceptionally broad bands. The frequencies $\nu_{\mathrm{c}}^{\mathrm{HB}}$ of their centres strongly depend on the strength of the H-bond. For a medium-strength H-bond of the type –O–H···O–, it corresponds to a wavenumber around 3000 cm^{-1}, where, as defined in the appendix of this chapter, wavenumbers are equal to frequencies divided by c, the velocity of light. For nearly all free O–H groups with no H-bonds, stretching O–H vibrations are at the origin of relatively narrow bands in the vicinity of 3600 cm^{-1}. For a medium-strength H-bond, we therefore have

$$\bar{E}_q^{\mathrm{HB}} - \bar{E}_q^{\mathrm{free}} = \frac{3000-3600}{2}\,\frac{96.3}{8054} = -3.6\ \mathrm{kJ\ mol^{-1}} = -0.85\ \mathrm{kcal\ mol^{-1}} \tag{1.7}$$

As defined below, the enthalpy of such an intermediate-strength H-bond is of about -20 to $-40\,\mathrm{kJ\ mol^{-1}}$, out of which some -3.6 come from vibrations. Without being preponderant this contribution due to stretching vibrations of H-atoms is consequently not negligible. The contribution of intermolecular vibrations, represented by the last two terms of eq. (1.5), is markedly smaller and in this rapid evaluation may be neglected. This is not possible in the case of Van der Waals complexes where, in opposition to H-bonds, the third and fourth terms of eq. (1.5) are negligible, whereas the last two terms, which contribute less than the first two at low temperature, become preponderant above a certain temperature, implying disruption of the complex.

Some confusion exists in the literature between the various energetic quantities we have defined above for H-bonds. This is because different experimental or theoretical methods often measure or calculate different quantities. Thus, theoretical methods most often consider the binding energy D_e. It is not equal to the enthalpy of formation of H-bonds, as it neglects the vibrational non-negligible contribution. Also, thermodynamic methods, such as for instance calorimetry, measure values of ΔG of a set of H-bonds that differ from ΔH by the presence of an entropy term $-T\Delta S$. This last term is often small in solids, somewhat greater in liquids, but may become important in gases where the disorder that is characterized by this term is great. We shall see in this book other methods, such as IR spectroscopy (Chs. 4 and 5), X-ray (Ch. 3) spectroscopy, etc. from which thermodynamic quantities can be extracted, which may be ΔH, ΔG or other quantities. All this means that great care should be taken when one compares values of thermodynamic quantities obtained by various methods. A typical example is the enthalpy of H-bonds in ice, which has been found to vary (10) between -3 and $-6\,\mathrm{kcal\,mol^{-1}}$. All measurements were precise, at least precise enough that this discrepancy could not be assigned to errors. But not all referred to ΔH, and this is still the case with many thermodynamic quantities in the literature.

We may conclude from this section on thermodynamics of H-bonds that the enthalpy of formation, ΔH, of an H-bond, which is in principle the quantity that characterizes an H-bond and which has been the object of many early measurements (10, 18), is in practice a quantity that is not so well defined and measured. It exhibits sufficiently great imprecision that it is not a quantity we shall retain to characterize an H-bond. We shall see that geometrical parameters we describe in Ch. 2 or, better, spectroscopic parameters we describe in Ch. 4, provide much more accurate characterizations of H-bonds.

EXAMPLES OF WEAK, INTERMEDIATE AND STRONG H-BONDS

Following the values of their enthalpies of formation, ΔH, H-bonds are roughly classified into three categories: weak, intermediate (or medium-strength) and strong H-bonds. We illustrate this classification by few selected examples.

Weak H-bonds

Weak H-bonds are for instance "π hydrogen bonds" that have acceptors that are not atoms with nonbonding orbitals, but a set of atoms with polarizable orbitals such as π-orbitals

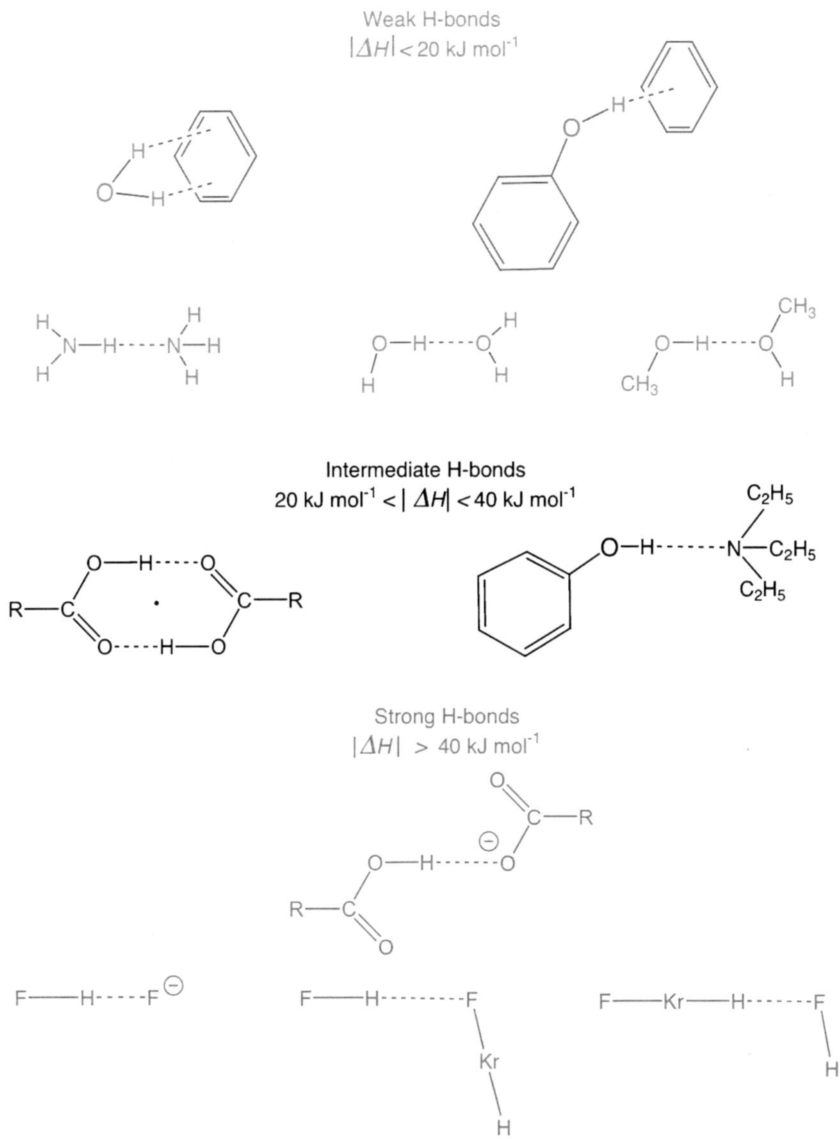

Figure 1.6 Examples of weak, intermediate and strong H-bonds. The strong H-bonds that involve F–H and a noble gas can adopt the two forms drawn in the bottom, middle and right-hand side (27).

extending, for instance, over aromatic systems. Such an H-bond is drawn in Figure 1.6 in the form of a water molecule, H_2O, as a donor and a benzene ring as an acceptor. The binding energy D_e of such a complex trapped in Ar solid matrices at very low temperatures has been evaluated to be around 2 kcal mol^{-1} (about 8 kJ mol^{-1}) by microwave spectroscopy (19). The distance between the O-atom of the H_2O molecule and the centre of the benzene ring has been measured equal to about 3.3 Å, with both H-atoms of the H_2O molecule pointing towards the benzene ring. By photoionization of a mixture of benzene and water molecules in a supersonic expansion beam, a value for D_e equal to about 3.3 kcal mol^{-1} (13.8 kJ mol^{-1}) (20) has been measured. Other authors (21) calculated a value of about 3.9 kcal mol^{-1} (16.3 kJ mol^{-1}) for this complex. Mass spectrometry of a cooled beam of a similar system where the H_2O molecule has been replaced by a phenol molecule gives a binding energy D_e of about 4 kcal mol^{-1} (16.7 kJ mol^{-1}) (22). IR spectra of similar systems reveal that this energy remains the same when the benzene acceptor is replaced by other molecules that also exhibit π-orbitals (23), such as an alkyl molecule containing a CH=CH group or an acetylenic molecule containing C≡C group.

More common weak H-bonds are found in simple molecules such as ammonia, NH_3; water, H_2O; or methanol, CH_3OH. The enthalpies ΔH of homogeneous H-bonds formed with these molecules are thus of about -3.5 kcal mol^{-1} (-14.6 kJ mol^{-1}) in the ammonia dimer, of about -5 kcal mol^{-1} (-21 kJ mol^{-1}) in the water dimer and of about -4 kcal mol^{-1} (-16.7 kJ mol^{-1}) for the methanol dimer (10, p. 212). These H-bonded dimers are shown in Figure 1.6. Heterogeneous H-bonded dimers formed by mixtures of these various molecules, for instance H_2O as donor and NH_3 as acceptor, have enthalpies of comparable magnitudes. The absolute values of these enthalpies are of the order of $5kT$ to $10kT$ at 300 K. As a consequence, the corresponding H-bonded complexes are stable at room temperature. They can, however, easily be disrupted or transformed, and their formation is reversible. It is therefore not so very surprising that we shall have to deal with weak H-bonds in liquid water, where their extremely great density is at the origin of exceptional physical and chemical properties, or in aqueous media that also displays particularly original chemical properties. In biomedia, which are a very special type of aqueous media, they are at the origin of the precisely defined sophisticated molecular complexes that we examine in Ch. 2. These complexes are stable but at the same time they are flexible, evolutive and adaptable, a set of properties that covalent bonds are unable to provide. It is not a surprise then if weak H-bonds are often encountered in biomedia.

Medium-strength H-bonds

The absolute values of enthalpies of medium-strength H-bonds extend from 5 to 10 kcal mol^{-1}, that is from 20 to 40 kJ mol^{-1}. Two typical examples of such intermediate H-bonds are drawn in Figure 1.6. The first one consists of the centrosymmetric cyclic dimers of carboxylic acids that are found in vapours of these acids, where they are in equilibrium with monomers with no H-bonds. These particular dimers are stable, with a well-defined structure, and are moreover easy to manipulate. All throughout this book we shall encounter them. They are excellent models of H-bonds. The enthalpies of formation of acetic acid dimers are found equal to $\Delta H = -7.5$ kcal mol^{-1} $= -31.4$ kJ mol^{-1} for one H-bond, and also for one D-bond, as deduced from analyses of IR spectra (24, 25). Corresponding values for formic acid are slightly smaller. A second example is given by phenol-amine bonds φ–O–H···N–. For phenol triethylamine, ΔH has been found in the vicinity of -9 kcal mol^{-1} (-38 kJ mol^{-1}) (18). These are bonds that may be qualified as strong H-bonds by biochemists, who compare these bonds to the average H-bond strength of the hyperdeveloped

H-bond network they are working with, that is the H-bond network of biomedia strongly developed by H_2O molecules. The average strength ΔH of this H-bond network falls around $-5\,\text{kcal mol}^{-1}$ ($-21\,\text{kJ mol}^{-1}$), as mentioned above. Such an H-bond with $\Delta H = -9\,\text{kcal mol}^{-1}$ is consequently stronger. For chemists it nevertheless remains an intermediate H-bond, because stronger bonds have been isolated.

Strong H-bonds

Strong H-bonds with $|\Delta H|$ greater than $10\,\text{kcal mol}^{-1}$ ($40\,\text{kJ mol}^{-1}$) are scarce. A few examples are nevertheless known. They often imply charged acceptor groups and have been observed in crystals. Thus acid salts of either carboxylic or carbonic acids have $|\Delta H|$ that varies between 10 and $20\,\text{kcal mol}^{-1}$, that is between 40 and $80\,\text{kJ mol}^{-1}$. Their structure is drawn in Figure 1.6. A neutral form of trichloracetic acid as a donor and trioctylphosphine oxide as an acceptor has also been reported to display an enthalpy of formation of $-29\,\text{kcal mol}^{-1}$ ($-120\,\text{kJ mol}^{-1}$) (26). Other strong H-bonds imply F–H in various forms. The strongest H-bond ever detected is $F\text{–}H\cdots F^-$ which displays a ΔH equal to about $-37\,\text{kcal mol}^{-1}$ ($-150\,\text{kJ mol}^{-1}$) (2). It is the energy of a covalent bond and the electronic structure of this complex that make it a molecule where the H-atom is indeed divalent. More recently, strong H-bonds implying F–H and some noble gases such as Ar, Kr or Xe have been identified. Their structure is also drawn in Figure 1.6. Their ΔH has been calculated to fall between -13 and $-18\,\text{kcal mol}^{-1}$ (between -54 and $-75\,\text{kJ mol}^{-1}$) (27). In opposition to preceding strong H-bonds they are not charged complexes.

NONCONVENTIONAL H-BONDS

We have seen above that H-bonds are formed between a molecular donor group, typically –O–H or –N–H, which displays an appreciable electric dipole moment, and an acceptor group that displays a polarizable part, most of the time a lone pair of electrons in a nonbonding orbital. C–H groups exhibit in their great majority no dipole moment and no nonbonding orbitals. They are consequently thought to be neither H-bond donors nor H-bond acceptors. There exist, however, molecules where these C–H groups display an appreciable dipole moment. This is the case, for instance, of hydrogen cyanide H–C≡N, where the difference of electronegativity of atoms N and C is at the origin of this important electric dipole moment. It makes H-bonds established by such C–H groups possible, and dimers of hydrogen cyanide of the form N≡C–H···N≡C–H are known that display a H···N distance of about 2.1 Å, smaller than the sum of Van der Waals radii of H- and N-atoms. This is also the case of some molecules such as chloroform CCl_3H or chlorofluoroforms with same formula with one or several Cl-atoms replaced by F-atoms. Upon addition of molecules such as cetones or diesters that exhibit acceptor groups of H-bonds, a clear deviation from the "ideal behaviour" can be detected. The ideal behaviour is that of a mixture of different gases that obeys the perfect gas law. More recently, H-bonds with calculated binding energies D_e ranging around somewhat less than $4\,\text{kcal mol}^{-1}$ ($16\,\text{kJ mol}^{-1}$) (28, 29) have been detected by IR spectroscopy with C–H groups of these chlorofluoroforms as H-bond donors and the O-atom of either dimethyl ester CH_3–O–CH_3 or of acetone CH_3–CO–CH_3 as acceptors, or with pentachlorocyclopropane as donor (30). Also the splitting into two bands of the C=O stretching band in cyclopentanone in its liquid or solid state has recently (31) been reassigned to the existence of a weak C–H···O=C interaction in these condensed states with a ΔH of about $-1.4\,\mathrm{kcal\,mol^{-1}}$ ($-6\,\mathrm{kJ\,mol^{-1}}$). It explains the difference with gaseous cyclopentanone where the C=O stretching band exhibits a single component.

There is consequently no doubt that C–H groups may interact with usual H-bond receptors such as O-atoms of esters or carbonyl groups. Should such an interaction be called an H-bond? The answer is not straightforward and is the object of many discussions that started about 35 years ago (32). The corresponding interactions are therefore (momentarily?) called "unconventional H-bonds" to distinguish them from usual H-bonds X–H···Y with X–H, a classical polar group. Except in the particular cases where the C-atom of these CH groups is covalently bound to a strong electron-withdrawing atom such as N in hydrogen cyanide, or Cl and F in chlorofluoroforms or other compounds, these bonds have energies weaker than any H-bond. Such weak energies are very hard to measure, and only IR spectroscopy has enough sensitivity to H-bonds to allow deciding whether a weak interaction exists or not. Unfortunately, these unconventional H-bonds affect IR spectra in a way most of the time opposite to the way *all* conventional H-bonds display, which is characterized by a dramatic increase of the integrated intensity of the stretching band ν_s of the H-atom upon establishment of an H-bond, an increase that is always accompanied, as described in Ch. 4, by an important shift of the centre of this band towards lower wavenumbers. In molecular complexes with unconventional H-bonds, integrated intensities of stretching C–H bands are smaller than in the absence of interaction and most of the time the shift of their centre is towards higher wavenumbers. Such unexpected shifts towards higher wavenumbers occur even with most (but not all) chlorofluorocarbons we have seen to clearly reveal the presence of an interaction. Furthermore, the integrated intensities of the CH stretching bands increase for some of them upon formation of an unconventional H-bond while it decreases for others, so that no correlation between shift and integrated intensity of this band (29) can be put into evidence as in the case of classical H-bonds.

Beyond the existence of a well-established interaction between polar C–H groups and H-bond receptor groups (33), one may wonder whether such an interaction established by C–H groups that are not especially polar is anecdotic or commonly encountered. Thus some crystallographers consider that it exists for amino acids where many C···O distances of 2.6 Å are measured, just at the edge of the sum of Van der Waals radii and may consequently reveal a systematic interaction. Also, quantum chemists (16, 34) explain the shift towards higher wavenumbers as having the same origin as that towards lower wavenumbers in classical H-bonds, with nevertheless a supplementary electronic rearrangement in the case of C–H groups that cause this inverse trend. *Ab initio* calculations of the energy of formation of formamide (see Figure 2.4) and N-methyl acetamide dimers reveal the formation of a classical N–H···O=C H-bond of about $-7\,\mathrm{kcal\,mol^{-1}}$ ($-29\,\mathrm{kJ\,mol^{-1}}$) to which about -1 to $-2.5\,\mathrm{kcal\,mol^{-1}}$ (-4 to $-10\,\mathrm{kJ\,mol^{-1}}$) should be added that comes from a C–H···O=C interaction (35). Other crystallographers conversely consider such a nondirectional interaction as nonexistent. It does not exist for many spectroscopists either: classical X–H···Y H-bonds exhibit dramatic effects in IR spectroscopy, as described in Ch. 4. Their absence in the case of unconventional C–H··· interactions, or their weak amplitude most often in an opposite direction, is a sufficiently clear criterion to exclude them from H-bonds. The problem of the existence of unconventional H-bonds remains consequently open.

H/D SUBSTITUTIONS IN H-BONDS

The H-atom may easily be replaced by its isotopic equivalent, the deuterium D-atom, sometime written ^{2}H. The latter, whereas scarcely encountered, has the advantage of clearly indicating that the only difference between H- and D-atoms lies in their nuclei: whereas the nucleus of the H-atom is composed of a single proton, that of the D-atom has a neutron in addition to the proton. The two nuclei have accordingly a different mass: that of D is twice that of H. But they have the same charge. In consequence, the electronic structures of both H- and D-atoms are exactly the same. This remains true when these atoms are parts of molecules and their electrons are found in σ-orbitals. As we have seen that the origin of H-bonds has a pure electric character, only ruled by Coulomb interactions between various charges, we see that a "deuterium" bond (D-bond) X–D···Y has the same electronic structure as its homologous H-bond X–H···Y, with same energy $V(q_0,Q_0)$, defined in eq. (1.3). This is also true of the electronic structures of X–H and X–D that have same energy $V(q_0,\infty)$. The binding energy defined above, $D_e = -\{V(q_0,Q_0) - V(q_0,\infty)\}$ of a D-bond is consequently the same as that of its homologous H-bond. It stresses the interest of this quantity that is otherwise the quantity that easily comes out of quantum calculations of electronic structures.

These binding energies being the preponderant quantities that enter values of enthalpies of H-bonds, we may deduce that enthalpies of H-bonds are nearly the same as those of their homologous D-bonds—nearly the same, but not identical, because the terms due to stretching vibrations of the H-atom, $\bar{E}_q^{\mathrm{HB}} - \bar{E}_q^{\mathrm{free}}$, in eq. (1.5) are not the same in an H-bond and in a D-bond. These differences between H-bonds and D-bonds are examined in more detail in Ch. 7, as a H/D substitution often conveys original information on H(D)-bonds. We do not reproduce them here, but we only indicate that for an intermediate-strength H-bond with total enthalpy of about -20 to $-40\,\mathrm{kJ\,mol^{-1}}$, the contribution of this stretching vibration of the H-atom amounts to about $-3.6\,\mathrm{kJ\,mol^{-1}}$ (eq. (1.7)). In a D-bond frequencies of stretching vibrations of D-atoms are, in a first approximation, equal to those of an H-bond divided by $\sqrt{2}$, the square root of the ratio of D to H masses. As a consequence, the contribution from these vibrations is also divided by $\sqrt{2}$. The difference of enthalpies between an H-bond of medium strength and the corresponding D-bond is therefore:

$$\Delta H^{\mathrm{H}} - \Delta H^{\mathrm{D}} = -7.2\,(1-\tfrac{1}{\sqrt{2}}) \simeq -1\ \mathrm{kJ\,mol^{-1}} \tag{1.8}$$

It therefore represents a percentage of the enthalpy of the H- or D-bond. This quantity is still reduced for a weak H-bond as found in most aqueous media for which the centre of the O–H stretching band falls now around $3400\,\mathrm{cm^{-1}}$. We may conclude that isotopic effects displayed by H-bonds are small for what concerns thermodynamics and geometry. As shown in more detail in Ch. 7, this is not so for dynamical properties of H-bonds, particularly transfers of protons or of H-atoms examined in Ch. 6, which drastically vary when a D-atom replaces the H-atom of an H-bond. It implies that if a H/D substitution has limited effects in physics and chemistry, this is not at all so in biology, where such a substitution has nearly always lethal consequences.

APPENDIX: ENERGIES AND RELATED QUANTITIES

In the international unit systems energies are expressed in joules (J). This is not a practical unit for molecular systems. We shall therefore encounter in this book other units, such as the electron-volt (eV), which is related to joule by the equation

$$E = 1\,\text{eV} = 1.60 \times 10^{-19}\,\text{J} \tag{1.A1}$$

The eV is well adapted to electronic transitions between outer orbitals of atoms or molecules that correspond to energies of the order of a few eV. For H-bonds that are weaker bonds than covalent bonds, other units, such the kilocalorie per mole or kilojoule per mole, are often encountered. They are related to eV by the equations:

$$E = 1\,\text{eV} = 23.04\,\text{kcal mol}^{-1} = 96.3\,\text{kJ mol}^{-1} \tag{1.A2}$$

Other quantities of a different nature and consequently with different units, such as temperature, frequencies or wavenumbers, may also be proportional to exchanges of energies. Thus energies of thermal fluctuations are expressed in terms of kT, with the Boltzmann constant $k = 1.38 \times 10^{-23}\,\text{J K}^{-1}$. Quanta of energies carried by an electromagnetic wave of frequency ν are equal to $h\nu$ with the Planck's constant $h = 6.626 \times 10^{-34}\,\text{J s}$. They are by definition the energies of photons, the names of these quanta. In this book, we do not characterize electromagnetic waves by their frequencies, expressed in Hz, but rather characterize them by their wavenumbers $\tilde{\nu}$ expressed in cm^{-1}. One wavenumber is the number of wavelengths in the unit distance $l_0 = 1\,\text{cm} = 10^{-2}\,\text{m}$. It is defined as

$$\tilde{\nu} = \frac{10^{-2}}{\lambda} = \frac{\nu}{10^2 c} \tag{1.A3}$$

where λ is the wavelength of the electromagnetic wave, expressed in meters, and $c = 2.998 \times 10^8\,\text{m s}^{-1}$ the velocity of light. Let us note that eq. (1.A3) may be simply written $\tilde{\nu} = \nu/c$ if we express c in cm s^{-1}. This is true for subsequent equations that exhibit this factor 10^2. This formulation is used in figures of various chapters of this book.

From all these definitions we conclude that an energy $E = 1\,\text{eV}$ has equivalent temperature T_{eV}, equivalent frequency ν_{eV} and equivalent wavenumber $\tilde{\nu}_{\text{eV}}$ that verify the equation:

$$E = 1\,\text{eV} = kT_{\text{eV}} = h\nu_{\text{eV}} = 10^2 hc\tilde{\nu}_{\text{eV}} \tag{1.A4}$$

with $T_{\text{eV}} = 11{,}600\,\text{K}$, $\tilde{\nu}_{\text{eV}} = 8054\,\text{cm}^{-1}$ and $\nu_{\text{eV}} = 241\,\text{THz}$ ($1\,\text{THz} = 10^{12}\,\text{Hz}$). It defines the correspondence between eV, K and cm^{-1}. We shall not use in this book the correspondence with THz, which may nevertheless be encountered in microwave experiments. It should be, however, kept in mind that correspondence does not mean equality, as these various quantities have not the same dimensions and are consequently not expressed in the same units.

From the definition of wavenumbers in eq. (1.A3) we deduce another correspondence that, although neither used in this book, may be useful having in mind. Thus

$$\tilde{\nu} = 1\,\mathrm{cm}^{-1} = \frac{\nu}{10^2\,c} \tag{1.A5}$$

implies

$$\nu \simeq 3 \times 10^{10}\,\mathrm{Hz} = 30\,\mathrm{GHz} \tag{1.A6}$$

which means that an electromagnetic wave with wavenumber $1\,\mathrm{cm}^{-1}$ has a frequency in a close vicinity of 30 GHz. We have already seen the photon emitted or absorbed by such a wave has an energy of 1/8054 eV. In the following, we shall often encounter in vibrational potentials the quantity $\omega = 2\pi\nu$ instead of ν. The quantum of vibrational energy in the harmonic approximation is then $h\nu = \hbar\omega$ with $\hbar = h/2\pi = 1.055 \times 10^{-34}\,\mathrm{J\,s}$.

REFERENCES

1. E. Lippert, in P. Schuster and G. Zundel (Eds.), The Hydrogen Bond: Recent Developments in Theory and Experiments, Vol. I, Ch. 1, North Holland, Amsterdam, 1976.
2. G. A. Jeffrey, An Introduction to Hydrogen Bonding, Oxford University Press, Oxford, 1997.
3. A. Werner, Liebigs Ann, 322 (1902) 261.
4. A. Hantzsch, Berichte, 43 (1910) 3049.
5. T. S. Moore and T. F. Winmill, J. Chem. Soc., 101 (1912) 1635.
6. P. Pfeiffer, Berichte, 47 (1914) 1580.
7. W. M. Latimer and W. H. Rodebush, J. Am. Chem. Soc., 42 (1920) 1419.
8. L. Pauling, The Nature of the Chemical Bond, University Press, Ithaca, NY, 1939.
9. J. Maddox, Nature, 332 (1988) 677.
10. G. C. Pimentel and A. L. McClellan, The Hydrogen Bond, Ch. 3, W.H. Freeman, San Francisco, 1960.
11. G. A. Jeffrey and W. Saenger, Hydrogen Bonding in Biological Structures, Springer Verlag, Berlin, 1994.
12. F. Fillaux, N. Leygue, J. Tomkinson, A. Cousson and W. Paulus, Chem. Phys., 244 (1999) 387.
13. S. Tsuzuki, T. Uchimaru, K. Matsumara, M. Mikami and K. Tanabe, J. Chem. Phys., 110 (1999) 11906.
14. C. A. Coulson, Research, 10 (1957) 149.
15. A. C. Legon and D. J. Millen, Chem. Soc. Rev., 16 (1987) 467.
16. P. Hobza, Phys. Chem. Chem. Phys., 3 (2001) 2555.
17. A. Nilsson, H. Ogasawara, M. Cavalleri, D. Nordlung, M. Nyberg, Ph. Wernet and L. G. M. Pettersson, J. Chem. Phys., 122 (2005) 154505.
18. H. Ratajczak, W. J. Orville-Thomas and C. N. R. Rao, Chem. Phys., 17 (1976) 197.
19. S. Suzuki, P. G. Green, R. E. Bumgarner, S. Dasgupta, W. A. Goddard III and G. A. Blake, Science, 257 (1992) 942.
20. A. Courty, M. Mons, I. Dimicoli, F. Piuzzi, M. P. Gaigeot, V. Brenner, P. de Pujo and P. Millié, J. Phys. Chem. A, 102 (1998) 6590.
21. D. Feller, J. Phys. Chem. A, 103 (1999) 7558.
22. J. L. Knee, L. R. Khundkar and A. H. Zewail, J. Chem. Phys., 87 (1987) 115.

23. A. Fujii, T. Ebata and N. Mikami, J. Phys. Chem. A, 106 (2002) 8554.
24. J. Bournay and Y. Maréchal, J. Chem. Phys., 59 (1973) 5077.
25. J. C. Lassègues and J. Lascombe, in J. R. Durig (Ed.), Vibrational Spectra and Structure, Vol. 11, Ch. 2, Elsevier, Amsterdam, 1982.
26. D. Hadzi, Chimia, 26 (1972) 7.
27. S.-Y. Yen, C.-H. Mou and W.-P. Hu, Chem. Phys. Lett., 383 (2004) 606.
28. S. N. Delanoye, W. A. Herrebout and B. J. van der Veken, J. Am. Chem. Soc., 124 (2002) 7490.
29. S. N. Delanoye, W. A. Herrebout and B. J. van der Veken, J. Am. Chem. Soc., 124 (2002) 11854.
30. A. B. Baker, C. Samet, J. T. Lyon and L. Andrews, J. Phys. Chem. A, 109 (2005) 8280.
31. P. D. Vaz and P. J. A. Ribeiro-Claro, J. Phys. Chem. A, 107 (2003) 6301.
32. G. A. Jeffrey, J. Mol. Struct., 485 (1999) 293.
33. R. Taylor and O. Kennard, J. Am. Chem. Soc., 104 (1982) 5063.
34. K. Hermansson, J. Phys. Chem. A, 106 (2002) 4695.
35. R. Vargas, J. Garza, R. A. Friesner, H. Stern, B. P. Hay and D. A. Dixon, J. Phys. Chem. A, 105 (2001) 4963.

– 2 –

Geometrical Properties of H-Bonds and H-Bonded Organized Supramolecular Structures

GEOMETRIES OF H-BONDS AT EQUILIBRIUM

In the preceding chapter thermodynamic properties of H-bonds have been examined. In this chapter their geometrical properties are described and related to thermodynamic properties. The appearance of well-defined fundamental supramolecular structures of macromolecules, a consequence of the special geometrical and thermodynamic properties of H-bonds, is then examined in the form of a description of some illustrative examples. The relative positions of X–H and Y in an H-bond X–H···Y are defined by three "intermonomer coordinates", labelled Q throughout this book. Q encompasses Q_s, the X···Y distance when the H-bond is in its linear conformation, and two angular coordinates θ and φ (Figure 2.1) for the relative orientation of the acceptor group Y with respect to X–H. When the donor group of X is, for instance, a –C–O–H group that establishes an H-bond –C–O–H···Y, θ is the angle of projection of the –O–H···Y group on the C–O–H plane and φ is the angle of the projection perpendicular to this plane (Figure 2.1). In some cases it is preferable to substitute Cartesian coordinates Q_θ and Q_φ to these angular coordinates. They are equal to these angles multiplied by a distance, for instance the H···Y distance, equal to $(Q_s - q)$ when θ and φ are small. When θ and φ are no longer small, as is often the case

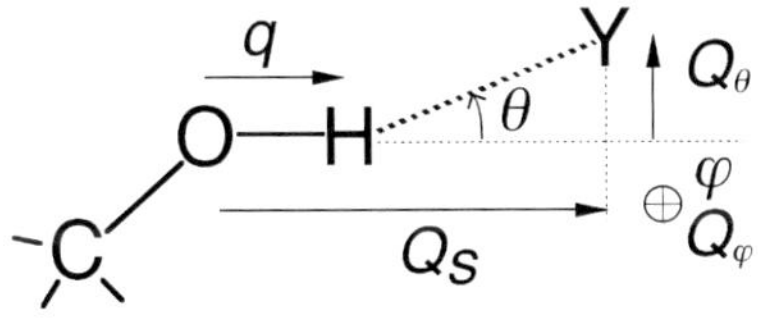

Figure 2.1 Definitions of intermonomer coordinates of an H-bond –C–O–H···Y. The C–O–H group defines the plane of the diagram. The in-plane angle θ is that of the O–H direction with the projection of H···Y on this plane. The out-of-plane angle φ is that of the O–H direction with the projection of H···Y perpendicular to this plane. Q_θ and Q_φ are Cartesian coordinates equal to angles θ and φ multiplied by a distance (see text, or see Ch. 4).

for such a small molecule as H_2O, for instance, several choices for this distance, and consequently for Q_θ and Q_φ, are possible.

In this chapter we focus on the equilibrium values Q_0, θ_0 and φ_0 of the three coordinates Q_s, θ and φ, respectively. They define the geometries of H-bonds X–H···Y at equilibrium. Departing from these equilibrium positions provokes the appearance of "intermonomer" vibrations that are examined in Ch. 4. The coordinate of the "stretching intermonomer" vibration is then $Q_s - Q_0$ and those of the two "bending intermonomer" vibrations are $\theta - \theta_0$ and $\varphi - \varphi_0$.

Equilibrium angles θ_0 and φ_0

These angles have been measured by X-ray diffraction for H-bonds found in crystals. The histograms of a series of several hundreds of such H-bonds (1) that had been examined around 30 years ago show that the distribution of equilibrium "solid angles" is a Gaussian distribution centred around 0 and having a width of about 15°. In other words, the average H-bond examined is directional and linear within about 15°, that is the three atoms A, H and B of the –A–H···B– complex are on an average positioned on a straight line at their equilibrium position, with a mean square deviation of 15°. This directionality is a fundamental property of H-bonds. It differentiates H-bonds from Van der Waals interactions that do not exhibit this property, at least at room temperature. It allows for building well-defined supramolecular structures. Some important structures are described later in this chapter. The H-bonds involved in these special structures are weak H-bonds that we have seen in the preceding chapter to have energies of formation of about 5–10 kT at room temperature. These supramolecular structures are consequently stable at room temperature but exhibit flexibility, capacities of evolution and adaptability because H-bonds, their cement, can be easily disrupted at the cost of energy that can be provided by thermal fluctuations. Covalent bonds are also directional and offer an even greater stability. They are, however, very stiff, offering only possibilities of rotations around σ-bonds and are consequently unable to provide enough flexibility and capacities of evolution or adaptability at room temperature. Van der Waals interactions offer no stability, as they are too easily disrupted at room temperature. With their directionality and their enthalpies of about 5–10 kT at room temperatures, H-bonds are consequently the only type of chemical interaction that can offer both stability and flexibility, two properties that are necessary in molecular biology. We begin to see the origin of their importance. We shall also see later in Ch. 6 that another property of their own strongly enhances their importance.

Some authors have postulated the existence, beside ordinary H-bonds, of "bifurcated H-bonds" or "three-centre H-bonds" that can be occasionally encountered and consist of a single X–H donor group that simultaneously interacts with two acceptor groups Y and Y′. These multicentred H-bonds lose the properties of directionality of ordinary H-bonds. These authors are mainly crystallographers (2; pp. 136–146) who measured, by X-ray or neutron diffraction, distances X···Y and X···Y′ that are nearly the same in an appreciable number of systems. The existence of such H-bonds has never been experimentally put into evidence by other techniques, particularly IR spectroscopy, the most sensitive method to observe H-bonds (Chs. 4 and 5). It has been occasionally used as an *ad hoc* explanation of some otherwise poorly understood results, especially in the case of the exceptional physical properties of liquid water. Bifurcated H-bonds are consequently a concept that has not conveyed the proof of its validity and the arguments given by X-rays and neutron diffraction experiments are weaker than they may appear, owing to the difficulty X-ray encounters in detecting H-atoms and other types of difficulties neutrons encounter with the relatively small coherent scattering factor of these H-atoms with respect to their great incoherent ones (see Ch. 3). This finding of equivalent X···Y and X···Y′ distances may indeed be due to a pure dynamic effect: the X–H group alternatively establishes H-bonds with neighbouring Y and Y′ molecules, making a kind of flip-flop motion between the two equivalent H-bonds X–H···Y and X–H···Y′. We expect such a behaviour to be quite natural in the case of a molecule such as H_2O that has such a small moment of inertia that it exhibits an especially rapid rotational dynamics, with characteristic time of the order of 1 psec (10^{-12} sec) (Chs. 4 and 9). In these conditions, scattering techniques see the average structure. Such a situation would then correspond to the existence of two energetically equivalent possibilities of establishing H-bonds, with two equivalent equilibrium positions. These equilibrium positions would, however, each correspond to two directional H-bonds between which the X–H group would oscillate. It is not consequently a bifurcated H-bond where neither the X–H···Y nor the X–H···Y′ angle is zero.

Equilibrium distances Q_0

For all H-bonds the X···Y equilibrium distances Q_0 decrease with the strengths of the H-bonds as defined by their enthalpies ΔH. For an –O–H···O– bond this distance varies between 2.95 Å for a very weak H-bond to 2.5 Å for a very strong H-bond. For an –N–H···O– bond these distances become 3.1 Å for a weak H-bond and 2.6 Å for a strong H-bond. For an –N–H···N– bond these distances are 3.2 and 2.7 Å, respectively. Distances of other H-bonds X–H···B– with X, a halogen, and the receptor atom B, either an O- or N-atom, have been precisely measured in gases by microwave or IR spectroscopy, as described in Ch. 3. In this gas state these H-bonds are isolated. Some typical values for X···B distances are given in Table 2.1, together with corresponding values for ΔH. For H-bonds in solution with such an apolar solvent as CCl_4, relatively precise values of Q_0 can be indirectly deduced from IR spectra, as shown in Ch. 4. Some of these distances determined using this method are displayed in Table 2.2, together with corresponding ΔH values measured by calorimetry or by IR. In both Tables 2.1 and 2.2 it clearly appears that the stronger the H-bond, the shorter is its equilibrium distance Q_0. It implies that a short Q_0 corresponds to a strong H-bond. This is, however, true for intermolecular H-bonds. In the case of intramolecular H-bonds, short Q_0s may also be the consequence of strong sterical constraints of the covalent bonds. The consequence is that for a given value of Q_0 one may find H-bonds with various ΔH values, which extend from weak to strong H-bonds. Shortening of H-bonds is consequently not correlated with increasing enthalpy of formation in these intramolecular H-bonds.

In the preceding chapter, it has been shown that characterizing H-bonds by their ΔH values conveyed many uncertainties. In some cases, for instance crystals with H-bonded molecules, Q_0 distances are relatively easy to measure. The same is true of H-bonds in

Table 2.1

Enthalpies ΔH and equilibrium distances Q_0 of some isolated H-bonds in gases

H-bond X–H···B	ΔH (kJ mol^{-1})	Q_0 (Å)
Cl–H···NH_3	–6.8	3.1
Br–H···NH_3	–7.3	3.25
F–H···NH_3	–13	2.7
F–H···N≡C–R	–21	2.8
F–H···OH_2	–26	2.7
F–H···$O(CH_3)_2$	–43	2.6
$(HCOOH)_2$	–2 × 30	2.7
$(CH_3COOH)_2$	–2 × 31.4	2.7

For H-bonds with NH_3 as the acceptor, values for ΔH are calculated (14) while values for Q_0 are experimental values deduced from microwave spectroscopy (14–16), as described in Ch. 3. For H-bonds with F–H as donor and various acceptors other than NH_3, and also for carboxylic acid dimers, ΔH comes out of the measurement of the equilibrium constant between separated monomers and H-bonded dimers by IR or microwave spectroscopy, and Q_0 is directly measured using these methods (17) otherwise described in Chs. 3–5. Acetic and formic acid cyclic dimers are shown in Figures 1.6 and 4.4.

Table 2.2

Enthalpies of formation ΔH and equilibrium distances Q_0 of a selection of H-bonded dimers in solution in CCl_4

H-bonded dimer	H-bond formula	ΔH (kJ mol^{-1})	Q_0 (Å)
Ethanol···dioxane	CH_3–O–H···O–$(CH_2$–$CH_2)_2$–O	–13	2.95
Methanol···acetone	CH_3–O–H···O=C–$(CH_3)_2$	–15	2.95
Ethanol···acetone	C_2H_5–O–H···O=C–$(CH_3)_2$	–15	2.95
Ethanol···dimethylformamide	C_2H_5–O–H···O=CH–N–$(CH_3)_2$	–16	2.85
Phenol···acetonitrile	C_6H_5–O–H···N≡C–CH_3	–18	2.8
Phenol···dioxane	C_6H_5–O–H···O–$(CH_2$–$CH_2)_2$–O	–20	2.75
Phenol···diethylether	C_6H_5–O–H···O–$(C_2H_5)_2$	–21	2.75
Phenol···pyridine	C_6H_5–O–H···NC_5H_5	–29	2.7
Phenol···triethylamine	C_6H_5–O–H···$N(C_2H_5)_3$	–36	2.65

Values for ΔH are those given by Ratajczak *et al.* (18). Values of Q_0 are calculated from centres of ν_s IR bands given in the same reference (18) using Novak's curve for –O–H···O– H-bonds (Figure 4.5).

gases where microwave spectroscopy allows relatively precise measurements of moments of inertia of H-bonded complexes from which fairly precise values of Q_0 can be deduced. It is then tempting to better characterize H-bonds by their Q_0 distances, a quantity measured with a better accuracy. We shall furthermore see in Ch. 4 that one of the most powerful and general method to observe H-bonds is IR spectroscopy, a method that can be routinely used in most laboratories. It allows us to rapidly deduce the value of Q_0 from the wavenumber of the centre of the υ_s band of this H-bond. Q_0 is consequently a better characteristic quantity for an H-bond than is its ΔH value.

Equilibrium distances q_0

These are the O–H or N–H equilibrium distances in H-bonds –O–H···Y or –N–H···Y. This distance is of 0.95 Å (1, 2; p. 95) for a free O–H group that does not establish an H-bond and of somewhat less than 1 Å for a free N–H group (1). This distance is slightly elongated by the establishment of an H-bond, thus reaching a value of 1 Å for a medium-strength –O–H···O– bond with corresponding O···O distance of 2.6 Å and reaching a value of 1.1 Å for a strong –O–H···O– bond such as that encountered in acid salts shown in Figure 1.6 with corresponding O···O distance of 2.5 Å. In this latter case the H-atom comes close to the middle point between the two O-atoms. This increase of the X–H distance with the strengths of the H-bonds X–H···Y is quite general. It is very well taken into account by the theory that considers the H-bond as mainly resulting from a transfer of electronic density from the nonbonding orbital of Y to the antibonding orbital σ^* of the X–H group (3).

ORGANIZED SUPRAMOLECULAR STRUCTURES OF MACROMOLECULES

We have seen above that the conjugation of structural and thermodynamic properties of weak H-bonds, namely their directionality combined with their enthalpies of about 5–10 kT at room temperature, allows building well-defined molecular structures that combine stability with flexibility and adaptability or evolution. Such structures are most important in the organization of macromolecules, especially biomacromolecules for which the appearance or disappearance of such supramolecular organization may completely change their properties, making for instance a living protein irreversibly becomes a bio-inert polypeptide.

The primary interaction that defines the structure of a molecule or a macromolecule is covalence, the strongest interaction in molecules. H-bonds intervene in their structures only inside the few degrees of freedom that may be left after covalent interactions have been established. In molecules, most of these degrees of freedom left by covalent bonds come from free rotations around bonds such as C–C, or other σ-bonds that are not part of a cyclic structure. They make these macromolecules free to move inside a limited range of various conformations with nearly the same electronic energies. Within this space the possibility of establishing H-bonds strongly favours some conformations that then become more stable than other conformations with no H-bonds. These special H-bonded conformations define the "secondary structure" of these macromolecules. This is what happens in the case of cellulose or proteins. Another possibility is establishing a very great number of intermolecular H-bonds within a special conformation of both constituent macromolecules, as in DNA. We select these three examples—cellulose, proteins and DNA—to illustrate the role of H-bonded biological supramolecular structures that are in a much greater number than these three examples, as shown in books on biochemistry (4, 5). In cellulose, the H-bond network has such an extension that cellulose is one of the most stable and resistant macromolecule encountered in biology. It is indeed a crystal, but a crystal that exhibits much more flexibility than ordinary mineral crystals. Proteins and DNA are two typical types of stable biomacromolecules where flexibility and the possibilities of evolution that are given by H-bonds are of fundamental importance. The appearance of several types of secondary structures of proteins that can be found throughout their very great varieties illustrates the subtle properties that H-bonds are able to convey. DNA is a structure of elementary components, the nucleotides, that is sufficiently stable but at the same time sufficiently evolutive that it has been able to keep and develop the memory of life for at least 3.6 billion years, the age of the oldest fossils of blue-green algae ever retrieved (4, 5).

Cellulose and amylose

In order to understand how covalent bonds may leave some room to build an H-bond network in macromolecules and how the establishment of this H-bond network modifies properties of macromolecules, we first examine cellulose, a particular carbohydrate. Carbohydrates form a simple class of biomacromolecules that contain only C-, O- and H-atoms, with the exclusion of N-atoms, and are directly obtained from atmospheric CO_2 and H_2O molecules during photosynthesis performed by plants. Cellulose is made of elementary repeating bricks that originate from a particular form of glucose, the β-D-glucopyranose linked in $1 \leftrightarrow 4$. Let us explain what all this is, noting that a minimum amount of vocabulary is regretfully unavoidable to understand the great wealth of "saccharide chemistry" (6) for which molecules with same atomic constituents adopting same arrangements but with different stereoconformations have completely different chemical properties. Small differences in these stereoconformations may induce great differences in the possibilities of establishing H-bonds. This is at the origin of great differences of macroscopic properties. D-Glucopyranose is thus a cyclic molecule made of six C- and O-atoms linked by σ-type covalent bonds, a "hexose" shown in Figure 2.2. Like practically all hexoses found in biology, they adopt the stereostable "chair form". With this form there exist $2^5 = 32$ different isomers that come from two possible stereoconformations for each of the five C-atoms of the ring. They are divided into 16 D "enantiomers" that provoke rotation of a linearly polarized optical beam towards the right (dextrogyre). The mirror image of these D enantiomers are 16 L (levogyre) enantiomers. Only the D-enantiomers are found in the living world. These 16 D enantiomers differ by permutations of −H and −OH groups on carbons 1, 2, 3 and 4, written as C1, C2, C3 and C4, and by permutation of H and CH_2OH groups on C5. They may be grouped into eight pairs that differ by the relative positions of these groups on C2, C3, C4 and C5. One of them, pyranose, is represented in Figure 2.2. It is also called glucopyranose, as glucose encompasses pyranose and another not represented form, a hexose with a five-atom cycle called furanose. Two forms of glucopyranose are shown in Figure 2.2, the α and β forms that differ by the stereoconformations of their C1-atom: the C5O5C1O1 dihedral angle is 60° for the α form, also called the axial form, and 180° for the β form, also called the equatorial form.

Upon a $1 \leftrightarrow 4$ condensation, shortly written as:

$$-\mathrm{C1{-}OH} + -\mathrm{C4'{-}OH} \rightarrow -\mathrm{C1{-}O{-}C4'}- + \mathrm{H_2O} \tag{2.1}$$

elimination of a H_2O molecule occurs. It is accompanied by condensation between the C1−O1−H group of a D-glucopyranose molecule and the C4′−O4′−H group of another identical molecule that establish a C1−O4′C4′ covalent bond between the two cycles. We obtain accordingly two different types of dimers depending on whether we use pyranoses

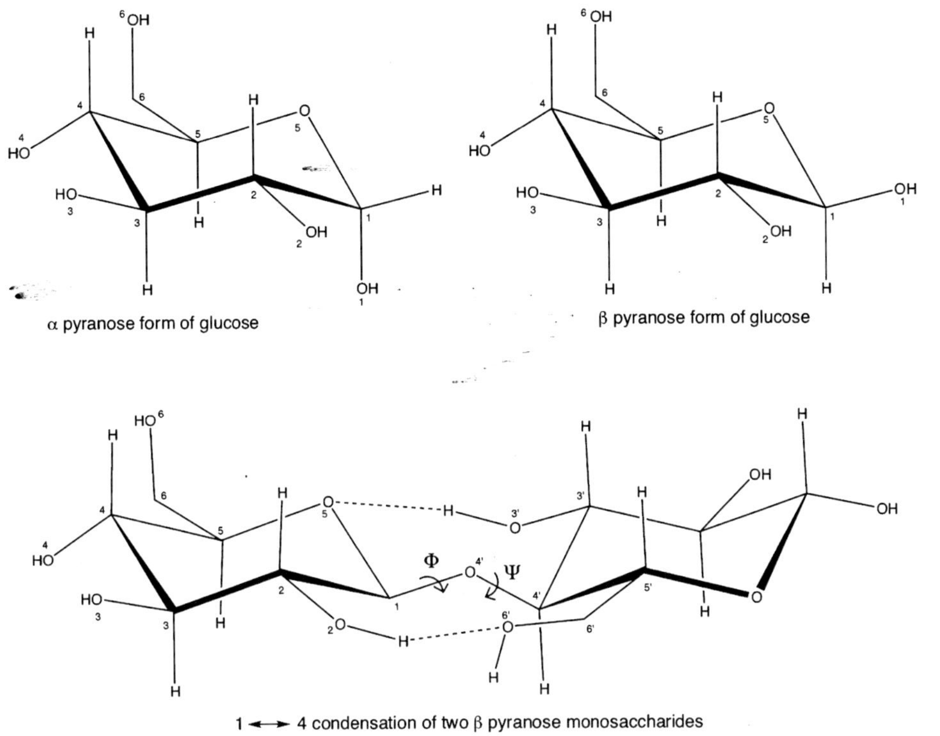

Figure 2.2 Upper diagram: α-D-glucopyranose and β-D-glucopyranose monosaccharides. Lower diagram: β-D-glucopyranose disaccharide after a 1↔4 condensation of two β-D-glucopyranose monosaccharides. Values of angles Φ and Ψ are such that the establishment of two collateral H-bonds C3′–O3′–H···O5 and C2–O2–H···O6′ are possible. This latter H-bond may be replaced by another H-bond C6′–O6′–H···O2.

of α or β forms. As can be seen at the bottom of Figure 2.2, condensation of two β-D-glucopyranoses allows the establishment of two collateral H-bonds that greatly stabilize this dimer. Such H-bonds are obtained for precisely defined values of the rotation angles Φ and Ψ around σ-bonds C1–O4′ and O4′–C4′ of this dimer. These rotation angles define the two degrees of freedom that are left by covalent bonds. In the optimum configurations that allow establishments of collateral H-bonds, the O3′···O5 and O6′···O2 distances fall around 2.7–2.8 Å and the establishment of nearly linear H-bonds that consequently suffer almost no constraint from covalent bonds is possible. Formation of such H-bonds cannot occur after condensation of two α-D-glucopyranoses because other covalent bonds do not then offer the possibilities of having such O3′···O5 or O6′···O2 distances, whatever be the values of Φ and Ψ.

Similar 1↔4 condensations of glucopyranoses are possible once these dimers are formed. We thus obtain trisaccharides and higher polymers named polysaccharides that are widespread forms of biological carbohydrates. Those obtained from the α and β forms of D-glucopyranose are most common but have, however, completely different macroscopic properties. Thus, α-D-glucopyranose is at the origin of polysaccharide macromolecules made of long chains with no possibilities of establishing intrachain H-bonds; in the same way no intramonomer H-bonds could be established in the disaccharide obtained after condensation of two α-D-glucopyranoses. A great number of alcohol sites of these macromolecules do not consequently establish H-bonds in the dried state. These are sites where H_2O molecules can easily do so and become consequently embedded within the macromolecule. Rotations around C1–O4 and O4–C4′ bonds, defined by angles Φ and Ψ in the lower diagram of Figure 2.2, are possible and are not at the origin of preferred conformations with lower energies that would appear if H-bonds could be established. Such a macromolecule consequently exhibits a great flexibility due to these two degrees of freedom that do not favour any conformation, and can be easily hydrated. It can also be easily digested as it can fix H_2O molecules and can therefore be easily hydrolysed, hydrolysis being the reverse reaction of condensation. Such a typical macromolecule is amylose, one of the two main components of starch. The energy of light captured during photosynthesis can thus be stored in the form of such polysaccharides that can be later digested. This is the way vegetal seeds store reserve of food.

In opposition, the 1 ↔ 4 polycondensation of β-D-glucopyranose is at the origin of a polysaccharide with completely different properties: cellulose. Several types of cellulose are known but they are all made of different relative arrangements of chains similar to that shown in Figure 2.3. They are all crystalline, insoluble, scarcely reactive species. All these properties are due to the well-developed H-bond network, part of which appears in Figure 2.3. The H-bonds of this network are of two types: intrachain H-bonds established by secondary alcohols of type C3′–O3′–H···O5 that are completed by H-bonds of type C2–O2–H···O6′, as shown at the bottom of Figure 2.2, and interchain H-bonds that are

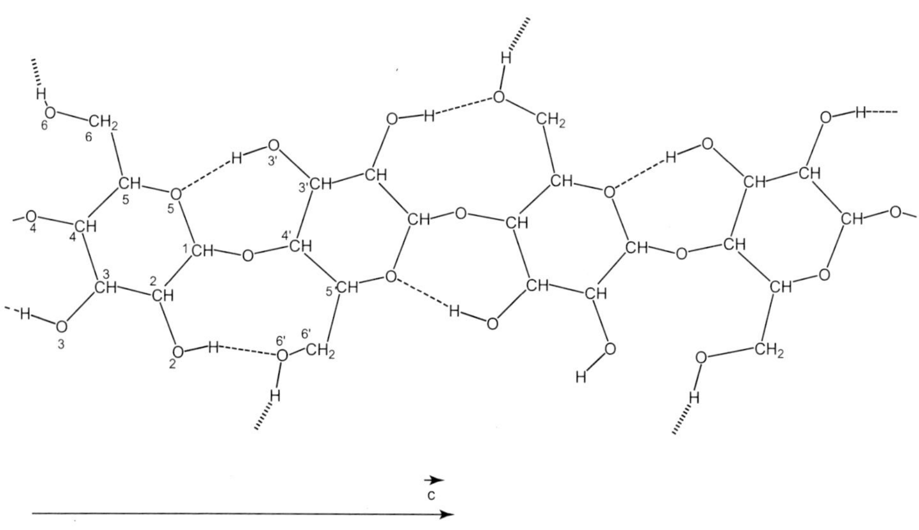

Figure 2.3 Projection of two repeat units of a single chain of cellulose I on a plane that contains the $\vec{c}$ axis that defines identical translations along the chain. Intrachain H-bonds are shown as thin dotted lines whereas interchain H-bonds are shown as thick dotted lines.

established by primary alcohols and are of type C6–O6–H···O where this latter O-atom is an H-bond receptor situated on another chain. Various receptors of these latter H-bonds are possible that allow the appearance of various crystalline forms that have all the chains shown in Figure 2.3, but where the relative positions of the various chains are different. In many native states these various chains are parallel. This type of cellulose is known as cellulose I. Note that many types of cellulose are native cellulose, as it is the most common component of many vegetal species. Thus, one type of cellulose known as *Tunicine* appears in cotton wool, another one known as *Valonia* appears in algae, etc. These parallel arrangements of chains are not the most stable ones. Consequently, chemical treatments of these native forms can produce crystalline forms, known as cellulose II, where chains are antiparallel. When treated by supercritical ammonia, cellulose III is obtained where the C2–O2–H··· hydrogen bond is perpendicular to axis $\vec{c}$ (7) and is consequently an interchain H-bond. This is different from cellulose I represented in Figure 2.3. The precise structures of the extended H-bond networks that give celluloses their macroscopic properties are still incompletely known (8).

The stability of cellulose(s) that is due to this particularly well-developed H-bond network makes it one of the most widespread biopolymers that exhibits particularly interesting mechanical and chemical properties. With this dense H-bond network, it is thus insoluble in water, it is very hard to hydrolyse and consequently not easily digested by many living organisms, among them human beings. They may nevertheless be digested by herbivores that possess developed batteries of special gastric juices to do so. This H-bond network also provides mechanical properties of a crystal. Cellulose thus forms cellular walls of many vegetal species. The dramatic difference of properties of cellulose and of amylose, obtained from a slightly different hexose, a α-D-glucopyranose instead of a β-D-glucopyranose, illustrates the effects of H-bonds in macromolecules and the different macroscopic properties that are related to the existence of supramolecular structures these H-bonds allow to appear. Before looking at H-bond networks in other types of biomacromolecules, let us note that cellulose, even if it is the most common biopolymer, is not the only one to display such exceptional mechanical and chemical properties due to its extended H-bond network. Thus, chitin and chitosan, polymers built from the same β 1↔4-D-glucopyranose where, however, the C2–O2H alcohol has been replaced by a C2–$NHCOCH_3$ (acetamido) group or by a C2–NH_2 (amino) group, exhibit an as well extended H-bond network at the origin of mechanical and chemical properties that are similar to those of cellulose. They form the carapaces of crabs, lobsters and of most crustaceans, as they are, like cellulose, relatively hard, insoluble and chemically resistant materials.

Proteins

Composition: polypeptide backbone

Proteins form another class of biomaterials within which most N-containing macromolecules of the living world are found. They form the basic components of many organs and are deeply involved in bioreactivity. Enzymes, for instance, are proteins. Proteins are also responsible for the structures and metabolisms of many tissues. Their hydrolyses, which either can be performed chemically or can occur during such metabolism processes as digestion, revealed, more than hundred years ago, that they are obtained through "polycondensation" of amino acids. They are consequently "polypeptides". Let us explain these terms looking at the diagrams in Figure 2.4 of an amino acid, an amide and a peptide. α-Amino acids are simple molecules that are made of a central C-atom, labelled C_α, that is bound to a primary amine group NH_2 and to a carboxylic group COOH. It is also bound to an H-atom and to a chemical group R that characterizes the various amino acids. Other amino acids exist where the amine and carboxylic groups are not bound to the same C-atom. We disregard these other amino acids and, for simplicity, drop the α-characterization of the amino acids in the following, calling them simply amino acids. Nearly all proteins are made of 20 various amino acids. All amino acids exhibit two enantiomeric forms, levogyre (L) and dextrogyre (D; we have already defined these terms in the case of carbohydrates), due to the stereoasymmetry of the C_α carbon that carries an H-atom and a C-atom that belongs to the R group. The exception is glycine that is optically inactive, as its two enantiomeric forms are identical, R being in this case an H-atom. In the living world only levogyre amino acids are found, with the exception, however, of optically inactive glycine, of some scarce bacterial amino acids and of some amino acids found in proteins of amphibian skin that are dextrogyre (9). This is also true of their derivatives, proteins or polypeptides. A dextrogyre amino acid is obtained from the levogyre one represented on top of Figure 2.4 by a permutation of H and R groups situated on the C_α-atom, all other atoms keeping their positions unchanged.

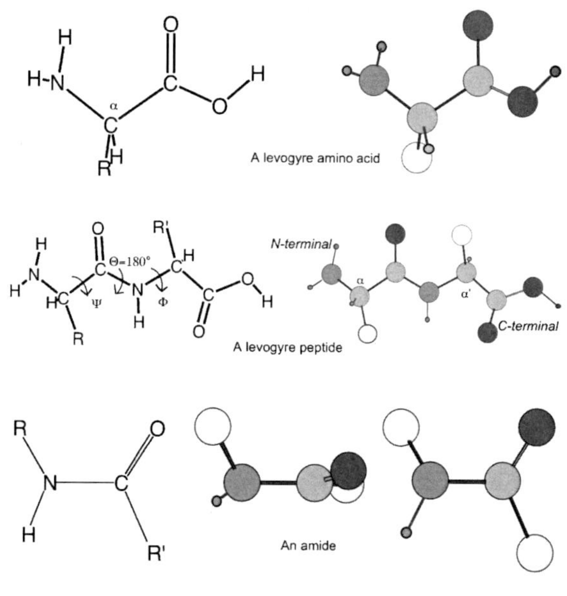

Figure 2.4 An amino acid, a peptide and an amide. All diagrams are projections of molecules on the plane of the figure. These projections are shown with conventional chemical symbols in the left part. The same projections are shown in the right part with spheres that make the stereostructures appear more clearly. For the amide in the bottom diagram, the chemical symbols correspond to the projection with spheres on the right, while the central projection (with spheres) is obtained from the other two projections by a rotation around the N–C axis of the molecule of about 90°. In all three diagrams of this amide, the N–C axis is the *X*-axis of the figure.

We have seen in the case of cellulose and carbohydrates that condensation consists of the elimination of a H_2O molecule between two groups that belong to two molecules and this elimination goes on with the simultaneous formation of a covalent bond between these two molecules. In the case of amino acids it can be written as:

$$-NH_2 + -COOH \rightleftarrows -NH-CO- + H_2O \tag{2.2}$$

with the reverse reaction being "hydrolysis". In eq. (2.2) $-NH_2$ is the amino group of the first amino acid and –COOH the carboxylic group of the second amino acid. The result is the formation of a "peptide" shown in the central part of Figure 2.4. It is a molecule that has an amine NH_2 group on one of its end, called "N-terminal" and a carboxylic group on its other end, called "C-terminal". Between them, two C_α type carbon atoms labelled C_α and $C_{\alpha'}$ carry levogyre groups R and R′ of the initial amino acids. Between these two C_α- and $C_{\alpha'}$-atoms, an "amide bond" is formed that is called a "peptide" bond in that case where we have an amine and a carboxylic acid at both ends of the dimer. An amide group that has particularly well-defined chemical properties is shown at the bottom of Figure 2.4. Its characteristic HNCO group is planar in the first approximation, which means the dihedral angle H–N–C=O falls in the vicinity of 180°, with the H- and O-atoms being in a "*trans* conformation". This *trans* conformation is more stable by about 90 kJ mol^{-1} than the corresponding *cis* conformation with dihedral angle H–N–C=O in the vicinity of 0°. At room temperature this *trans* conformation is consequently by far predominant.

This stereoconformation is also found in peptides we have seen to be amides with amine and carboxylic end groups. In the case of the peptide shown in the central part of Figure 2.4, the angle Θ is consequently fixed at a value in the vicinity of 180°. Two other angles, Φ and Ψ often called "Ramachandran angles" (2), equal to the dihedral angles C–N–$C_{\alpha'}$–C and N–C_α–C=O remain free. These two degrees of freedom are of central importance in the properties of polypeptides that are obtained by having the reaction of condensation described by eq. (2.2) proceed beyond the formation of monopeptides. The peptide shown in Figure 2.4 exhibits an amine group at its N-terminal and a carboxylic group at its C-terminal. It can consequently perform a condensation reaction with any other amino acid chosen among the 20 possible ones, leading to the formation of a dipeptide made of two amide groups between an N-terminal and a C-terminal. This dipeptide can proceed on with condensation and we finally obtain a polypeptide. Such a polypeptide is entirely defined by the succession of its amino acids that compose it, starting from the N-terminal and ending at the C-terminal. Let us note that hydrolysis of such a polypeptide leads to reformation of its constituent amino acids. At the end of the 19th century, such constituents were called "residues", as they were the results of the decomposition of polypeptides or proteins. This term "residue" is sometimes found in the literature and is used instead of "amino acids".

Primary, secondary and tertiary structures of a protein

Proteins are polypeptides where the degrees of freedom are restrained by the establishment of H-bonds. In a polypeptide the Ramachandran angles Φ and Ψ of each peptide (amide) group, shown in Figure 2.4, adopt any value. In a protein they adopt well-defined values that allow the establishment of intrachain H-bonds. These H-bonds make the corresponding conformations more stable than other ones. Such stabilized conformations constitute the "secondary structures" of the protein, the primary structures being defined as the succession of the constituent amino acids of the "polypeptide backbone" of the protein. Starting from the amino acid that contains the N-terminal and is given number 1, the successive amino acids, or residues, that are encountered along the sequence of the protein receive numbers each time increased by 1. The residue "Serine51" of a given protein, for instance, means that when starting from the N-terminal of this protein, defined in Figure 2.4, the R group found on the 51st C_α-atom is $-CH_2OH$, characteristic of serine, one of the about 20 amino acids found in proteins. The number of amino acids in a given protein varies from 129 in lysozyme, the smallest protein, to about 1000 in big proteins.

Two characteristic secondary conformations, α-helices and pleated β-sheets we describe below in more detail, were defined in 1951 by Pauling *et al.* (10, 11) after they had determined the structure of a particular protein, keratine, found in wool, silk and other natural materials. These secondary conformations have been later found in almost all other proteins. Parts of amino acids of a protein adopt these two ordered conformations. Between them, amino acids that adopt neither of these conformations form "random coils" within which the values of Φ and Ψ of amino acids (or of peptide groups) are not defined as in α-helices and β-sheets and may take a much wider range of values. The presence along the succession of amino acids of ordered α-helices and β-sheets is the criterion that allows us to distinguish a protein from a polypeptide. It is acquired during the synthesis of proteins, following mechanisms that are not yet entirely understood. A given protein has a well-defined secondary structure. Breaking the secondary structures of a living protein, a process called *denaturation*, which may be provoked by changing the neutral pH of the biomedium, heating it, dehydrating it, etc., transforms this protein into a bio-inert polypeptide. Such a polypeptide is an entirely random coil, which is equivalent to having no secondary structure.

Pleated β-sheets

They are of two kinds: parallel and antiparallel. An example of a parallel β-sheet is shown in Figure 2.5, where the amino acids of two tetrapeptides adopt values for Φ and Ψ in the vicinity of $-110°$ and $+120°$, respectively (2). These two tetrapeptides belong to a same protein but are separated along the sequence of the protein by an appreciable number of amino acids that do not belong to the β-sheet. They are held together by bent H-bonds. A third or even a fourth tetrapeptide sequence with same values for Φ and Ψ and positioned on another separated part along the sequence of amino acids of the protein may extend this sheet along the *X*-axis of the figure. This conformation is a parallel one because walking along the succession of amino acids makes us travel twice in the same direction (from top to bottom in Figure 2.5) when passing through these two tetrapeptides. Such a conformation is not limited to tetrapeptides but may be established by any set of *n*-peptides. The projection on the right side of Figure 2.5 clearly shows the pleated character of this conformation. An antiparallel β-sheet is shown in Figure 2.6. The antiparallel character stems from the 180° change of direction taken when one travels along the succession of peptides within this conformation. Thus, starting from the N-terminal of the sheet shown in Figure 2.6, one first goes downwards in that β-sheet structure, meets a "turn" that does not belong to this conformation and then meets again peptides that are part of this antiparallel β-sheet. But one then travels upwards before meeting another "turn" and then new peptides that belong

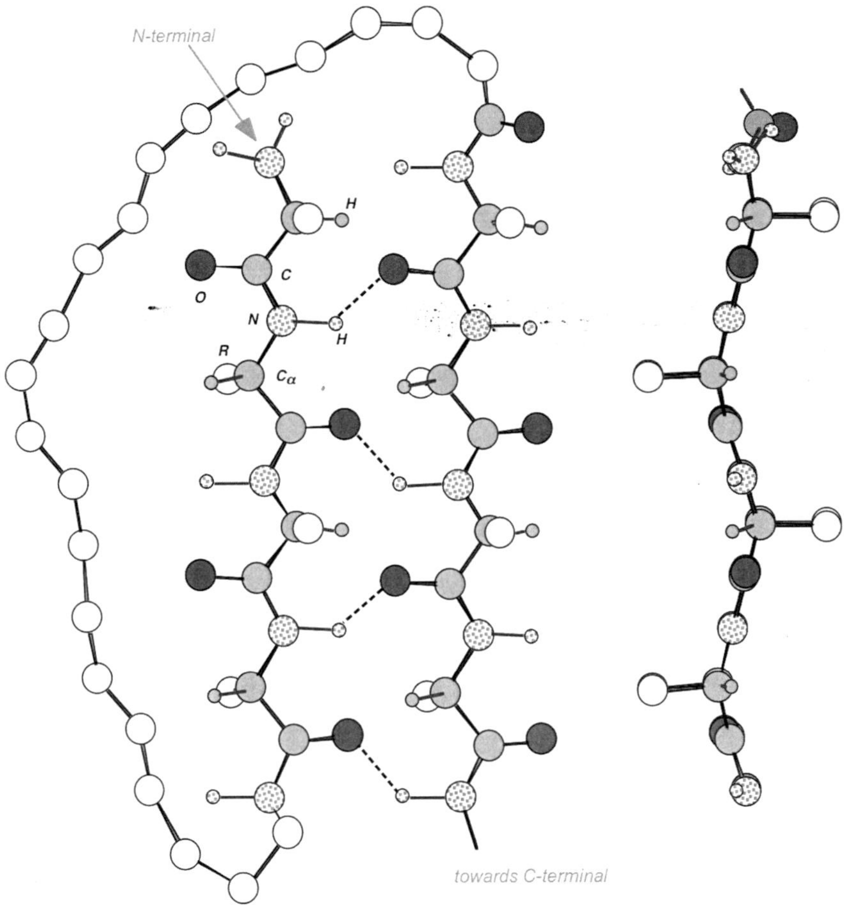

Figure 2.5 Two projections of a parallel pleated β-sheet established by two sets of four peptidic groups in a protein. The left diagram is a projection on the plane of the H-bonded N–H···O=C groups of the β-sheet, and the right diagram is a projection of the same part of the protein rotated by about 90° along the *Y*-axis of the figure. The first set of four peptides is shown starting at the N-terminal of the protein. The second set appears further along the succession of peptides of the protein. The sequence of peptides that link these two peptidic groups are shown as white circles in the left diagram and are not represented in the right diagram for clarity.

to this sheet and travelling again downwards. We may see that H-bonds are in less constrained configurations in these antiparallel β-sheets than the H-bonds of parallel β-sheets, as they can be in their preferred linear configuration with bending angles θ and φ of Figure 2.1 equal to 0°. Furthermore, the "turns" that separate two sequences of such a conformation can be much shorter than the sequences that separate the sets of peptides that establish parallel β-sheets in Figure 2.5. These antiparallel β-sheets have a tendency to be consequently more stable and more easily formed than parallel ones. They are encountered with Φ and Ψ values of the peptides in the vicinity of $-130°$ and $+145°$, respectively (2).

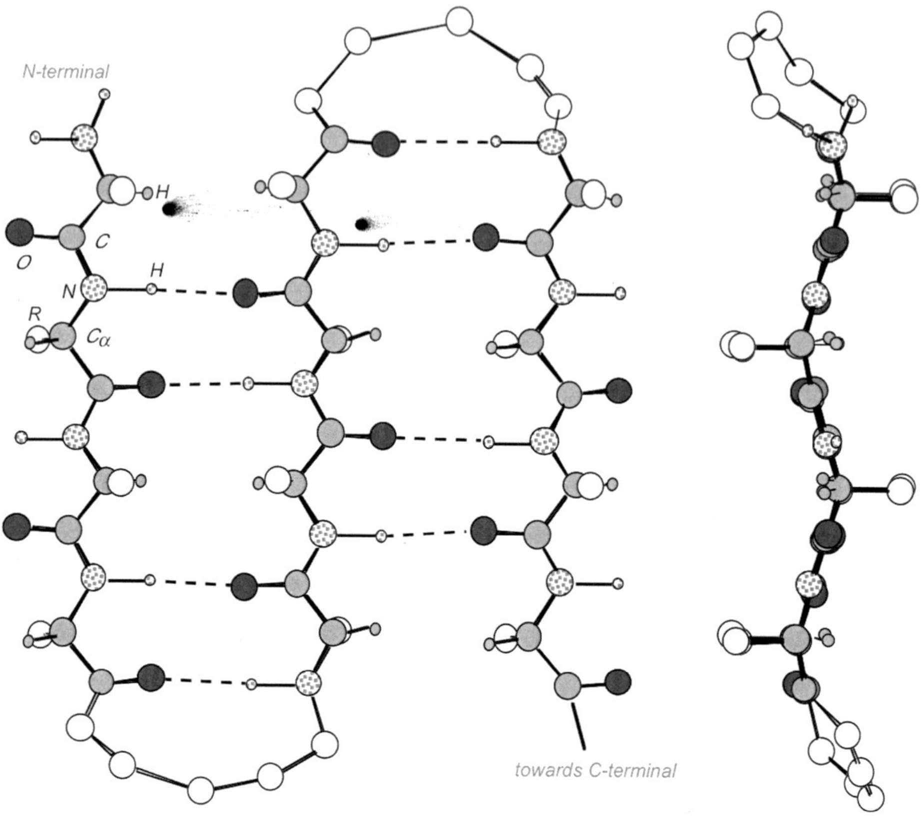

Figure 2.6 Two projections of an antiparallel pleated β-sheet established by three sets of four peptidic groups in a protein. The left diagram is a projection on the plane of the H-bonded N–H···O=C groups of the β-sheet and the right diagram is a projection of the same part of the protein rotated by about 90° along the *Y*-axis of the figure. The first set of four peptides is shown starting at the N-terminal of the protein. The two other sets appear further along the succession of peptides of the protein. The sequence of peptides that link these three peptidic groups are shown as white circles.

They are somewhat rigid but mechanically resistant conformations. They are typically found in silk proteins.

α-Helices

The systematic establishment of H-bonds between N–H and C=O parts of amino acids that are separated by a few repetitive number of amino acids along the succession of the primary structure is at the origin of helices, called α-helices, that have their axes roughly parallel to all these H-bonds. Nearly all constituent amino acids of all proteins are levogyre (L), as already seen. It favours right-handed helix structures that the great majority of α-helices adopt. Artificially substituting D-amino acids to the naturally occurring L-amino acids destabilizes these α-helices (9). These α-helices are not easy to visualize and require the expert eyes of a crystallographer to be detected. Projections of such helices with various

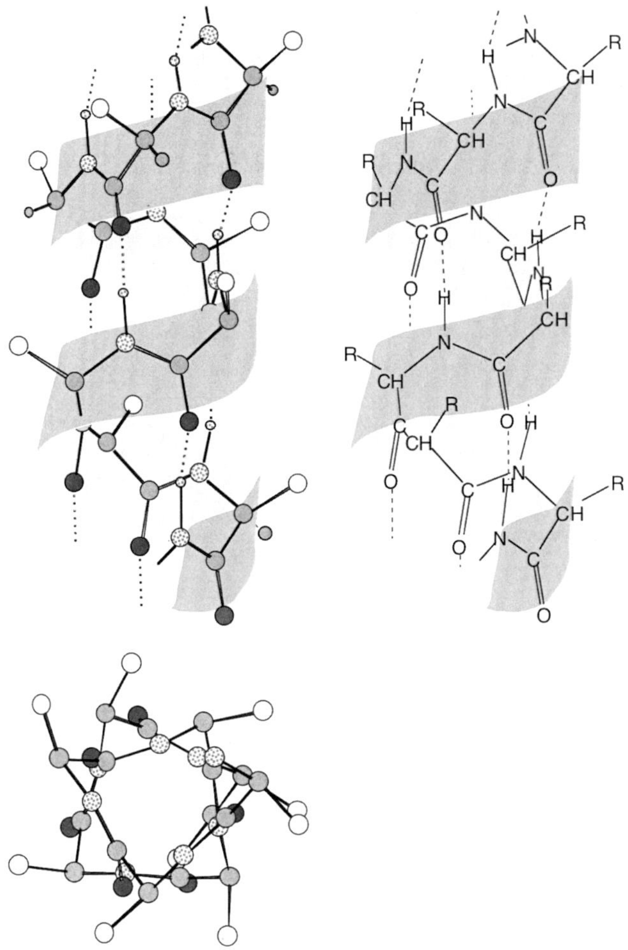

Figure 2.7 A projection of a α-helix on a plane that contains the axis of the helix in a ball and sticks representation (left) and in a conventional representation (right), with a ribbon added to mimic the helix (upper diagrams). The bottom diagram (balls and sticks) is the same helix as above but projected on a plane perpendicular to the axis of the helix.

representations are given in Figure 2.7, where it is intended to make them appear as clearly as possible. The helix is symbolized as a ribbon in the upper diagrams. The lower diagram, which is a projection along the axis of the helix, clearly shows we have to do with a helix. We may see on this figure that various geometrical possibilities exist for the establishment of H-bonds. N–H donor groups are thus all parallel. The H-bonds these N–H groups establish may be those represented in Figure 2.7, but might also be H-bonds established instead on neighbour C=O groups. In that case, the pitch of the helix and its average diameter are different from that represented. In order to characterize these various possible helices, crystallographers assign them two numbers X and n (2) written as X_n, with X being the number of amino acids in one turn of the helix and n being equal to 1 (H-bond) plus the number of covalent bonds in a cyclic structure such as N–H···O=C where the N- and C-atoms of this H-bond are otherwise linked by a chain of (not shown) covalent bonds. The α-helix shown in Figure 2.7 is accordingly a 3.6_{13} helix, a typical representative of all α-helices.

These α-helix structures are more flexible than β-sheets encountered above but mechanically less resistant. They exhibit the elasticity of a spring, while β-sheets are much stiffer. Thus, part of α-helices of the protein that form the thread extruded from sericigen glands of a spider are replaced by β-sheets in the definitive cobweb. These β-sheets give this definitive thread its solidity, while the α-helices of the forming thread give it the flexibility required to be extruded. Myoglobin, a globular protein, exhibits nearly only α-helices (4; p. 70).

Tertiary structures and beyond

Secondary structures are stabilized by H-bonds that require the angles Φ and Ψ of each peptide, which are part of this structure, to take on special values. Most of the time, secondary structures are not independent from other ones in an active protein. It means that α-helices and β-sheets should be arranged in some definite order for the protein to be active. This ordering of secondary structures is known as tertiary structure of proteins. H-bonds play no direct role in this ordering, in opposition to secondary structures where they are directly involved. They may nevertheless intervene in the building of these tertiary structures through the presence of H_2O molecules that may establish H-bonds on sites other than those that are responsible for the establishment of secondary structures. These tertiary structures define the global forms of proteins. An example is globular form, for instance, that is a roughly spherical form for the whole protein. The tertiary structure is particularly important for enzymes, which are proteins that act as biocatalysts, and have to take their optimal form so as to have their active site sterically accessible to the protein it acts upon. Finally, some proteins require the presence of another one to play its role. These two proteins should also adopt some relative order to be active. This is called the "quaternary" structure of proteins.

Stability of secondary structures and denaturation of proteins

From the conventional view of a chemist, a protein is entirely defined by its primary structure that defines the succession of peptides that compose it when starting from its N-terminal, the remaining NH_2 group of the head peptide. In other words, a protein is not different from a polypeptide with this conventional point of view. In biology, proteins and corresponding polypeptides are two different entities: a protein is bio-active whereas the corresponding polypeptide is bio-inactive. This crucial difference physically appears with the existence in the protein of secondary structures stabilized by H-bonds and sometime tertiary structures. A given protein always has a well-defined secondary structure that may sometimes display some evolution during its lifetime. For most proteins, the secondary structure consists of the appearance along the succession of peptides of various conformations we have seen above, namely α-helices and β-sheets, plus random coils where a definite secondary structure is absent. The formation of these secondary structures may depend on the stereoproperties of the various groups, R, of the peptides that adopt such a conformation. It is thus established that the presence of such amino acids as glutamic acid, leucine or alanine favours the establishment of α-helices, whereas methionine, valine or isoleucine rather favour the establishment of β-sheets. This is, however, no well-established law, just a tendency to make guesses, and presently no predictions of the formation of secondary structures can be made knowing only the primary structure.

In living organisms, the secondary structure is adopted during the synthesis of the protein following mechanisms that are not yet fully understood. The consequence is we do not know how to transform a polypeptide into a protein. We know how to do the inverse, that is "denaturating" a protein, which is transforming it into its corresponding polypeptide, destroying at the same time all its biological properties. It consists of modifying the H-bond network established by water molecules around the protein, which makes the secondary structures stable. This can be done by changing for instance temperature, a familiar effect that transforms the transparent albumin of an egg into a denatured white polypeptide. The secondary configuration of a protein represents only a few configurations, among myriads of possible configurations of the corresponding polypeptide. It is consequently a low-entropy state as compared to the set of all other possible configurations of the polypeptide. Increasing temperature will consequently favour these less-ordered states, as it enhances their weight characterized by a greater entropy. This is the mechanism of irreversible denaturation by heat. Denaturation can also occur by changing the (neutral) pH of the aqueous biomedium, or adding denaturating agents such as urea, NH_2–CO–NH_2. Some chemical agents such as trimethylamine-N-oxide (TMAO) have a tendency to restore this H-bond network and consequently may, in some conditions, induce renaturation of the protein (12). Destroying these secondary structures is consequently much easier to do than destroying the primary structure, which requires breaking covalent bonds. It illustrates how bioreactions are delicate as compared to reactions in ordinary chemistry. Polypeptides that are the result of denaturation of proteins are most often easier to hydrolyse than native proteins. Denaturation is consequently often the first step of digestion that results in breaking a protein into its constituent amino acids by hydrolysis. These secondary structures therefore provide the protein some stability in its aqueous medium.

Denaturation of a protein may also be induced by a genetic mutation, where by accident an amino acid at a given position in the primary structure of a protein is replaced by another one. If this other one hinders the protein from taking its programmed secondary structure, this mutation is likely to be lethal. If this other amino acid is not very different, and does not hinder the protein to take on its secondary structure, it may be viable. It may be one of the mechanisms at the origin of evolution of living species. It may also be at the origin of diseases, such as Parkinson's, Alzheimer's or Creutzfeldt–Jacob diseases, that are suspected to be induced by a mutation that favours several β-sheets of some cerebral proteins to coalesce into a single much stiffer and more extended β-sheet, called an "amyloid fibrillation". It hinders these proteins from correctly transmitting neural signals.

DNA

Deoxyribonucleic acid, or DNA, is the macromolecule inside which all genetic information of a living body is encoded. It preserves and transmits all genetic characters of all known living organisms. DNA is thus the universal system that keeps the memory of life. By genetic engineering it may be exchanged between various living organisms without any consequent damage. The particular H-bond network it possesses combines stability and flexibility in such a way that it allows full transmission of this genetic character with an exceptionally low rate of errors. At the same time, DNA displays a sufficient reactivity that it allows transcription of parts of this genetic code into another kind of nucleic acid, ribonucleic acid or RNA, that is able to carry this genetic information to definite places in the cell, called ribosomes, where synthesis of proteins proceeds. It also allows for duplication of the whole DNA molecules in new forming cells. The "double-helix" structure of DNA, described later, was discovered by J. D. Watson and F. H. C. Crick (13) in 1953. This discovery, together with that of the structure of proteins by L. Pauling some years earlier, has been the starting point of the explosive development of molecular biology. Let us examine this structure of DNA in more detail. As established by German and American biochemists at the beginning of the 20th century, the basic bricks of DNA are "nucleotides", in some way the corresponding equivalent of amino acids for proteins. These bricks bind together to form very long chains. Nucleotides are an association of three chemical groups: a phosphate group, a pentose and a base, as shown in Figure 2.8. The pentose, deoxyribose, is a cyclic saccharide of five atoms. The backbone of DNA relies on the phosphoribose group,

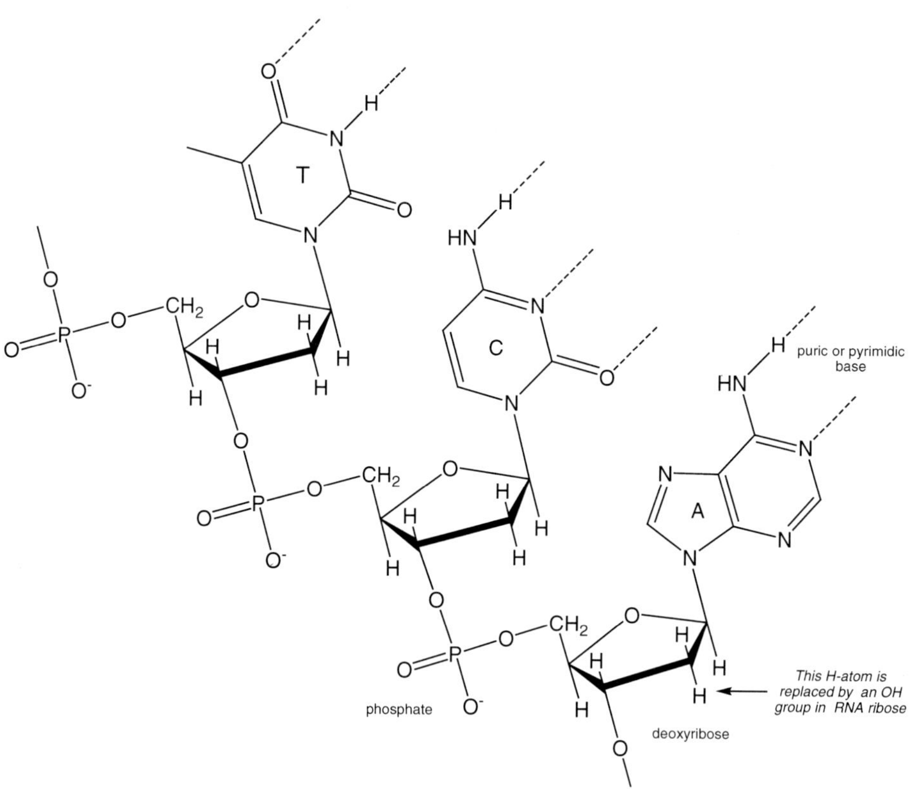

Figure 2.8 Part (three nucleotides) of a single strand of DNA.

composed of phosphate and pentose, which constitutes the repeat unit of the chain. Each base is hanging on the deoxyribose part of this phosphoribose repeat unit and constitutes a side group of the nucleotide. There are four different bases, two "puric" bases, adenine and guanine, and two "pyrimidic" bases, thymine and cytosine that form H-bonds from there. These H-bonds are of central importance.

The four bases adenine, guanine, thymine and cytosine, represented by letters A, G, T and C, respectively, have different sites to establish H-bonds. They, however, establish particularly stable H-bond configurations with "complementary" bases. Thus, adenine establishes two H-bonds with its complementary base thymine and vice versa, while guanine establishes three H-bonds with cytosine and vice versa. These H-bond configurations of bases are the only ones in DNA. They are shown in Figure 2.9. They allow the formation of double strands of DNA that adopt a helicoidal conformation, as schematized in Figure 2.10. The two strands are complementary because the succession of bases along one strand exactly matches that on the other strand by replacing any base by its complementary one: A by T, G by C and vice versa for the two pairs. The structure of DNA is therefore that of a double

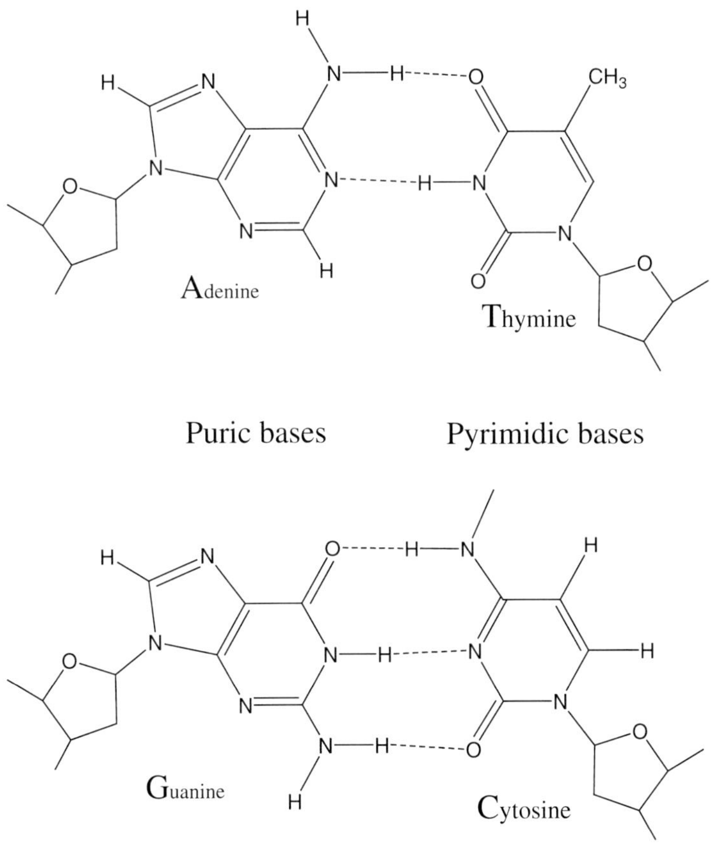

Figure 2.9 The two H-bond configurations between A–T and C–G base pairs in DNA.

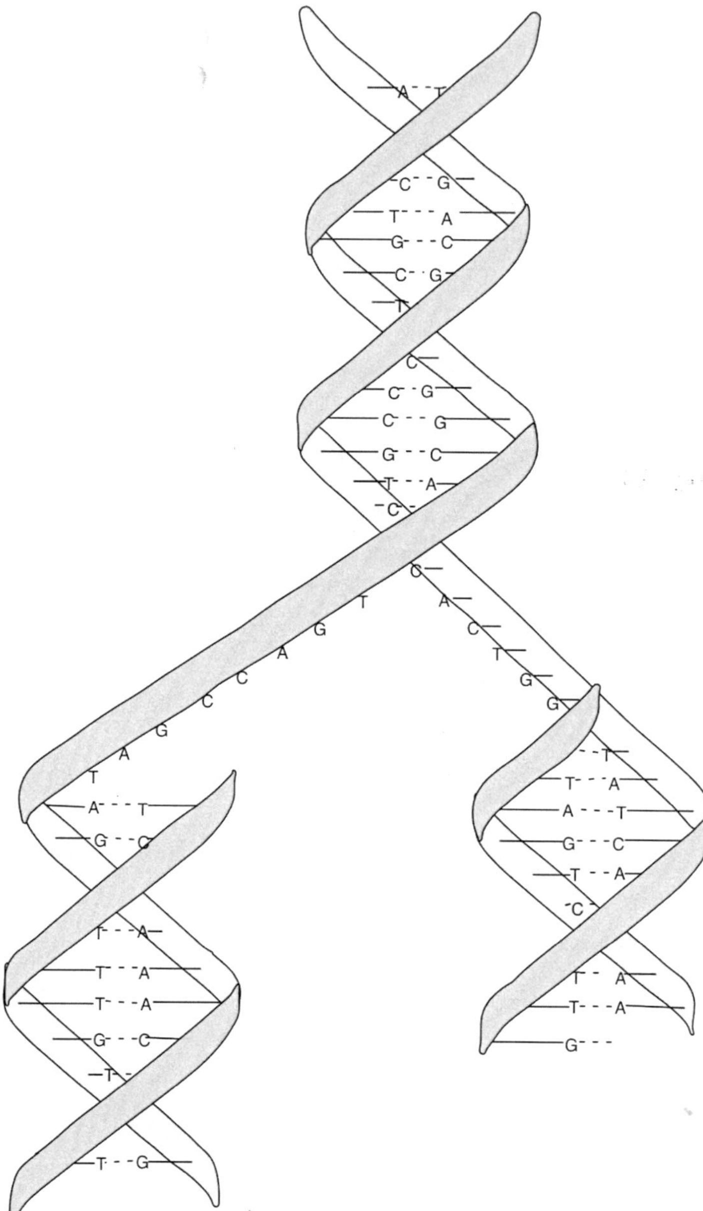

Figure 2.10 The helicoidal double strand of DNA. In the lower part of the figure replication into two daughter double strands proceeds.

strand with the two strands being linked by the H-bonds represented in Figure 2.9 between the two sets of complementary bases. This was in 1953 a discovery that reveals an extraordinary intuition from Watson and Cricks, especially when one considers the techniques that were available at that period of time, in particular the absence of informatics tools.

This conformation of DNA in a helical double strand appears as the optimum conformation to have both an exceptional stability and at the same time enough flexibility to exhibit some reactivity, in particular for reproducing itself with nearly no error. The stability of DNA is remarkable. In particular, its interstrand H-bond network represented in Figure 2.10 allows it to be stable in an aqueous biomedium, as well as in a waterless, or nearly so, medium. Thus, DNAs of organisms that have lived several thousands or tens of thousands years ago are still enough preserved that their analysis can be performed! In this sense, their stability is much greater than that of proteins. It owes much to its helicoidal double strand conformation. A protein is denaturated by a lack of water molecules, which destabilizes its secondary structure. DNA is kept intact because the presence or absence of H_2O molecules does not modify its interstrand H-bond configuration shown in Figures 2.9 and 2.10. Denaturation of DNA consists of separating the two strands. It requires a much stronger chemical action than that required for proteins, such as putting DNA into a strong acid or base that makes the pH of the surrounding medium completely different from that of a neutral one and consequently modifies the H-bond network of DNA. Denaturation is also obtained in nearly boiling water, an extreme condition that also offers a completely different H-bond network of the surrounding medium. The H-bond network of DNA furthermore allows it to be duplicated during transmission of the genetic code or to be partly copied in view of synthesizing programmed proteins. These reactions, duplication and copy, occur with a minimum of energy in a biomedium essentially made of liquid water. It requires disrupting a limited number of interstrand H-bonds, following the way depicted in Figure 2.10. The cost of such a disruption is somewhat low because during this process disrupted H-bonds can be replaced by H-bonds with water molecules, liquid water playing in this case the role of an infinite reservoir of H-bonds. Copying a part of DNA is somewhat different. As in duplication it requires disrupting all H-bonds between the two strands along a limited portion of DNA. Only one of the two strands is copied in the region of disrupted H-bonds. It gives a single strand, not of DNA, but of a most similar nucleic acid: RNA, or ribonucleic acid. This RNA has the same backbone as DNA with, however, two modifications. The first one is the substitution of deoxyribose by ribose. Ribose is the same pentose as deoxyribose shown in Figure 2.8 where one H-atom of the CH_2 group of the five-atom cycle is replaced by an OH group. This substitution has important biological effects: the lifetime of an RNA strand is limited because of the possibility of hydrolysis this OH groups provides, whereas that of DNA is very long. The second modification is the replacement of thymine (T) by uracil (U), another pyrimidic base. The role of RNA is to transfer the information copied on DNA to ribosomes, the sites where proteins are synthesized in the cell. This synthesis requires several steps that are now rather fairly well understood, at least in their gross features. Once this synthesis has been carried out, RNA is destroyed by hydrolysis, but the memory of the way this synthesis is to be performed is kept intact in DNA.

The four bases A, T (replaced by U in RNA), C and G are the four letters of an alphabet that allows encoding the genetic code of all living species along sequences of DNA that are also read sequentially. The encoding unit, or codon, is based on a set of three letters that materialize by three nucleotides. Each set of such three letters encodes for a particular amino acid that belongs to the set of about 20 amino acids that are at the origin of all proteins. There is some redundancy in this encoding, as a particular amino acid can be defined by several sets of three bases. The significance of this redundancy is not known. This is not surprising: even though we know the principle of the genetic code, as described above and which is in itself the result of a huge interest that has made biology progress significantly during these last tens of years, we still completely ignore many points. The genetic code of some living organisms, among them humans, has thus been recently read. It means that the whole set of genes of these organisms has been transcripted into sequences of three letters. We still ignore, however, the meaning of the great majority of these sequences. It has furthermore been established that only a part of DNA encodes the genetics of a species. The role of the nonencoding part is still obscure. In other words, we are still far from having understood the universal code of life.

CONCLUSION

H-bonds are directional. They share this property with covalent bonds that are at the origin of molecules with well-defined geometries, that is well-defined arrangements of the atoms that form the molecule. In a similar way, the directionality of H-bonds is at the origin of supramolecular arrangements that can also adopt well-defined geometries. H-bonds, however, differ from covalent bonds by their enthalpies of formation, which we have seen in the preceding chapter to be one order of magnitude weaker than those of covalent bonds. The consequence is that the energies required to form these supramolecular arrangements are this same order of magnitude weaker than those required to form molecules. It falls then in the vicinity of about $10\,kT$ at room temperature. As a consequence, these supramolecular arrangements held by H-bonds can be easily modified, as the energy of such modifications can be easily provided by thermal fluctuations. No surprise then that H-bonds are central in molecular biology: they are at the origin of stable but nevertheless flexible, adaptable and evolutive supramolecular structures. In this chapter we examined three of such supramolecular structures: cellulose, proteins and DNA, which represent three classes of macromolecules that are central in biology. These are not the only ones found in nature, as thousands of other structures could have been mentioned. We have thus not mentioned another class of biomaterials: lipids. They are not macromolecules. They are for the most part "amphiphile" molecules, that is molecules that have one or several sites to establish H-bonds on one of their sides and no sites to establish H-bonds on their other side. They do not develop a well-defined H-bond network like the structures we have examined previously in this chapter, but they nevertheless adopt typical conformations described in Ch. 9 when they are immersed in the huge H-bond network of liquid water.

The three kinds of structures described in this chapter are representative of the central structuring role of H-bonds, which takes on a special importance in molecular biology. The helicoidal double strand of DNA thus displays such a stability while at the same time allowing a sufficient reactivity that it appears certainly the best compromise to build a material capable of keeping the genetic memory of all living species and transmitting it to all newly born organisms. Its universality, together with that of the amino acids that compose proteins, is one of the strongest argument that firmly establishes the foundation of the theory of evolution proposed in the 19th century by C. Darwin, at an epoch when DNA was unknown. It strongly suggests that all living organisms have a unique and common ancestor cell. This compromise is so good that it has ruled life for more than 3.6 billion years. The central biomaterials, proteins, also owe very much to H-bonds that allow them to adopt, within the degrees of freedom that are left by the succession of peptides (or amino acids) along their chains and that are ruled by laws of covalence, particular conformations, called their secondary structure. The conformations adopted in these secondary structures allow them to be biologically active and distinguish them from biologically inert polypeptides. Two main conformations compose these secondary structures: α-helices and β-sheets, in addition to a nonorganized structure called random coils. Along its sequence of amino acids, a protein exhibits various parts that adopt these conformations. Finally, a rather simple H-bond network allows a particular carbohydrate, cellulose, to be a widely spread polymer that gives many materials exceptional mechanical properties. Carbohydrates are molecules that are formed during photosynthesis, from atmospheric CO_2 and H_2O molecules. They are found in many bioenergetical processes.

The two properties of H-bonds, directionality and enthalpies of formation of about $10\,kT$ at room temperatures for weak H-bonds, are thus fundamental and explain the central role they play in molecular biology. We shall, however see, in Ch. 6 and the following ones, that they have a third property that makes them even more crucial: they allow transfer of protons or of H-atoms between the two molecules they link. This third property of H-bonds is at the origin of the reactivity of aqueous media, particularly biomedia.

REFERENCES

1. I. Olovsson and P.-G. Jönsson, in P. Schuster and G. Zundel (Eds.), The Hydrogen Bond—Recent Developments in Theory and Experiments, Vol. II, Ch. 8, North-Holland, Amsterdam, 1976.
2. G. A. Jeffrey and W. Saenger, Hydrogen Bonding in Biological Structures, Springer Verlag, Berlin, 1994.
3. P. Hobza, Phys. Chem. Chem. Phys., 3 (2001) 2555.
4. D. E. Metzler, Biochemistry. The Chemical Reactions of Living Cells, Academic Press, New York, 1977.
5. R. H. Garret and C. Grisham, Biochemistry, 2nd ed., Saunders College Publishing, 1999.
6. S. David, Chimie Moléculaire et Supramoléculaire des Sucres, Inter Editions et CNRS Editions, Paris, 1995.
7. M. Wada, L. Heux, A. Isogai, Y. Nishiyama, H. Chanzy and J. Sugiyama, Macromolecules, 34 (2001) 1237.
8. Y. Maréchal and H. Chanzy, J. Mol. Struct., 523 (2000) 183.
9. R. Sudha and M. F. Jarrold, J. Phys. Chem. B, 109 (2005) 11777.
10. L. Pauling, R. B. Corey and H. R. Branson, Proc. Natl. Acad. Sci., 37 (1951) 205.
11. L. Pauling and R. B. Corey, Proc. Natl. Acad. Sci., 37 (1951) 729.
12. A. Di Michele, M. Freda, G. Onori and A. Santucci, J. Phys. Chem. A, 108 (2004) 6145.
13. J. D. Watson and F. H. C. Crick, Nature, 171 (1953) 737.
14. M. Biczysko and Z. Latajka, Chem. Phys. Lett., 313 (1999) 366.
15. N. W. Howard and A. C. Legon, J. Chem. Phys., 86 (1987) 6722.
16. N. W. Howard and A. C. Legon, J. Chem. Phys., 88 (1988) 4694.
17. J. C. Lassègues and J. Lascombe, in J. R. Durig (Ed.), Vibrational Spectra and Structure, Vol. 11, Ch. 2, Elsevier, Amsterdam, 1982.
18. H. Ratajczak, W. J. Orville-Thomas and C. N. R. Rao, Chem. Phys., 17 (1976) 197.

– 3 –

Methods to Observe and Describe H-Bonds

In this chapter various methods to observe or describe H-bonds are examined. Among them are experimental methods that allow the measuring of various quantities that are related to these H-bonds. These are thermodynamic quantities, such as their enthalpies of formation we have already discussed in Ch. 1 or their free energies. They may also be geometrical quantities, already seen in Ch. 2, or spectroscopic quantities. Spectroscopic methods may be encountered in the present chapter, but the specific signatures of H-bonds, as they appear in IR spectroscopy, which is the most sensitive and precise method to observe H-bonds, are the subject of two special chapters, Chs. 4 and 5. In this chapter, we examine various other experimental and theoretical methods, indicating the quantities they allow to measure, together with their advantages and limitations. We concentrate on H-bonds in general. The case of H-bonds established by H_2O molecules, which is particularly important in all biomedia, may occasionally be mentioned in this chapter, but the specific methods to detect and observe these water molecules, which include the determination of the important H-bond network they develop around them, is also the subject of a special chapter, Ch. 11. In the summary at the end of this chapter, a reverse attitude is adopted that consists of indicating which method should be used when one has to measure a definite quantity related to H-bonds.

CALORIMETRY

This is a traditional method. The concept of H-bonds emerged, at the beginning of the 20th century (Ch. 1), from the analysis of calorimetric measurements. Calorimetry aims at measuring temperatures where such phase transitions as melting or boiling occur and also aims at measuring corresponding latent heats, that is differences of enthalpies that appear during these phase transitions. As argued by Pauling (1), considerable variations of latent heats appear between liquids formed from molecules that are mainly linked by H-bonds and liquids where H-bonds are in a smaller number, and the interactions between molecules are mainly of Van der Waals type. In the case of H-bonded liquids, evaporation requires breaking H-bonds, a process that implies more energy than evaporating molecules bound by Van der Waals interactions. Figure 3.1 illustrates this point. It represents the evolutions of

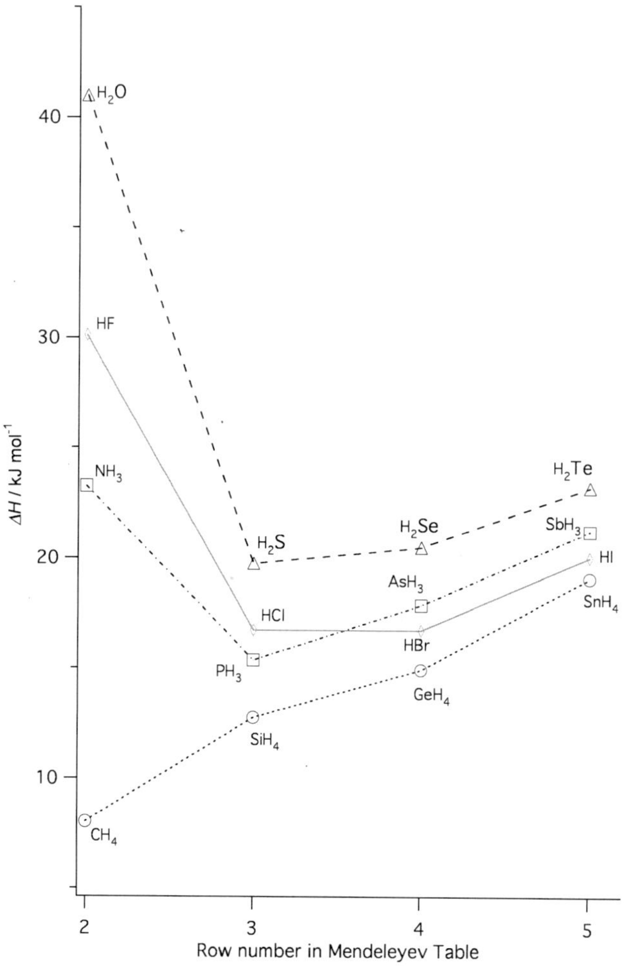

Figure 3.1 Latent heats of evaporation ΔH for various hydrides H_nA_m in their liquid state, with A being a single atom. Lines join hydrides with their A's that belong to same columns in the Mendeleyev periodic table.

the latent heats ΔH of simple hydrides H_nA_m of the atomic elements A up to 53 in the Mendeleyev periodic table. The elements C, Si, Ge and Sn of the column of carbon in the Mendeleyev table display a monotonic variation. The molecules in these liquids, CH_4, SiH_4, etc., are apolar and unable to establish H-bonds. They consequently form the series with the lowest values of ΔH. The hydrides with elements of the column of nitrogen, N, P, As, Sb, show a roughly similar behaviour, with the exception, however, of NH_3, a molecule that is able to accept and donate H-bonds. It consequently shows a marked departure from the monotonous behaviour displayed by the other elements of this column. It indicates that PH_3, AsH_3 and SbH_3 are molecules that hardly establish H-bonds. The same trend appears for the elements F, Cl, Br, I of the halogen column: liquids made of HI and HBr molecules have difficulties establishing a great number of H-bonds and consequently display relatively small values for ΔH. HCl has less difficulties doing so and starts departing from this behaviour illustrated by hydrides of elements of the C column. HF, which establishes strong H-bonds, displays an anomaly for halogens. The same trend is found for elements of the O column. The exception, liquid water, is the most marked of all. We shall see in Ch. 9 that liquid water is characterized by weak H-bonds that are, however, so numerous that their numbers equal that of covalent bonds, an exceptional and unique situation. It is no surprise then that H_2O is the molecule that displays the most exceptional behaviour in Figure 3.1. It also shows that H_2S is a poor molecule to establish H-bonds with similar H_2S molecules, especially if compared to its analogue molecule H_2O. Boiling and melting temperatures (1; 2, p. 214) of these same hydrides exhibit mostly similar behaviours. It clearly shows the influence of H-bonds on these quantities. The boiling temperature of liquid water, for instance, is higher by more than 150° over that which it would have if it were a normal liquid with no H-bonds such as H_2S, H_2Se or H_2Te.

From measurements of latent heat of vapourization of these liquids, enthalpies ΔH of formation of H-bonds can be deduced. Measurements similar to those described in Ch. 8 give the enthalpy of H-bonds in ice. These calorimetric measurements work particularly well in those cases where we have to do with homogeneous mixtures of small molecules, as those of Figure 3.1. In the case of more complex molecules, Van der Waals interactions between molecules may not be negligible, as is in the case of these simple molecules, and deducing values of ΔH from calorimetric measurements often requires supplementary measurements. When they can be used, calorimetric measurements are relatively precise, especially with modern techniques, such as differential scanning calorimetry (DSC) or thermogravimetry that measures both mass and heat exchange. Calorimetric measurements are therefore often used in conjunction with other methods, providing complementary data.

MODERN EXPERIMENTAL METHODS

Most modern experimental methods to study molecules or molecular complexes are based on interactions of these molecules with a wave that acts as a probe for this molecule or molecular complex. Most of the time this wave is an electromagnetic wave that consists of coupled oscillating electric and magnetic fields that propagate along some direction. Such a wave is defined by its frequency or its wavenumber, a proportional quantity as defined in the appendix of Ch. 1. The studied molecule has various ways to interact with it. It may, for instance, interact with its electric field or with its magnetic field. This wave may also not be that of an electromagnetic field, but may be, for instance, one associated with particles such as neutrons or electrons. It means that a lot of different methods enter this kind of experiments based on the interaction of the molecular H-bonded complex with a wave. We classify these methods into two categories: absorption methods, which imply a single photon, or occasionally a single particle, and are consequently a first-order process, and scattering methods that imply two photons or particles and are second-order processes. In each category several subcategories may be found.

Absorption of an electromagnetic wave

Optical spectroscopy is typical of the kind of methods in this category. This is a "one-photon method" where a photon may be either absorbed or emitted. Absorption of a single photon of an electromagnetic wave provokes excitation of the probed molecule or molecular complex. Emission of a single photon is a consequence of the de-excitation of a molecule or molecular complex initially in an excited state. These two processes, absorption or emission, are thus the same processes with, however, magnitudes of opposite sign. In the following we only consider absorption, the predominant method, but all that is said can be applied to emission. Absorption is a resonant process: the energy difference ΔE of the molecule after and before absorption of the photon is equal to the energy of this photon, which is itself equal to $h\nu$, with ν being the frequency of the electromagnetic wave and h the Planck's constant, equal to 6.626×10^{-34} J s. Absorption consequently strongly depends on the frequency of the electromagnetic wave. When this frequency is that of a UV or visible wave, electrons are excited, because the energies of the photons of such waves correspond to differences of energies of electronic states of molecules. In the case of H-bonds this kind of spectroscopy is scarcely used, because it allows detection of a difference of electronic energy levels between H-bonded and non-H-bonded molecules but does not otherwise convey much information on H-bonds themselves. By contrast, spectroscopy in the IR and far-IR (FIR) regions conveys a lot of information on H-bonds themselves, as they correspond to vibrational transitions that are strongly affected by the establishment of H-bonds, as described in detail in Chs. 4 and 5. In this chapter, we mainly consider a particular type of oscillations that has much to do with vibrations: rotations, and we briefly mention IR spectroscopy only as a tool to determine rotational spectroscopic signatures of H-bonded complexes. We concentrate on two kinds of absorption spectroscopy that operate at frequencies different from optical spectroscopy: microwave spectroscopy that involves electromagnetic waves with frequencies ν in the range 3–300 GHz, that is wavenumbers ν in the range 0.1–10 cm^{-1} (wavenumbers are defined in the appendix of Ch. 1), and nuclear magnetic resonance (NMR) spectroscopy that involves radiowaves with frequencies in the range 100–1000 MHz (= 1 GHz). We also briefly mention X-ray absorption spectroscopy, which operates at the other extreme end of electromagnetic radiation, with frequencies in the range of 10^{20} Hz.

Microwave spectroscopy

The energy levels that absorb photons with wavenumbers typically of 1 cm^{-1} have energies typically of 10^{-4} eV, as seen in the appendix of Ch. 1. These energies are those of rotations of isolated molecules or of isolated H-bonded complexes *in gases*. The techniques used to emit and detect radiation in such a range are microwave techniques, and are completely different from the optical techniques used in the FIR neighbouring region. Thus, even if they share a common spectroscopic region and look at identical physical mechanisms, microwave and FIR spectroscopies are considered as two different methods. Microwave spectroscopy of H-bonds consists of inducing transitions between rotational levels of isolated H-bonded molecular complexes found in the gas phase. Such transitions occur when these H-bonded complexes have permanent electric dipole moments. This condition excludes the study of such model systems as cyclic dimers of carboxylic acids

$(RCOOH)_2$ that we have already mentioned in Ch. 1 and shall encounter later in this book. They are shown in Figure 4.4, where it clearly appears that the presence of an inversion centre prohibits the existence of such a permanent dipole moment. Most H-bonded complexes display no such inversion centres and consequently exhibit permanent electric dipole moments. Rotations of such complexes induce variations of the dipole moment in the laboratory setting within which the electromagnetic wave propagates. These variations couple to the electric field of the wave, inducing absorption of photons one by one when the energies of these photons are equal to energy differences of rotational levels. These rotational levels depend on two quantum numbers: j that defines the magnitude of the total kinetic moment and m that defines the magnitude of one component of the rotational moment along a chosen axis Z. j and m are integer numbers, j being positive or zero and m taking on integer values between $-j$ and $+j$. Transitions induced by absorption of a photon can only occur between levels that have j and m each differing by -1, 0 or 1. These rather simple "selection rules" are of a great help for the interpretation of spectra, particularly the assignments of m and j of the initial and final state of all experimental bands.

The Hamiltonian H_R that governs rotations of isolated H-bonded molecules in the gas phase is particularly simple when, in a first approximation, this H-bonded molecule is considered as rigid, as no potential energy is present. It therefore consists of only a kinetic energy term that is written as

$$H_R = \frac{J_X^2}{2I_X} + \frac{J_Y^2}{2I_Y} + \frac{J_Z^2}{2I_Z} \tag{3.1}$$

where J_X, J_Y and J_Z are the components along axes X, Y and Z of the molecular complex of the total angular momentum $\vec{J}$ and I_X, I_Y and I_Z are the moments of inertia of this molecular complex with respect to these three axes. The moment of inertia I_X, for instance, is equal to the sum over all particles (electrons and nuclei) that compose this molecular complex of mass m_i multiplied by the square of its distance $r_{i,X}$ to the X axis, the three axes X, Y and Z having as a common origin the centre of gravity of the molecular complex:

$$I_X = \sum_i m_i r_{i,X}^2 \tag{3.2}$$

We thus see that quantities that are measured by such a technique are magnitudes of moments of inertia. As the number of rotational bands that enter such microwave spectra is large, determination of the geometry of the H-bonded molecule can be performed with great accuracy. Microwave spectroscopy is consequently a very precise tool to determine the geometry of simple H-bonded dimers, sometimes of trimers, etc. The first prerequisite, however, is that these dimers exist in the gas phase, because molecular free rotations that appear in gases and are characterized by the absence of potential energy terms in eq. (3.1) are replaced in liquids and solids by hindered rotations that are characterized by the appearance of additive potential terms in eq. (3.1). These hindered rotations appear therefore at higher frequencies in the FIR region, sometimes in the conventional IR region as in the case of liquid water. They are vibrations of rotational type. The second prerequisite we have already mentioned: the H-bonded molecular complex carries a permanent electric dipole moment.

A typical microwave spectrum thus consists of numerous narrow bands. Separating them requires having a great resolution. With modern techniques, the resolution at which frequencies of bands are determined may indeed be great, of the order of 1 Hz in the GHz region. However, this great resolution is useless if too many overlapping bands are present, as assigning them then becomes most difficult. This is the case at room temperature, because the separation of rotational levels is of the order of some cm^{-1}. Their energy separation is consequently a small quantity with respect to kT at 300 K (see the appendix of Ch. 1, where the ratio $hc\tilde{\nu}/kT$ of the energy of a wave with wavenumber $\tilde{\nu} = 1\,cm^{-1}$ to thermal energy kT is equal to 1 for $T = 1.44$ K and consequently equal to about 5×10^{-3} at room temperature), which implies that a great number of rotational levels are populated at this temperature. In consequence, many transitions can be induced from these populated levels. Having a small number of populated rotational levels only, and consequently a smaller number of transitions in the spectra, requires lowering the temperature down to some K. This cannot be simply done, because all gases at these temperatures condense into solids with vapour pressures nearly equal to 0. The solution consists then of observing gases at such temperatures when they are out of equilibrium. One of the most used techniques to do this consists of provoking an adiabatic cooling of this gas diluted in an inert gas such as Ar. This adiabatic cooling is obtained by expanding the gas, through a nozzle, from a cell at a temperature where the vapour tension is still appreciable (3), into an evacuated chamber that forms the microwave cavity. Before this gas condenses when hitting the walls of the chamber, it acquires a supersonic speed and no collision or exchange of energy between the various molecular complexes that compose it occurs. This is equivalent to having an effective temperature of this expanding gas of some K (4) only. This expansion is not continuous but pulsed, as is the microwave beam. As the Fourier transform (see eq. (5.A17) of the appendix in Ch. 5 for the definition of a "Fourier transform") of a pulsed wave that lasts a very short time is an everlasting wave made of a superposition of a very great number of waves of almost all frequencies from 0 to ∞, this way of operating with short pulses instead of monochromatic waves is called "Fourier transform microwave spectroscopy".

Systematic studies of simple H-bonded dimers such as F–H···N≡C–H have been initiated by Legon and Millen (3, 5) and have been later extended to many other simple H-bonded dimers. These authors could precisely measure the moments of inertia of these complexes in their equilibrium positions. They were then able to determine their geometries. These measurements also conveyed a wealth of information on the couplings of vibrations with these rotations, which implies going beyond the approximation of a rigid H-bonded complex. Information on charge redistribution during dimer formation could also be collected. Thus precise structures of such H-bonded dimers as Cl–H···OH_2 (6), Cl–H···NH_3 and various analogous complexes where the H-, N- and Cl-atoms are replaced by various isotopes (7) could be precisely determined. H-bonds of the type X–H···$N(CH_3)_3$ and various isotopic analogues, where X is a halogen atom Cl (8), Br (9), or I (10), have also been precisely studied. These results constitute invaluable basic data on the structural properties of H-bonds that we have examined in Ch. 2. Information on transfers of protons, a fundamental mechanism in H-bonds that we examine in Ch. 6, could also be obtained from the precise analysis of microwave spectra of these species. It could thus be shown that no proton transfer occurs in Cl–H···OH_2, Cl–H···NH_3, but some ionic character starts appearing in Cl–H···$N(CH_3)_3$ that contains some amount of Cl^-···$N(CH_3)_3^+$.

It clearly indicates that trimethylammonium $N(CH_3)_3$ is a stronger base than ammonia NH_3. This ionic character becomes more pronounced in Br–H···$N(CH_3)_3$ and complete when the halogen atom is I (iodine), so that the H-bond established between hydrogen iodide and trimethylammonium is I^-···$N(CH_3)_3^+$ instead of the above written initial neutral species I–H···$N(CH_3)_3$.

Microwave spectroscopy therefore appears to be a highly precise technique that conveys a lot of fundamental data not only on the geometry of H-bonds, but also on transfers of protons through H-bonds that are mechanisms at the origin of all acid–base chemistry. The requirement that such measurements can only be performed in gases limits its interest. The studied systems are consequently made of simple molecules only. The data obtained using these techniques nevertheless constitute highly useful starting points for more complicated systems that are not observable in the gas phase. They allowed understanding many properties of H-bonds. The second limitation of such methods is that the studied H-bonded complex has a permanent electric dipole moment. A small dipole moment gives weak microwave bands that disappear when this dipole moment becomes zero. This is not a really strict condition as most H-bonded systems exhibit such a permanent dipole moment.

IR spectra: rotational bands

Rotational features of almost all H-bonded complexes in the gaseous phase appear in the microwave region, with wavenumbers less than $10\,cm^{-1}$. They correspond to transitions between "pure" rotational levels, "pure" meaning that vibrations remain unchanged, or no vibrational transition accompanies such rotational transitions. Rotational features, however, also appear in the IR spectra of these H-bonded complexes. IR bands correspond to transitions between various vibrational levels of a molecule. When this molecule is isolated, as in the gas phase, these transitions are always accompanied by transitions between rotational levels that obey the same selection rules as pure rotational transitions detected in microwave spectroscopy. The information conveyed by these rotational features in IR spectra are therefore most similar to those conveyed by microwave spectra, even if the mechanism at the origin of their appearance is different. Their interests lie in the use of an IR spectrometer, a common instrument in many laboratories, instead of a microwave spectrometer, which is a much more specialized instrument. However, the resolution of usual IR spectrometers are lower than that of microwave spectrometers that use Fabry–Perot cavities. This IR technique has been used in the case of simple H-bonded dimers with relatively small moments of inertia, such as, for instance, F–H···N≡C–H (3). Such complexes are far from simple to manipulate, but provide particularly simple IR spectra with a limited number of bands that do not show any overlap.

NMR spectroscopy

Principles

In NMR spectroscopy a radiowave induces transitions between spin levels of the nuclei of the molecules that compose the sample that is immersed in a static magnetic field. The frequencies of these radio waves fall below 1 GHz. They consequently induce transitions between spin levels that are separated by energies always smaller than kT, even at such low temperatures as 1 K, as can be seen from values of constants h and k defined in the appendix of Ch. 1. It implies that the difference of population of these levels is always very small at equilibrium. It is, however, compensated by a very great number of absorbing nuclei, all protons of a sample, for instance. Furthermore the electronics technology that operates in this range of frequencies has reached such a precision and sensitivity that precise analyses of the shapes of bands that appear in NMR spectra can be performed. It is at the origin of precious information related to the dynamics of nuclear spins. It is an advantage of NMR as compared to optical spectroscopy where mostly intensities and positions of bands are taken into account. Their shapes may be occasionally analysed, in the case of particularly simple spectra, or in the case of the ν_s band of H-bonded systems described in Ch. 5, but are most often disregarded, except in quite recent methods such as time-resolved nonlinear IR spectroscopy sketched in Ch. 4.

In a typical NMR experiment the sample is positioned in a strong and homogeneous static magnetic field, now delivered by a superconducting coil that operates at liquid He temperature around 4 K. This static magnetic field separates the various nuclear spin levels that have same energies in the absence of this magnetic field. This separation of spin levels in a magnetic field is called "Zeeman splitting". Radiofrequency waves with their magnetic field perpendicular to the static magnetic field induce transitions between these different nuclear spin levels. In a first approximation, the Zeeman splitting between these spin levels is proportional to the amplitude B of this magnetic field. In order to induce transitions of spin levels, the frequency ν of the radiowave should also be proportional to B and equal to ν_n, the "Larmor frequency" of nucleus n, with

$$\nu_n = \frac{\gamma_n B}{2\pi} \tag{3.3}$$

Larmor frequencies ν_n are typically of 500 MHz for a proton immersed in a static magnetic field of amplitude $B = 12$ T, a very strong, but nowadays current magnetic field. In the same magnetic field, the Larmor frequency of a fluorine nucleus F is a bit smaller, that of a ^{13}C carbon atom (^{12}C that represents more than 99% of all C-atoms in natural conditions has no spin and is consequently not directly seen in NMR) is four times smaller and that of a ^{14}N about 0.07 that of the proton. This first approximation consequently allows us to define various ranges of radiofrequencies to be used to induce transitions between spin levels of a definite nucleus, H, F, ^{13}C, ^{14}N, ^{15}N, etc.

The absolute resolution of spectra in this range of frequencies around 500 MHz can be smaller than 1 Hz. It makes NMR a method with an exceptionally great relative resolution that approaches 10^{-9}. It implies that in a usual experiment, only one type of nuclei is looked at because even the limited range of frequencies of a single type of atom scanned with such a resolution requires recording a very great number of data, and studying with the same resolution the same species around the Larmor frequency of another nucleus is most of the time worthless. The frequency of the radio wave is then set around the Larmor frequency of this nucleus and absorption of this wave is recorded by scanning a limited range around this frequency. It then appears that for a given atom, H for instance, the most interesting atom in the case of H-bonds, resonance does not occur at quite the same frequency when this H-atom is part of a CH_2 group, or of an OH group, or of a NH group, etc. Figure 3.2 illustrates this point in the case of the NMR spectrum of ethanol CH_3–CH_2–OH in its liquid

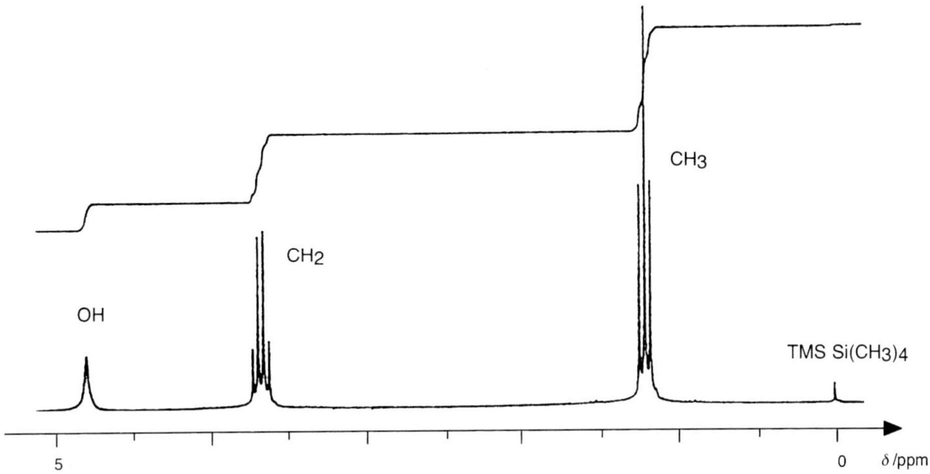

Figure 3.2 The NMR spectrum of ethanol C_2H_5OH (lower diagram) in its liquid phase. The upper curve in the form of steps that appear at same δ's as bands in the lower spectrum is the integrated spectrum, offset for clarity. δ is the chemical shift in "parts per million". The (not labelled) ordinate is the absorbed power of the radio wave that defines the NMR intensity. From Canet (42) with permission.

phase. The resonance frequencies of all protons are found in the vicinity of the Larmor frequency defined in eq. (3.3), but nevertheless fall at slightly different values when the H-atom is that of an OH group, or one of the CH_2 group, or one of the CH_3 group. Generalizing eq. (3.3), one can account for this point by writing, in a second approximation, that a transition among spin levels occurs when the frequency ν of the radiowave is equal to

$$\nu = \nu_H(1-\sigma_G) = \frac{\gamma_H B(1-\sigma_G)}{2\pi} \tag{3.4}$$

where the index G stands for the group CH_2, CH_3, etc., and ν_H is the Larmor frequency of the proton defined in eq. (3.3). This formula, written in the case of a proton, is valid for any other nucleus n, provided γ_H is replaced by the γ_n of this other nucleus. The coefficients σ_G account for the screenings of the static magnetic field by electrons of the molecule. These electrons are not in the same orbital in an O–H group or in a CH_2 group, or in a CH_3 group, and consequently provide different screening coefficients σ_G that characterize these various groups. The magnitudes of these screening coefficients, which are dimensionless quantities, are of the order of about 10^{-6}. They are very small quantities that can nevertheless easily be put into evidence, thanks to the high resolution encountered in this radiofrequency range, and are fully reproducible. The quantity represented in abscissa of Figure 3.2 is not σ_G, but the "chemical shift" δ_G used in practice, which is a measurement of the relative shift of frequency due to this screening. The following equation relates these two quantities

$$\delta_G = 10^6 \frac{\Delta\nu_G}{\nu} = 10^6(\sigma_{ref} - \sigma_G) \tag{3.5}$$

In this equation $\Delta\nu_G$ is the difference of frequencies $\nu_G - \nu_{ref}$ of the proton in the chemical group G and in a reference chemical group, here that of tetramethylsilane $Si(CH_3)_4$, or TMS, at the origin of a thin line that clearly appears in Figure 3.2. δ_G is defined for each chemical group with accuracy of the order of 10^{-2}. It justifies the approximation implicitly made in the second part of eq. (3.5) where $1 - \sigma_{ref}$ is written equal to 1 with σ_{ref} of the order of 10^{-6}. δ_G is usually expressed in "parts per million" or ppm. In Figure 3.2 the proton of the OH group exhibits a chemical shift δ_{OH} equal to about 4.8 ppm, which becomes 3.7 for δ_{CH_2} and 1.2 for δ_{CH_3}. These are values that are retrieved when these groups are parts of other molecules and are immersed in any magnetic field, provided, however, the same reference, TMS, is used. A change of reference is equivalent to making a δ_G change of a constant quantity when expressed in ppm. The upper integral curve in Figure 3.2 allows us to define the number of H-atoms that are implied in each group: the height of the step of the CH_2 group is twice that of the OH group while that of the CH_3 group is three times that of the OH group. When dealing with an unknown species this might be valuable information.

NMR of H-bonds established by small or medium-size molecules

Various groups of molecules are characterized by their various chemical shifts δ_G. Chemical shifts allow us to characterize H-bonds in NMR spectroscopy, because the screening of the spin of the proton of an O–H group is not the same when this O–H group is free or when it establishes an H-bond. Thus the chemical shift of the O–H group of liquid ethanol is equal to 4.8 ppm (Figure 3.2) and shifts to a value of −0.7 ppm for ethanol diluted in CCl_4. In its liquid form at room temperature, most O–H groups are H-bonded, while they are free when ethanol is diluted in CCl_4. The CH_2 and CH_3 bands hardly move between these two cases (11), which means that their chemical shift does not vary when a H-bond on the OH group is established. It shows that chemical shifts are sensitive to the establishment of H-bonds and NMR spectroscopy has been for some time thought to be an interesting tool to observe H-bonds (12). Thus correlations between chemical shifts $\delta_{X\text{–}H\cdots Y}$ due to H-bonds and X···Y distances (the equilibrium distance Q_0 of Q_s defined in Figure 2.1) that were obtained from X-ray diffraction of crystals could be established. However, these results did not suggest any systematic trend that could have been used to characterize H-bonds in NMR spectra, as shifts of ν_s bands do in IR spectra. This assertion should nevertheless be somewhat moderated, as the deuteron quadruple coupling constant, a quantity that can be measured in NMR experiments of D-bonds, has been shown to correlate quite well with the wavenumbers of the centres of some ν_s bands in O–D···O bonds of some solid hydrates (13). It nevertheless remains a quantity of a limited use. We may thus conclude that NMR is a method that allows us to observe H-bonds in solution.

Let us note that these latter correlations between $\delta_{X-H\cdots Y}$ and Q_0 in crystals could not have been obtained before the problem of dipole–dipole interactions has been solved. This problem is the following: two nuclear spins, situated on two separated nuclei interact when a small distance separates these nuclei. This interaction is of a classical dipole–dipole type, with these dipoles being magnetic dipoles. It is a somewhat long-range interaction and its magnitude exceeds the chemical shift by some orders of magnitude. It depends on the relative orientation of the two spins with the orientation of the line that joins these two spins. It is consequently at the origin of a most important broadening of most bands that completely masks the effects due to chemical shifts. This is what happens in solids but it is, however, absent in the case of liquids, where rotations of individual molecules are sufficiently rapid to average out this dipole–dipole interaction, transforming these broad bands into narrow Lorentzian bands centred at frequencies equal to those of the centres of these broad bands. Chemical shifts can consequently be observed, as appears clearly in Figure 3.2, despite the presence of dipole–dipole interactions of much greater amplitudes. This averaging-out effect is known as "motional narrowing": a rapid intermixing of levels among some distribution, due to some external factor such as rotations of molecules in a liquid, leads to observing only the average value of this distribution with a band centred at this average value and having a Lorentzian shape (14). It is much narrower than the band that would be obtained with a slow intermixing, which would have a much broader shape that directly reflects this distribution. A rapid rotation of a solid sample at a frequency of about 20–30 kHz around an axis that makes an angle of 54.4° with the magnetic field, called the "magic angle", has the same effect of averaging out this dipole–dipole interaction. After the discovery of this method NMR spectroscopy of solids has been possible, replacing bands that are broadened by dipole–dipole interactions by narrow bands, as in liquids. It means that NMR spectroscopy always works with conditions where these dipole–dipole interactions are averaged out, either by rapid molecular fluctuations in a liquid or by artificially averaging them out by magic angle spinning. It allows obtaining narrow bands within which chemical shifts may be seen.

In Figure 3.2 another interaction established by nuclear spins also clearly appears. It is an interaction through valence electrons between the H-atoms of the CH_2 groups and those of the CH_3 groups. It is responsible for the appearance of four lines in the band due to CH_2 group and of three lines in the band due to CH_3 groups. In this case of ethanol the splitting of these bands is appreciably smaller than the splitting due to chemical shifts, which is responsible for the separation of the bands due to CH_2 and CH_3 groups. In general this latter interaction, called "scalar coupling", has a magnitude comparable to that of chemical shifts.

NMR of macromolecules

When applied to macromolecules, NMR faces another problem, which is that the number of protons is great and the overlapping of bands consequently important, making the interpretation of NMR spectra a difficult problem (15, p. 280). This is true, even with macromolecules in liquids, at the origin of narrow bands. Due to their great number these narrow bands remain strongly entangled. They appear at slightly different frequencies because they correspond to different nuclei that suffer slightly different chemical shifts or scalar couplings. Information can nevertheless still be extracted, thanks to the high resolution of NMR (16) that allows taking into account the effects of the environment on the studied nuclear spin system. The main effect of the environment is to limit the lifetime of this system, which appears in the form of "relaxation times" for this system. These relaxation times convey statistical information on this environment. Sending thus a time-limited pulse of radiowaves with their magnetic fields perpendicular to the static one B provokes transitions of spins of the nuclei. Observing the relaxations of the various spins that interact through what is left of the dipole–dipole interactions after their main parts are averaged out by rotations of the liquid within which the macromolecule lies, allows us to observe correlations between spins of various nuclei. These spins should be separated by distances smaller than 4–5 Å (15, 17, 18) for these remaining interactions to be efficient. For longer distances no correlation is observable. These correlations consequently convey important structural information in the form of "structural constraints": they only appear when two protons are positioned at relative distances smaller than 4–5 Å. It is widely used in structural studies of proteins. Needless to say, in that case its implementation and exploitation could only be performed thanks to the advent of powerful informatics.

The establishment of these structural constraints makes NMR spectroscopy a basic method to determine the structures of many macromolecules. It is then often used in conjunction with diffraction (coherent scattering by ordered systems) methods described below and theoretical methods. It is invaluable to determine the structure of many proteins, where H-bonds are central for the formation of secondary structures, as described in Ch. 2. The information NMR conveys about H-bonds nevertheless remains indirect as it only identifies H-atoms that are positioned within close distances. It does not consequently directly identify H-bonds themselves in macromolecules, as IR spectroscopy naturally does (see Ch. 4), but only suggests the possibility of the establishment of such H-bonds. Progress in this way is nevertheless observable with the increase of resolution conveyed by improvements of the magic angle spinning that allowed, for instance, correlations of relaxation of spins of the H- and C-atoms of N–H···O=C H-bonds of proteins (19) to become evident. It showed that scalar coupling is not limited to covalency, but may also appear through H-bonds, with smaller but now detectable amplitude. It may lead to the possibility of directly observing H-bonds with NMR in the near future.

Conclusion on NMR

H-bonds established by small or medium size molecules can be characterized by their chemical shifts. These do not, however, convey enough original information on H-bonds to become a general method to observe H-bonds. In that sense, NMR spectroscopy did not materialize the hopes that were put into it during the 1970s (12). Nevertheless, NMR spectroscopy remains one of the most precise, basic methods in chemical physics and analytical chemistry, due to its ultrahigh spectral resolution and the precisely known spin–spin interaction dependence on distances. In these domains it has for long been a pioneering method. It introduced the use of precise time-resolved experimental protocols to fully exploit its ultrahigh resolution capacities that we have seen to allow precise correlations between nuclear spins. These powerful protocols are now adapted in IR spectroscopy under the name of nonlinear time-resolved IR spectroscopy which, as described in Ch. 4, is a promising method to study H-bonds. In NMR spectroscopy they allow to establish precise correlations between various groups of a macromolecule, leading to "structural constraints" that convey useful complementary information to data obtained by other structural methods such as X-ray or neutron scattering methods, described below. Even put together, these structural methods do not constitute, however, direct methods to study H-bonds in macromolecules. NMR spectroscopy may become one if the recent observation of scalar couplings through H-bonds (19) becomes exploitable.

X-ray absorption spectroscopy

This is a relatively recent method that should be clearly differentiated from the usual scattering of X-rays, a fundamental and very old method for the study of H-bonds we describe below. X-ray absorption experiments require exceptional qualities of the X-ray emitting source and have consequently much benefited of the implementation of high-quality X-ray beams produced by synchrotrons. Although a method that is now fully operative in chemical physics, it has yet scarcely been used in the case of H-bonds, and we consequently only sketch the potential interest of such a method. It is often labelled XAS (X-ray absorption spectroscopy), XANES or NEXAFS (near-edge X-ray absorption fine structure). It consists of absorbing an X-ray photon that sends an electron of the core of an atom such as O or N of a molecule into an empty electronic orbital of the molecule of high energy or into a "free-electron" orbital. The band that corresponds to such a transition is centred around 540 eV in the case this core electron belongs to the 1s electron of an O-atom (K-edge X-ray absorption zone). Around this value some fine structure may appear that falls either in the pre-edge region with energies between 535 and 540 eV or in the post-edge with energies greater than 540 eV. A modification of H-bonds, due to passing for instance from ice to liquid water, or due to a temperature change, or due to solvation of ions, changes this fine structure. It indicates that absorption of X-ray photons is sensitive to H-bonds and may consequently be an interesting source of data on them, provided the X-ray beam is stable enough to make possible the detection of the relatively small changes of this absorption band. It could thus be shown (20), using this method, that solvation of ions in liquid water scarcely modifies its H-bond network. The only H-bonds that suffer some change are those of the first sphere of H_2O molecules around each ion, a result in agreement with recent results of nonlinear time-resolved IR spectroscopy (21), described in Ch. 4. It strongly questions the older concept of "structure breaker" or "structure maker" ions for which ions were supposed to appreciably modify the H-bond network of bulk water, which in turn was supposed to be at the origin of the modifications of macroscopic properties of liquid water, such as its viscosity.

These XAS results however still require some clarification before being operational as, for instance, the comparison of XAS spectra of ice Ih and of liquid water lead to the conclusion that an important proportion of H_2O molecules in liquid water have only two strong H-bonds, one accepted and the other one donated. This image conflicts with that given by IR spectroscopy, which, as described in Ch. 9, shows that each H_2O molecule of liquid water donates two H-bonds and accepts two other ones. These H-bonds are nevertheless somewhat different from those found in ice Ih, the ordinary form of ice described in Ch. 8. In this form of ice H_2O molecules adopt a unique and rigid tetrahedral arrangement, with four H-bonds around each O-atom. The tetrahedral arrangement in liquid water is flexible, but the coordination remains equal to four. The interpretation of XAS spectra, based on the close resemblance of bands of liquid water and of surface ice Ih where at least one half of H_2O molecules are thought to be free (not H-bonded), consequently certainly needs some revision. On the other hand, another interpretation of XAS spectra of liquid water at various temperatures from $-22°$ to 15 °C indicated the presence of two kinds of H-bonds in liquid water with enthalpies that differ by about 6.3 kJ mol^{-1} (1.5 kcal mol^{-1}) (22), a result that now agrees with those given by vibrational spectroscopy.

Scattering of electromagnetic waves or particles

In this section experiments are described where photons or particles, instead of being directly absorbed or emitted, are scattered. These are two-photon or second-order mechanisms, in opposition to absorption described previously, which is a one-photon or first-order mechanism. Some of these scattering methods, such as X-ray diffraction, are basic and often old routine tools in chemical physics. In all scattering methods, incident photons or particles propagate through a transparent medium that displays no absorption. In these conditions the great majority of these photons or particles travel through the sample and then travel out of it in the same state they entered it. A very few of them are nevertheless scattered in various directions, following two types of scattering: coherent or incoherent, which differ in the conservation or not by the scattered wave or particle of the phase of the incident wave or particle. Conservation of this phase allows waves scattered by various centres to interfere, providing precise information on the spatial distribution of these centres, in other words on the structure of the sample.

Coherent scattering (X-rays and neutrons)

This category has two main methods, X-ray and neutron diffractions, whereas other less used methods, such as diffraction of electrons or Rayleigh coherent scattering of light, can also be occasionally found. In these experiments an incident plane wave with frequency ν propagates through the sample. This wave is defined by a constant electric field $\vec{E}$ in the case of X-rays or light. We write all following equations for the case of X-rays. At spatial position $\vec{r}$ inside the sample and at time t the electric field of this plane waves can be expressed as

$$\vec{E}_{\vec{k}}(\vec{r},t) = \vec{E}\,\mathrm{e}^{\mathrm{i}(2\pi\nu t - \vec{k}\cdot\vec{r})} \tag{3.6}$$

with $\mathrm{i}^2 = -1$. Instead of the electric field, which describes X-ray diffraction, any other vector characteristic of the wave in other experiments may be considered. In eq. (3.6), $\vec{k}$ defines the direction of propagation of this wave. It is called the "wavevector" of this wave. Maxwell equations for a travelling wave show it is perpendicular to the electric field $\vec{E}$ and also perpendicular to the accompanying magnetic field. Eq. (3.6) means that the electric field oscillates everywhere as $\vec{E}\,\mathrm{e}^{\mathrm{i}2\pi\nu[t-\tau(\vec{r})]}$ where the time delay τ of this wave between positions $\vec{r}$ and 0, equal by definition to the time this wave takes to travel between these two points, is, following eq. (3.6),

$$\tau(\vec{r}) = \frac{\vec{k}\cdot\vec{r}}{2\pi\nu} = \frac{|\vec{k}|\,r_{//}}{2\pi\nu} \tag{3.7}$$

As τ is proportional to $r_{//}$ and independent of $r_{\perp}$, points $\vec{r}$ in a plane perpendicular to $\vec{k}$ have all same time delay τ with respect to $\vec{r} = 0$ as they all have along $\vec{k}$ the same component $r_{//}$ (Figure 3.3). This is the definition of a plane wave that propagates at a constant velocity in the direction defined by $\vec{k}$. For an electromagnetic wave, we have:

$$r_{//} = c\tau \tag{3.8}$$

with c the velocity of light. For a wave that accompanies a neutron, for instance, the same relation holds, with c replaced by the velocity of the neutron beam. By comparing eqs. (3.7) and (3.8), we deduce:

$$|\vec{k}| = \frac{2\pi\nu}{c} = \frac{2\pi}{\lambda} \tag{3.9}$$

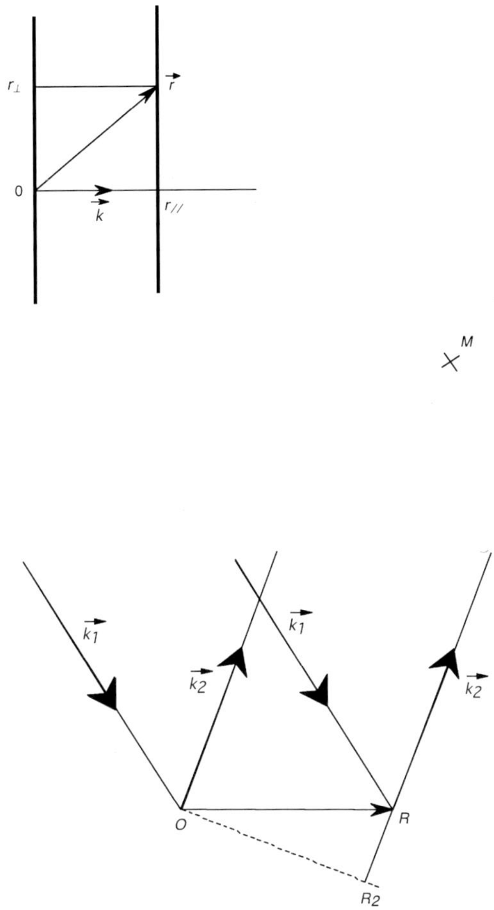

Figure 3.3 Geometry of a plane wave propagating along $\vec{k}$ (upper diagram) and scattering along $\vec{k}_2$ by two centres positioned at points O and R of an incident wave defined by $\vec{k}_1$ (lower diagram).

with λ the wavelength, equal to c/ν.

Let us now examine a wave that is scattered by two scattering centres situated at the origin O and at a point R, drawn in Figure 3.3. For X-rays such scattering centres are electrons. They are nuclei in the case of neutrons. The oscillation of the incident wave defined by wavevector $\vec{k}_1$ is written as $\vec{E}_{\vec{k}_1}(0,t) = \vec{E}\mathrm{e}^{\mathrm{i}2\pi\nu t}$ at the origin O if the origin of time is properly defined (in case the origin of t is defined differently, an additional constant phase factor $i\varphi$ appears in the exponential; this constant phase factor has no effect in the following). The wave scattered by scattering centre at O is defined by its wavevector $\vec{k}_2$. At a point M sufficiently far away from O, the oscillation due to this scattered wave is expressed as $s\vec{E}\,\mathrm{e}^{\mathrm{i}(2\pi\nu t - \overrightarrow{\mathrm{OM}}\cdot\vec{k}_2)}$ where s is the scattering factor of both centres at O and R (in the case of a particle s is called cross-section). On writing this equation, we implicitly suppose that the phase of the incident wave is kept unchanged during this (coherent) scattering. In an incoherent scattering a random (phase) factor appears within the parenthesis of the imaginary exponential. If a constant phase factor appears in this parenthesis, we have yet to do with coherent scattering. The incident wave arriving at point R, the second scattering centre, can be expressed as $\vec{E}\,\mathrm{e}^{\mathrm{i}(2\pi\nu t-\overline{\mathrm{OR}}\cdot\vec{k}_1)}$, so that the wave scattered by the centre situated at R with wavevector $\vec{k}_2$ becomes $s\vec{E}\,\mathrm{e}^{\mathrm{i}(2\pi\nu t-\overline{\mathrm{OR}}\cdot\vec{k}_1-\overline{\mathrm{RM}}\cdot\vec{k}_2)}$ when arriving at point M. With $\overrightarrow{\mathrm{RM}}\cdot\vec{k}_2=(\overrightarrow{\mathrm{OM}}-\overrightarrow{\mathrm{OR}})\cdot\vec{k}_2$, this latter scattered wave can then be expressed as $s\vec{E}\,\mathrm{e}^{\mathrm{i}(2\pi\nu t-\overline{\mathrm{OM}}\cdot\vec{k}_2-\overline{\mathrm{OR}}\cdot\vec{k}_1+\overline{\mathrm{OR}}\cdot\vec{k}_2)}$ so that the wave at M due to scattering of the incident wave by the two scattering centres in O and M is then equal to

$$\vec{E}_2(M,t)=s\vec{E}\,\mathrm{e}^{\mathrm{i}(2\pi\nu t-\overline{\mathrm{OM}}\cdot\vec{k}_2)}\{1+\mathrm{e}^{\mathrm{i}\overline{\mathrm{OR}}\cdot(\vec{k}_2-\vec{k}_1)}\} \tag{3.10}$$

In a coherent scattering experiment the frequency ν is the same for incident and scattered waves. Eq. (3.9) implies then that $|\vec{k}_1|=|\vec{k}_2|$, which means that $\vec{k}_1$ and $\vec{k}_2$ have same amplitudes whereas their directions are different. In the case of scattering of particles, incident and scattered particles also have same energies. Eq. (3.10) may be easily generalized to a set of numerous scattering centres with same scattering coefficient or cross-section. If these scattering centres are defined by their spatial density $\rho(\vec{r})$, the wave scattered by all scattering centres that are defined by the wavevector $\vec{k}_2$ can then be expressed as

$$\vec{E}_\rho(t)=s\vec{E}\,\mathrm{e}^{\mathrm{i}(2\pi\nu t-\overline{\mathrm{OM}}\cdot\vec{k}_2)}\int\rho(\vec{r})\,\mathrm{e}^{\mathrm{i}(\vec{k}_1-\vec{k}_2)\cdot\vec{r}}\,\mathrm{d}\vec{r} \tag{3.11}$$

This eq. (3.11) shows that the amplitude of the scattered wave is proportional to the Fourier transform, defined in eq. (5.A17) of the appendix of Ch. 5, of the density ρ of scattering centres. This is a particularly favourable situation, because the measurements of the amplitudes of the waves coherently scattered in all directions, that is with all possible wavevectors $\vec{k}_2$, allows in principle to calculate by an inverse Fourier transform, also defined by eq. (5.A17) of the appendix of Ch. 5, the density $\rho(\vec{r})$ of scattering centres. This last equation holds for a one-dimensional (1D) Fourier transform, that is a single coordinate x (or t) and consequently a single "conjugate" coordinate k (or ν in eq. (5.A17)). The extension to a 3D space defined by $\vec{r}$ and $\vec{k}_1$, $\vec{k}_2$ is trivial and can be found in any textbook on X-ray. Without entering the details of calculations of inverse 3D Fourier transforms, we nevertheless have to note that in a coherent scattering experiment the measured quantities are not amplitudes $\vec{E}_\rho$ (eq. (3.11)) of vectors, which are complex quantities, with real and imaginary parts, but their square amplitudes $|\vec{E}_\rho|^2$, called intensities, which no longer convey information on imaginary parts of $\vec{E}_\rho$. As a consequence, we cannot exactly calculate by inverse Fourier transforms the densities $\rho(\vec{r})$ of scattering centres but we can calculate their correlation functions $\int\rho(\vec{r})\rho(\delta\vec{r}-\vec{r})\mathrm{d}\vec{r}$ as a function of $\delta\vec{r}$.

This inverse Fourier transform calculation of the correlations of density of scattering centres of the sample gives particularly precise results when this sample is a crystal. In this case $\rho(\vec{r})$ is periodic. The scattered intensities are then "δ functions", or Dirac's functions, that are zero almost everywhere, except for well-defined values of $\vec{k}_2$ where they take on great amplitudes. They are known as "Bragg peaks" for which all scattered waves have the same phase. Interferences of all these waves are consequently constructive in the directions where Bragg's peaks appear. This is the consequence of the mathematical result that the

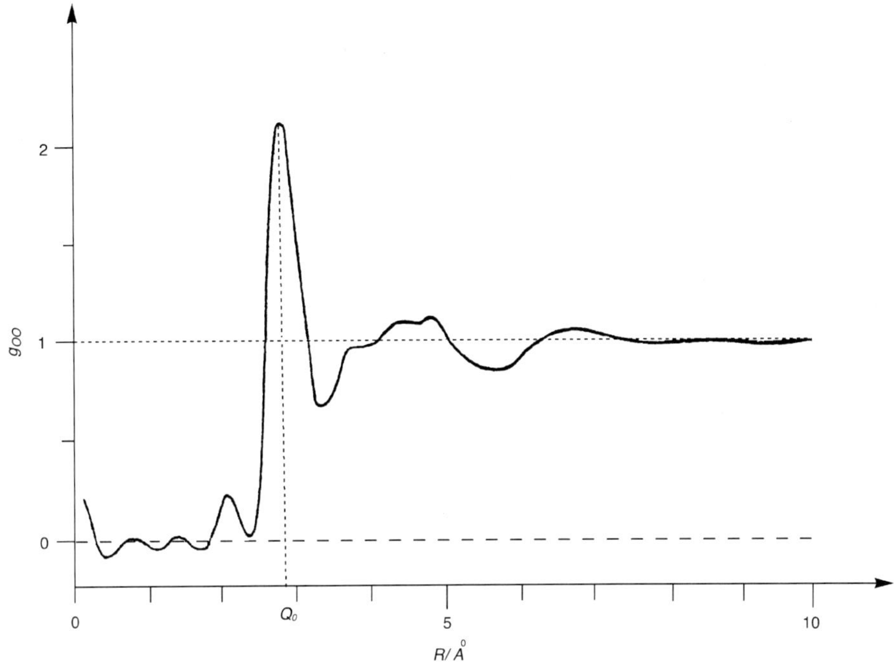

Figure 3.4 The autocorrelation function g_{OO} of liquid heavy water at 11 °C as a function of interatomic distance R. The value of R at maximum of g_{OO} is the value of Q_0, the average equilibrium O···O distance Q_s of O–D···O bonds in this liquid, defined in Figure 2.1. Reproduced from Figure 1b of reference (24), *with permission*. Copyright (2006) by the American Physical Society.

Fourier transform of a periodic sinus or cosine is a 1D δ function that is zero everywhere except at position 0 where it is infinite with, however, its integral between $-\infty$ and $+\infty$ equal to 1. It allows a precise determination of the structure of molecular complexes. The order in crystals is of long-range type because the periodicity extends very far. This particular case of scattering by a very great number of periodically ordered particles is called diffraction. It occurs in the case of X-rays, allowing us to map electronic densities in the sample, and it also occurs in the case of neutrons where it provides the densities of nuclei. Such a coherent scattering also occurs when ordinary light meets a grating made of periodically arranged lines.

Besides this ideal situation of a periodically ordered crystal, coherent scattering may still convey information in less ordered samples. Thus scattering by a liquid where some order remains at short distance but disappears at longer distances, provides, for instance, measurements of O···O distances in liquid water or liquid alcohols. This is reproduced in Figure 3.4 where the autocorrelation function g_{OO} is drawn as a function of distance R for liquid heavy water (23, 24). This function is proportional to the probability that when sitting on an O-atom of this liquid, one finds at distance R another O-atom (25). As the nearest O-atoms are those that establish H-bonds on this O-atom we are sitting on, we deduce that in this liquid, the value of Q_0, the equilibrium O···O distance Q_s of O–D···O bonds, defined in Figure 2.1, is equal to R at first significant maximum of g_{OO}, 2.75 Å in this case.

Second and third maxima, which occur at higher values of R, define average distances of second and third neighbours of this O-atom. At still higher values of R, g_{OO} takes on a nearly constant value that indicates a random distribution of these distances. This curve is typical of H-bonded liquids, which exhibit a well-marked order at short distances due to H-bonds. This order disappears at longer distances. It allows measuring O···O average distances Q_0 of these H-bonds. Other short distance correlation functions such as those of O–H or H···O can also be extracted in favourable cases from diffraction patterns in these liquids. In the experiments this diffraction pattern appears in the form of light cones centred on the sample and that all have for their axes the direction of the incident X-ray or neutron beam. Along these cones the diffracted beam is more intense than in other directions. The existence of these cones is a consequence of the isotropy of the liquid. They are the equivalents of Bragg's peaks in crystals. This type of coherent scattering of X-ray or neutrons has been used in the case of such chemically well-defined liquids as liquid water or alcohols. It has been scarcely used for gases, which are not dense enough to allow scattering intensities to be measured.

A very few highly diluted gases have, however, been the objects of coherent *electron* scattering experiments. In more dense gases, absorption of electrons rapidly becomes predominant and hinders measuring scattering intensities. Diffraction of electrons thus allowed determining with precision the structure of H-bonded cyclic dimers of carboxylic acids (26–28) that constitute model systems of H-bonds we have already seen in Ch. 2 and shall see later in this book. They nevertheless remain scarcely used methods.

Scattering of X-rays and of neutrons are basic structural methods. X-rays and neutrons are now delivered by special machines that provide intense fluxes of monochromatic beams. Thus X-rays delivered by synchrotron radiation have such high fluxes that recording scattered intensities in all directions can be made within 1 day on so small samples that they can be hardly visible. This is particularly interesting in the case of crystals of proteins, for instance, which are much harder to obtain than crystals of organic substances. Their sizes are consequently severely reduced. It allows extending the application of X-ray scattering to biomacromolecules. Determination of the structures of a lot of biological systems (29), which always remain partially disordered, has thus been carried out using these machines. It should be nevertheless kept in mind that exploitation of data for such complicated systems, which may be recorded within 1 day, may take months of complex calculations! The advent of especially designed nuclear reactors or accelerators of particles has also allowed precise neutron scattering experiments to be performed after 1970. The flux of neutrons in these machines is, however, weaker than that of X-ray photons by orders of magnitude. The acquisition of data takes consequently a much longer time, 1 week instead of 1 day, and only much more voluminous samples can be studied. Even with their weaker fluxes, neutron scattering experiments remain useful when X-ray scattering cannot be performed. This is, for instance, so when the sample can only be maintained in a metallic container, as in some high-pressure experiments. Metals absorb X-rays much more efficiently than organic substances and cannot be used in that case, whereas neutrons are weakly scattered by some light metals that can then be used in neutron diffraction experiments.

In the case of H-bonds, a great wealth of structural data has been obtained, making X-ray and neutron coherent scattering experiments basic methods to determine the geometries of a considerable number of H-bonded systems, as extensively described in excellent books (2, 29).

No point, then, to further stress the fundamental interest of such methods, and in this short chapter only rapid comments are added. Most structural parameters of H-bonds have been deduced from diffraction in crystals. The great majority of data concerns Q_0, the equilibrium X···Y distance Q_s defined in Figure 2.1. Measurements of these distances have most often been quite accurate. They may have the same level of accuracy as the distances provided by microwave spectroscopy, described previously and which are, however, limited to gases. The accuracy with which equilibrium angles θ_0 and φ_0 of H-bonds, also defined in Figure 2.1, have been measured is somewhat poorer than that for Q_0 for both these diffraction techniques. A lot of measurements concerning these angles have nevertheless been given by these diffraction methods. This relatively poorer accuracy finds its origin in the fact that measurements of such angles too much depend on the accuracy of the determination of the position of the H-atom of the H-bond X–H···Y. This position is determined with a relatively poor accuracy, because the coherent scattering factor of the H-atom is particularly small in both X-ray and neutron experiments. X-rays are scattered by electrons and the H-atom having a single electron around it has a weak scattering factor as compared to those of the other atoms of the sample. Neutrons are scattered by nuclei, but the positions of H-atoms are hardly observable by neutron diffraction, because the coherent scattering "cross-section" of H-nuclei is small in comparison to their especially great incoherent cross-section. It makes the coherent signal disappear into a much more intense incoherent one. Fortunately, however, this is not so for the isotopic companions, the D-nuclei, which have a much smaller incoherent cross-section and a coherent one that is of the same magnitude as other common atoms such as C, O or N (30). It allows, for instance, to observe with coherent neutron scattering (31) collective motions of D-atoms of different D-bonds in H-bonded ferroelectrics, described in Ch. 6. As H-atoms that establish H-bonds are in many materials relatively easily replaced by D-atoms, for instance by immerging the species into heavy water, structures of D-bonds may be directly observed. As described in Ch. 7, D-bonds have, in a first approximation that is often very good, same geometries as H-bonds. Structural properties of H-bonds can consequently be measured on equivalent D-bonds by neutron diffraction. It nevertheless requires supplementary preparations that may not always be easy. In complicated systems with numerous H-bonds, this indirect observation of H-bonds, due to the difficulty to detect the presence of the H-atom itself, may be a source of errors. Thus only part of H-bonds in nucleosides (32) is detected by X-ray diffraction, making their number estimated by this method only half the number determined using IR spectroscopy. In more complicated systems such as macromolecules, the spatial resolution attained by X-ray diffraction is about 1.5 Å (33, 34). It allows a fair determination of the structures of macromolecules, but is not sufficient to directly observe the H-bonds. In Ch. 11, it will be seen to be nevertheless enough to localize the O-atoms of some water molecules that are parts of the structures of proteins.

All these restrictions concerning these scattering methods have nevertheless not hindered them to be at the origin of banks of geometrical parameters of many H-bonds (29).

Incoherent scattering

The phase of the wave scattered by a particle at position $\vec{R}$ is random and has no relation with the phase of the incident wave or particle. It consequently gives no interference pattern between waves scattered by different scattering centres, and consequently no direct structural information. Each scattering centre scatters waves or particles independently from other centres. Incoherent scattering does not consequently give Bragg peaks in crystals. Ordinary light or neutrons can be incoherently scattered. In both cases, this incoherent scattering may be elastic or inelastic. Elastic scattering, which we have already encountered in the preceding case of coherent scattering, occurs when photons or particles of the scattered beam have same energies as those of the incident beam. In the case of a wave, same energy means same frequency of photons associated with this wave. Inelastic scattering conversely occurs when scattered photons or particles are less or more energetic than those of the incident beam.

Incoherent light scattering

In the case of light, incoherent scattering is called "Rayleigh" scattering when it is elastic. It is a common process that makes us see objects with our eyes, which detect light scattered by these objects. Thus leaves appear green as chlorophyll pigments scatter green light, while it absorbs blue and red lights and is transparent for green light. It is also a common process in the atmosphere that is responsible of the blue colour of a cloudless sky, that is of a completely transparent atmosphere. This blue colour is due to the fact that scattering of blue visible light is much more effective than that of red visible light. Incoherent Rayleigh scattering is consequently also responsible for the red colour of the sun at sunset, as direct light from the sun has travelled a long distance in the atmosphere and has scattered much of its blue light and relatively few of its red light. When inelastic, scattering of light is called Raman scattering. It occurs with incident photons of frequencies ν either in the visible region or in the NIR region. The frequencies of scattered photons are $\nu - \delta\nu$, and consequently their energies $h(\nu - \delta\nu)$ are not the same as those of incident photons. Let us note that scattering with $\delta\nu > 0$ has a greater amplitude than scattering with $\delta\nu < 0$, but both exist. $\delta\nu$, called Raman shift, is usually equal to frequencies of vibrations in the molecules. The energy loss or gain between incident and scattered photons in Raman spectroscopy therefore corresponds to excitation or de-excitation of vibrations in molecules. We shall see in Ch. 4 that we see in Raman spectra the same vibrational bands as in IR spectra. This does not mean that the Raman spectrum of a molecule is identical to its IR spectrum. Intensities of individual bands are not the same. Intensities of IR bands are proportional to the square of the variation of the dipole moment of the molecule induced by vibrations. Intensities of Raman bands are proportional to the square of the variation of the polarizability of the molecule induced by vibrations. In the case of centrosymmetric molecules, such as H-bonded cyclic dimers, only antisymmetric vibrations appear in IR spectra as symmetric ones do not induce any change of the dipole moment that remains equal to 0. The inverse holds for Raman bands, and the IR spectrum of these cyclic dimers differs of the Raman spectrum. IR and Raman spectroscopies, although observing the same phenomenon, vibrations of molecules or of molecular complexes, are thus complementary. In the absence of centres of symmetry transition probabilities remain different between IR and Raman bands, making intensities of common bands different.

As already mentioned, IR spectroscopy is a particularly powerful tool to study H-bonds, and both Chs. 4 and 5 are dedicated to this technique. This is not so of Raman spectra as, in opposition to IR bands, Raman bands are not especially sensitive to the presence of H-bonds. As a consequence, Raman spectroscopy, which gives most similar information, even if not identical, is a scarcely used method to study H-bonds, because of its small sensitivity to H-bonds, as compared to IR, and also because it is more sophisticated and harder to implement a technique than simple IR spectroscopy, a routine technique. It has for some time been used in cases where the hypersensitivity of IR spectroscopy to H-bonds is more an inconvenience than an asset. This has been the case for liquid water and for many aqueous media, where this hypersensitivity, combined with the extremely great number of H-bonds in these species lead to catastrophic saturation effects in IR spectra. Soon after these inconveniences have been overcome, as described in Ch. 11, Raman spectroscopy has been only scarcely used as a systematic method to study H-bonds, even those found in liquid water or aqueous media. On the other hand, it is still used in those spectroscopic regions where detection of IR waves becomes hard. This is the case of the FIR region, below 400 cm^{-1}, where Raman spectroscopy may then favourably replace IR spectroscopy in some cases. It is however a region that conveys much less information on H-bonds than the mid-IR region above 400 cm^{-1}.

Incoherent neutron scattering (INS)

This is a method that without being routine is regularly encountered in the study of H-bonds because, as already mentioned in the case of coherent scattering, the H-atom exhibits an exceptionally large incoherent cross-section, $\sigma_H^{inc} = 80$ barn (30), which is to be compared with its coherent cross-section, $\sigma_H^{coh} = 1.8$ barn. Let us recall that the scattering cross-section of a particle is equivalent to the scattering coefficient s of an electromagnetic wave, which is given in eqs. (3.10) and (3.11). This incoherent cross-section σ_H^{coh} of H-atoms is also far greater than coherent or incoherent cross-sections of any other atoms. It means that INS mainly sees H-atoms and is sensitive to transitions that imply H-atoms. These transitions are most often vibrational transitions. In that respect, it works the same way Raman spectroscopy does. Selection rules are nevertheless different between Raman and INS, so that both methods are also complementary. INS has, however, the same inconvenience as Raman spectroscopy with respect to IR spectra: it is not particularly sensitive to H-bonds in the region of conventional IR spectra, with wavenumbers greater than 400 cm^{-1}, and is much less easy to implement. It becomes interesting in the very far IR region where IR and also Raman spectra display a poor sensitivity. This is typically the case of vibrations or of any transitions that imply changes of wavenumbers smaller than 10 cm^{-1}, that is energy changes less than 1 meV. This region is known as the "quasi elastic region". In this region relative vibrations of heavy entities such as polymers, or great parts of macromolecules, can be observed. Transfers of protons or of H-atoms can also be observed (35) by INS. They are large amplitude motions of H-atoms or protons inside H-bonds that cannot easily be seen by IR or Raman methods. Their interest is fundamental for H-bonds and is described in more detail in Ch. 6. In this potentially interesting field, INS is a method that may prove of interest in the future.

THEORETICAL DESCRIPTIONS OF THE ELECTRONIC STRUCTURES OF H-BONDS

These are methods where description prevails over observation or measurement. Their fundamental interest is to provide or predict properties of H-bonded complexes in the case of a lack or absence of experimental data. They rely on an as precise as possible description of the electronic structure of these complexes. The point is that these electronic structures only slightly differ in X–H···Y from that of X–H and Y when they establish no H-bond. In other words, theoretical descriptions of the electronic structures of H-bonds should be precise to give exploitable results. For simple H-bonds established by very small molecules such as H_2O, NH_3, HF, etc., *ab initio* quantum methods that make in principle no approximation can satisfactorily account for the electronic structures of H-bonded dimers, or even trimers. Electronic structures of such gaseous simple complexes as Cl–H···O–$(CH_3)_2$ that we have seen have been the object of studies by microwave or IR spectroscopy in the gaseous phase, can also be satisfactorily computed. These computations allow testing theoretical methods against experimental data in view of extending them to the prediction of other properties of H-bonds or of properties of H-bonds established by bigger molecules that cannot be entirely computed without approximations. If correlation of electrons situated on the two parts of the H-bond, donor and acceptor parts, is correctly included (36), such *ab initio* methods allow to quite well reproduce the geometry of these complexes, in particular equilibrium distances Q_0 of Q_s (Figure 2.1). This is for instance the case of simple systems in the gas phase, where theoretical methods that have given correct values of the wavenumbers of bands in IR spectra are used to predict the geometries of H-bonded complexes (37). These theoretical methods can at the same time provide fair values of the enthalpies ΔH of formation of H-bonds and of wavenumbers of vibrational bands. In this case of vibrational spectra, only the wavenumber of the centre of the stretching X–H band ν_s in X–H···Y can be computed. The fundamental strong anharmonic coupling with intermonomer modes this band suffers, which is described in Ch. 5, has not up to now been reproduced by these *ab initio* methods, even if it is a characteristic feature of H-bonds that shows the peculiarity of the dynamics of H-atoms in H-bonds. As already seen in Ch. 2, for instance, these theoretical approaches of H-bonds allow establishing correlations between their enthalpies and values of Q_0, inside H-bonds of same types (38). These *ab initio* computations may also be applied to properties of H-bonds in solids, provided the hypothesis that H-bonded complexes weakly interact is valid, so that the solid may be thought of as an ordered assembly of weakly interacting H-bonds. Even in those cases, however, approximations should be made for calculating the electronic structures of H-bonded complexes with a greater number of atoms. This is because the number of atoms that can be fully taken into account with almost no approximation is as low as 4 (39). Even in the case of simple H-bonded complexes just mentioned above, a full *ab initio* treatment cannot consequently be performed, mainly because of the limitation of the number of orbitals that form the basis on which all wavefunctions are decomposed, and approximations are necessarily introduced. They often consist of exactly calculating some restricted parts of the H-bonded complex in a complete *ab initio* treatment and replacing the effects of the rest of the complex by simple approximate potentials. In the case of the intramolecular H-bond of, for instance, malonaldehyde drawn in Figure 6.2, or of similar molecule acetylacetone (40), where H-atoms are replaced by CH_3 groups, only the electronic structure of the C=O–H···O=C parts are computed. This is enough to compute the height of the potential barrier that governs the transfer of the H-atom between the two configurations C–O–H···O=C and C=O···H–O=C when O–H and C=O distances vary. It allows calculating such above described quantities of bigger H-bonded complexes and comparing them with experimental data.

The problem with these approximations is that all of them should be checked against experimental results and that one type of approximation may be the best one to describe a type of property, while another one may be the best one for another type of property. Optimizing the approximation is, therefore, often a hard task to perform. It is often performed within the frame of "density functional theory" methods, or DFT, which are often used for H-bonds. They consist of optimizing the electronic wavefunctions with respect to electronic densities, which is different of classical SCF (self-consistent field) approaches that optimize wavefunctions themselves. They exhibit a favourable scaling factor with the number of electrons (41), as the time required to calculate the electronic structure of a H-bonded complex of N atoms vary as N^4 or higher power using SCF methods, whereas it varies as N^3 with DFT. In addition, DFT methods take on correlations of electrons more easily, a mandatory requirement to have reliable descriptions of these electronic structures of H-bonds, which imply small energies as compared to covalent bonds, the main contributors to the total energy of the complex. Taking these correlations explicitly into account partially compensates for the errors introduced by the limitation of the number of basic orbitals on which all wavefunctions are decomposed. It nevertheless remains that even if they give good results for some H-bonded systems, they should be assessed in each particular case.

H-bonds are often found in more complicated systems than gas or solids as discussed above. They are thus most often found in liquids, aqueous solutions or solutions of other solvents. Reproducing experiments that have been performed in these conditions then require supplementary approximations in theoretical descriptions. It often requires a more drastic reduction of the number of parameters to be computed by *ab initio* methods while mimicking the effect of the surrounding molecules or complexes by a classical mechanical treatment. In the case this surrounding consists of identical complexes such an approximation is at the origin of an iterative procedure that falls in the same vein as "consistent-field theory" of solids. It can alternatively feed molecular dynamics (MD) computations of dynamic properties of the considered H-bonded complex. The potentials used in these MD computations are then potentials where the surrounding of the H-bond is treated as precisely as possible in this type of approach, and the H-bond itself is described in a full quantum frame. From these two relatively precise treatments, interactions between both parts are deduced in the form of simple approximate action and reaction potentials. The principle of MD computations is to calculate the evolution of a system made of many particles such as, for instance, H_2O molecules in liquid water. The dynamics of these particles is governed by a classical (not quantum) potential. The configuration of all these particles is set in a reasonable form at time $t = 0$ and its evolution is calculated applying this classical potential to these numerous particles when time is increased by small steps of about 10^{-15} s. The computations stop when a statistical stability is reached. The point is that, in order to be efficient in such a treatment of numerous particles, this potential has to be simple. Applied to H-bonds, this simplification hinders taking into account their directionality, one of their fundamental properties. This is at the origin of the failure of MD to reproduce the exceptional properties of liquid water described in Ch. 9, and more generally to predict how water molecules get embedded in macromolecules, where they often play important roles (Ch. 10). Other kinds of approximation can evidently be made, as these approximations depend on the kind of problem that is looked at. In all cases the theoretical results are to be precisely compared to experiments. It implies that predictive properties of such theoretical approaches are restrained. They can be only trusted when the approximations are precisely checked against well defined experimental results and predictions reduce to small extrapolations from them. They can still be used to predict entirely new properties from a model that has been shown to very well reproduce completely different data. Even with these restrictions, they are most useful to give preliminary estimations of the properties of hydrated forms of macromolecules, for instance, or give preliminary results to still poorly known effects such as transfers of protons (40). These are fundamental mechanisms of H-bonds described in Ch. 6, which may in the near future become important fields of research. Presently computational methods are the only ones to give estimates of relevant quantities for such mechanisms. They are also often valuable methods that are used as complementary methods in experiments where they allow the estimate of otherwise unknown quantities that are necessary to give more precise interpretations of these experiments. It explains why thousands of publications on the electronic structure of H-bonds are found in the literature, thus showing that despite their limitations stressed above, they are important methods in the study of H-bonds.

SUMMARY

Many methods have been used and will be used in the future to study H-bonds. In this chapter we described most of them, with the exception of IR spectroscopy, which is described in Chs. 4 and 5. We also left out of consideration specific methods that are used to observe H_2O molecules, as these methods are detailed in Ch. 11. Experimental methods, which include most of the methods described in this chapter, have been divided into calorimetry, an old but still interesting method, and "other" methods, which encompass most of more recent experimental methods. These other methods have been divided into "absorption", or first-order methods, and "scattering", or second-order methods. IR spectroscopy, not described here, is included in the category of absorption, first-order methods. In this chapter we have described these methods one by one. This summary is organized the other way: it considers the type of property that is intended to be looked at, and then lists the possible methods that may be used.

Three types of properties are particularly important for H-bonds: thermodynamics, geometry and dynamics.

(1) Thermodynamics quantities, such as enthalpies ΔH, may be directly measured by calorimetry on simple H-bonds found in systems that exhibit ideally a single type of H-bonds. These enthalpies ΔH may also be obtained from IR, Raman or NMR spectroscopy, in the case, for instance, where H-bonded complexes X–H···Y coexist with their free X–H and Y components and variation of their relative proportions can be induced by variations of temperature, of concentration, or of any other external parameter. Precise measurements of enthalpies ΔH of H-bonds of carboxylic acids from the variations of the equilibrium constants that defines the equilibrium

between monomers and dimers with temperature are thus described in Ch. 4. We have already seen in Ch. 1 that these ΔH may also be computed, which leads to relatively precise values if one can calibrate them against a limited number of calorimetric measurements.

(2) Geometrical quantities rely on three basic methods: rotational spectroscopy, X-ray or neutron diffraction, and IR spectroscopy. Rotational spectroscopy in the microwave region is a highly precise method, but it works only in gases at low pressures and low temperatures, which considerably reduces the number of H-bonds that can be studied. Furthermore, it requires the H-bonded complex to have a nonzero electric dipole moment. This is not a serious limitation, but it nevertheless excludes centrosymmetric complexes. Results obtained using these methods have been highly valuable when precise data on H-bonds started being collected, particularly all those examined in Ch. 2. Now relatively few H-bonds remain to be seen using this method, which requires the molecular components X–H and Y of the X–H···Y complexes to be simple molecules. It may remain interesting in astrophysics or physics of the atmosphere, but certainly not for aqueous or biological systems. We have mentioned that rotational constants are mainly obtained from microwave spectra, but can also be observed and exploited in IR spectra of similar gaseous H-bonded systems. These IR methods applied to rotational spectra suffer the same limitations just described above.

Diffraction methods, either employing X-rays or of neutrons, are basic general methods. Their main limitation is that they have difficulties seeing H-atoms, especially in the case of X-ray diffraction. This is strongly compensated by the possibility they offer to study complicated systems, such as macromolecules. They remain highly precise for measurements of such quantities as X···Y distances. They furthermore work in condensed states, mainly ordered crystals. Most X···Y distances of X–H···Y H-bonds have been measured using these methods, and banks of geometrical data concerning many H-bonded systems are thus proposed to the scientific community.

IR spectroscopy does not directly measure geometrical quantities. In Ch. 4 it is shown that measurements of the wavenumber of the centre of the most intense ν_s stretching band ($X-\vec{H}\cdots Y$) of a H-bond X–H···Y can be relatively precisely correlated to the X···Y equilibrium distance of this H-bond. IR measurements are most easy, in any conditions: gas, solid, liquid, solution, etc., and IR spectrometers are routine instruments in many laboratories. IR spectroscopy is consequently a basic general method to rapidly obtain geometrical parameters of H-bonds. Let us note that theoretical methods may also occasionally be used to estimate geometrical quantities of H-bonds made of small molecules.

(3) Dynamic quantities such as spectroscopic constants (frequencies or wavenumbers of various vibrations) are obtained from IR or Raman spectra. IR spectra are by far the most precise method to measure such quantities related to intramonomer vibrations (vibrations of X–H component, which are for some of them described in Chs. 4 and 5, are highly sensitive to the establishment of a H-bond X–H···Y), which contain a lot of information on H-bonds themselves. Raman spectra are more useful for intermonomer vibrations that are found in a region of smaller wavenumbers, where the quality of IR spectra becomes poorer. They also have been useful in the case of liquid water and aqueous media which give such intense signals in IR spectra that

saturation occurs and for a long time the H_2O molecule was considered as a poison in IR spectroscopy, as described in Chs. 9 and 11. Theoretical methods may also occasionally provide information on these quantities. It is thus likely that such dynamic quantities as transfers of protons or of H-atoms, described in Ch. 6, will be studied by these methods in the near future. The interest in such mechanisms is recent, and up to now theoretical methods have given the greater number of related data, even if they are still in a stage of evaluation. We have seen in this chapter that from microwave spectra of $X{-}H\cdots N^{-}R_3$ complexes, with X a halogen atom, the proportion of $X^{-}\cdots HN^{+}R_3$ complexes with transferred proton could be measured. Up to now they still did not convey information on the dynamics itself of these transfers. They nevertheless may appear most useful in this domain in the near future.

REFERENCES

1. L. Pauling, The Nature of the Chemical Bond, Ch. 12, Cornell University Press, Ithaca, NY, 1960.
2. G. A. Jeffrey, An Introduction to Hydrogen Bonding, Oxford University Press, Oxford, 1997.
3. A. C. Legon and D. J. Millen, Chem. Rev., 86 (1986) 635.
4. T. J. Balle, E. J. Campbell, M. R. Keenan and W. H. Flygare, J. Chem. Phys., 72 (1980) 922.
5. A. C. Legon and D. J. Millen, Faraday Discuss. Chem. Soc., 73 (1982) 71.
6. A. C. Legon and L. C. Willoughby, Chem. Phys. Lett., 95 (1983) 449.
7. N. W. Howard and A. C. Legon, J. Chem. Phys., 88 (1988) 4694.
8. A. C. Legon and C. A. Rego, J. Chem. Phys., 90 (1989) 6867.
9. A. C. Legon, A. L. Wallwork and C. A. Rego, J. Chem. Phys., 92 (1990) 6397.
10. A. C. Legon and C. A. Rego, J. Chem. Phys., 99 (1993) 1463.
11. F. A. Bovey, Nuclear Magnetic Resonance Spectroscopy, Academic Press, New York, London, 1969.
12. G. C. Pimentel and A. L. McClellan, The Hydrogen Bond, Ch. 3, W. H. Freeman, San Francisco, 1960.
13. B. Berglund, J. Lindgren and J. Tegenfeldt, J. Mol. Struct., 43 (1978) 179.
14. R. Kubo, in D. ter Haar (Ed.), Fluctuation, Relaxation and Resonance in Magnetic Systems, Oliver and Boyd, Edinburgh, 1962.
15. C. Branden and J. Tooze, Introduction to Protein Structure, Garland Publishing, New York, London, 1991.
16. K. Wüthrich, NMR of Proteins and Nucleic Acids, Wiley-Interscience, New York, 1986.
17. L. Stryer, Biochemistry, W. H. Freeman, San Francisco, 1995.
18. S. P. Brown and H. W. Spiess, Chem. Rev., 101 (2001) 4125.
19. S. Grzesiek, F. Cordier, V. Jaravine and M. Barfield, Progr. Nuc. Magn. Res. Spectrosc., 45 (2004) 275.
20. L.-A. Näslund, D. C. Edwards, P. Wernet, U. Bergmann, H. Ogasawara, L. G. M. Pettersson, S. Myneni and A. Nilsson, J. Phys. Chem. A, 109 (2005) 5995.
21. A. W. Omta, M. F. Kropman, S. Woutersen and H. J. Bakker, Science, 301 (2003) 347.
22. J. D. Smith, C. D. Cappa, K. R. Wilson, B. M. Messer, R. C. Cohen and R. J. Saykally, Science, 306 (2004) 851.
23. S. H. Chen and J. Teixeira, in I. Prigogine (Ed.), Advances in Chemical Physics, Vol. LXIV, John Wiley, 1986.
24. L. Bosio, S. H. Chen and J. Teixeira, Phys. Rev. A, 27 (1983) 1468.

25. G. W. Neilson, P. E. Mason, S. Ramos and D. Sullivan, Phil. Trans. R. Soc. Lond. A, 359 (2001) 1575.
26. A. Almenningen, O. Bastiansen and T. Motzfeldt, Act. Chem. Scand., 23 (1969) 2848.
27. J. L. Derissen, J. Mol. Struct., 7 (1971) 67.
28. J. L. Derissen, J. Mol. Struct., 7 (1971) 81.
29. G. A. Jeffrey and W. Saenger, Hydrogen Bonding in Biological Structures, Springer Verlag, Berlin, 1994.
30. P. Chieux and J. C. Dore, in John C. Dore (Ed.), Hydrogen Bonded Liquids, Vol. 329, Kluwer Academic, Dordrecht, 1991.
31. H. Stiller, in P. Schuster and G. Zundel (Eds.), The Hydrogen Bond: Recent Developments in Theory and Experiments, Vol. III, Ch. 24, North Holland, Amsterdam, 1976.
32. M. Rozenberg, C. Jung and G. Shoham, Phys. Chem. Chem. Phys., 5 (2003) 1533.
33. H. Lücke, B. Schobert, H. T. Richter, J. P. Cartailler and J. K. Lanyi, J. Mol. Biol., 291 (1999) 899.
34. H. Belrhali, P. Nollert, A. Royant, C. Menzel, J. P. Rosenbusch, E. M. Landau and E. Pebay-Peroula, Structure, 7 (1999) 909.
35. F. Fillaux, M. H. Limage and F. Romain, Chem. Phys., 276 (2002) 181.
36. S. Tsuzuki, T. Uchimaru, K. Matsumara, M. Mikami and K. Tanabe, J. Chem. Phys., 110 (1999) 11906.
37. A. B. Baker, C. Samet, J. T. Lyon and L. Andrews, J. Phys. Chem. A, 109 (2005) 8280.
38. H. Ratajczak, W. J. Orville-Thomas and C. N. R. Rao, Chem. Phys., 17 (1976) 197.
39. J. Mavri and J. Grdadolnik, J. Phys. Chem. A, 105 (2001) 2045.
40. J. Mavri and J. Grdadolnik, J. Phys. Chem. A, 105 (2001) 2039.
41. N. A. Benedek, I. K. Snook, K. Latham and I. Yarowsky, J. Chem. Phys., 122 (2005) 144102.
42. D. Canet, La RMN, Concepts et Méthodes, Interéditions, Paris, 1991.

– 4 –

Infrared and Related Spectroscopies of H-Bonded Systems: Experimental Point of View

IR SPECTROSCOPY AND H-BOND VIBRATIONS

Among the various methods that are used to observe H-bonds, infrared spectroscopy (IR in the following) plays a special role, due to its particularly great sensitivity to the presence of H-bonds. It therefore comes as no surprise if this whole chapter and the following one are devoted to it. Before examining the effects of H-bonds in IR spectra, let us, however, clarify some points concerning IR spectroscopy. First, its abbreviation, IR. This was the usual abbreviation before the appearance and generalization of "Fourier transform spectrometers", around 1970. We use this abbreviation, because the other one that appeared more recently in the literature, FTIR, refers to the technique used to obtain spectra in the IR region, not to the principle of the method itself, which remains spectroscopy in the IR region. The physical interaction at work in spectroscopy is the interaction between a sample and an electromagnetic wave in the spectroscopic region we are interested in and this interaction does not depend on the technique used to monitor it. Most of the time FTIR is based on a Michelson interferometer, which delivers an "interferogram" that represents the value of the IR intensity that comes out of this interferometer as a function of the optical path difference δ of the two beams that interfere at the output of the Michelson (1). The IR spectrum itself is equal to the Fourier transform of the interferogram, and is computed from the recorded interferogram (a definition of the Fourier transform of a distribution is given in eq. (5.A17) of the appendix of Ch. 5). This is presently the technique that gives the best signal-to-noise ratio in IR spectra, at least when considering broad spectroscopic regions. In consequence most IR spectrometers use this technique, and nearly all IR spectra presented in this book were obtained using it. This is, however, only a technique, and nothing tells us that in the future IR spectra will always be recorded using this technique. Thus modern nonlinear time-resolved IR spectra that we examine in this chapter do not rely on the FTIR technique. The second point to clarify is that most spectra called IR spectra are related to spectra in the mid-IR region, that is with wavenumbers $\tilde{\nu}$ ($= \nu/c$, with ν the frequency of the IR wave and c the velocity of light) which are in the region $400 < \tilde{\nu} < 4000\,\text{cm}^{-1}$. When another spectral region is considered, it will be specified as FIR (far infrared region) for $\tilde{\nu} < 400\,\text{cm}^{-1}$ or NIR (near infrared region) for $4000 < \tilde{\nu} < 12{,}500\,\text{cm}^{-1}$. For H-bonds, we shall be mainly interested in the mid-IR region where *intramonomer* vibrations appear, that is, vibrations that mainly occur either in the X–H molecule or in the Y molecule that constitute the X–H···Y, H-bonded complex. When the IR region is not specified, it implicitly refers to this mid-IR region, following the usual terminology in the literature.

In the case of H-bonded systems, two kinds of vibrations are encountered: the *intramonomer* vibrations, defined above, and the *intermonomer* ones. These latter vibrations are the relative vibrations of the two molecules X–H and Y that establish an H-bond and form the H-bonded complex X–H···Y. These relative vibrations are ruled by a force constant due to the presence of the H-bond itself. In its absence this force constant is zero, and the two molecules X–H and Y perform independent translations and rotations, with no relative vibrations or, equivalently with vibrations having frequencies equal to 0. The force constants that govern intramonomer vibrations are due to the much stronger covalent bonds established inside X–H and Y. We start examining *intermonomer* vibrations due to H-bonds that appear in the FIR region. Vibrations in this FIR region are scarcely observed and the next section devoted to these vibrations is consequently relatively short. This is not so in the case of the intramonomer vibrations invoked in a subsequent section. For some of these vibrations the establishment of an H-bond has spectacular consequences and these vibrations consequently display a lot of information on H-bonds. The theoretical descriptions of the particular features of these intramonomer vibrations are described in the next chapter where general theoretical aspects of spectroscopy are also developed in an appendix. Reference to this appendix is made all along the present chapter when necessary. In this chapter, we mainly consider isolated H-bonds, occasionally systems with two or more H-bonds. It allows us to put into evidence fundamental spectroscopic properties of H-bonds. It will be of a great help to later interpret IR spectra of "H-bond networks", assemblies of closely lying and interacting H-bonds, which naturally appear in the presence of H_2O molecules.

INTERMONOMER VIBRATIONS IN THE FIR REGION

Description

The three intermonomer vibrations of a single H-bond consist of small oscillations around the equilibrium positions Q_0, θ_0 and φ_0 of the X–H···Y complex we have defined in Ch. 2 (Figure 2.1). These oscillations are defined by coordinates $Q_s - Q_0$, for the intermonomer stretching vibration that is of a translational type, and by angles θ–θ_0 for in-plane intermonomer bending and $\varphi - \varphi_0$, for out-of-plane intermonomer bending vibrations that are of rotational type. These intermonomer bending vibrations may also be called torsional vibrations. As we are dealing with vibrations that are governed by a potential, it is easier to consider the intermonomer bending vibration coordinates Q_θ and Q_φ, equal to $\Delta Q(\theta - \theta_0)$ and $\Delta Q(\varphi - \varphi_0)$, respectively, instead of $\theta - \theta_0$ and $\varphi - \varphi_0$ themselves. The quantity ΔQ is a distance that is equal to the H···Y distance $(Q_0 - q_0)$ (see Figure 2.1) when the X–H molecule of the H-bonded complex X–H···Y cannot move or is much heavier than Y. It is equal to the X···Y Q_0 distance when Y cannot move or is much heavier than X–H, or to an intermediate value when both molecules X–H and Y have no other constraints than being H-bonded. The point is that when one chooses $\theta-\theta_0$ and $\varphi-\varphi_0$ as representative coordinates instead of Q_θ and Q_φ, the kinetic energy is expressed in terms of angular momenta, and the corresponding operators in quantum mechanics are more difficult to handle than operators of usual vibrations. These angular coordinates are consequently mainly used in case of free rotations in the gas phase, described in Ch. 3, where no potential energy term is present and where eigen functions and eigen values of angular momentum operators are known. The point to be noted is that quantities such as $M_\theta \Delta Q^2$ and $M_\varphi \Delta Q^2$ have the dimension of a moment of inertia. For simplification, we suppose in the following that at equilibrium the H-bond is linear, that is θ_0 and φ_0 are equal to 0. Changing these values at equilibrium often reduces to nothing except a longer writing. In the harmonic approximation, which is valid when the amplitudes of vibrations remain small, the potential that governs these intermonomer vibrations may be written, within a constant irrelevant term, as

$$U(Q_s, Q_\theta, Q_\varphi) = \frac{1}{2} M_s \Omega_s^2 (Q_s - Q_0)^2 + \frac{1}{2} M_\theta \Omega_\theta^2 Q_\theta^2 + \frac{1}{2} M_\varphi \Omega_\varphi^2 Q_\varphi^2 \tag{4.1}$$

so that the corresponding vibrational Hamiltonian is written as

$$H(Q_s, Q_\theta, Q_\varphi) = \frac{P_s^2}{2M_s} + \frac{P_\theta^2}{2M_\theta} + \frac{P_\varphi^2}{2M_\varphi} + U(Q_s, Q_\theta, Q_\varphi) \tag{4.2}$$

where M_s, M_θ and M_φ are the reduced masses (see appendix of Ch. 5) of the vibrations defined by coordinates Q_s, Q_θ and Q_θ and the conjugated momenta P_s, P_θ and P_φ are equal to

$$P_s = -i\hbar \frac{\partial}{\partial Q_s} \quad P_\theta = -i\hbar \frac{\partial}{\partial Q_\theta} \quad P_\varphi = -i\hbar \frac{\partial}{\partial Q_\varphi} \tag{4.3}$$

with $\hbar = h/2\pi$ (h = Planck's constant). In these equations, the masses M_s, M_θ and M_φ are reduced masses of molecules. They are bigger than the atomic masses encountered in the case of intramonomer vibrations of H-atoms by at least one order of magnitude. Furthermore, the force constants $M_s\Omega_s^2$, $M_\theta\Omega_\theta^2$ and $M_\varphi\Omega_\varphi^2$ that appear in eqs. (4.1) and (4.2) are those due to an H-bond. They are consequently weaker than the force constants that govern intramonomer vibrations due to covalent forces by more than one order of magnitude as shown in eq. (5.A27) of the appendix of Ch. 5. It therefore comes as no surprise that if intramonomer vibrations are found in the mid-IR region, these intermonomer vibrations appear in the FIR region. Within the harmonic approximation, that is with $U(Q_s, Q_\theta, Q_\varphi)$ defined by eq. (4.1), vibrational energy levels of these intermonomer vibrations are equal to

$$E_{N_s, N_\theta, N_\varphi} = \left(N_s + \frac{1}{2}\right)\hbar\Omega_s + \left(N_\theta + \frac{1}{2}\right)\hbar\Omega_\theta + \left(N_\varphi + \frac{1}{2}\right)\hbar\Omega_\varphi \tag{4.4}$$

where N_s, N_θ and N_φ are positive (or null) integers. They are consequently equidistant by energies $\hbar\Omega_s$, $\hbar\Omega_\theta$ and $\hbar\Omega_\varphi$. Let us note that the absence of terms that couple Q_s and Q_θ (Q_φ) in eq. (4.1) make these three vibrations the normal intermonomer modes (it is the definition of normal modes). Let us note that, as seen in the appendix of Ch. 5, the equilibrium position for the intermonomer stretching mode Q_s is not exactly Q_0, but $\bar{Q}$ (eq. (5.A44)), which differs from Q_0 by a correction factor due to the strong characteristic anharmonic coupling of q (the stretching vibration of the H-atom in X–H) with Q_s. We shall discuss this point in the next section related to intramonomer vibrations, but we presently disregard this effect, in a first approximation.

In this harmonic approximation only transitions appear between these various vibrational energy levels that correspond to a change ΔN_i of a single N_i of eq. (4.4) being equal to 1 or −1 ($i = s$, θ or φ) while the two other ΔN_j with $j \neq i$ are equal to 0. IR bands consequently appear at wavenumbers $\tilde{\nu}_i$ that are related to frequencies of vibrations ν_i and related quantities Ω_i by the equation

$$\hbar\Omega_i = h\nu_i = hc\tilde{\nu}_i \tag{4.5}$$

Experimental values of Ω_s and Ω_θ or Ω_φ measured on simple isolated H-bonded dimers in the gas phase are given in Table 4.1. In these experiments, the distinction between Ω_θ and Ω_φ was not always experimentally possible. Furthermore, these two bending vibrations are often nearly degenerated in isolated H-bonds, which means they have nearly equal energies. For these reasons, we give in this table a single value for most of them. We may see that the number of simple H-bonded systems where such measurements have been performed is limited. It is an indication of the difficulty to perform them. It nevertheless gives an indication concerning the wavenumbers of these intermonomer vibrations; the bending vibrations appear at wavenumbers that are nearly always smaller than stretching vibrations. They often fall at the limit of feasibility of most IR spectrometers, around 50 cm^{-1}. They are more visible in Raman spectra, which may be more adapted to this FIR region. The exceptions to this rule are ice and liquid water (Chs. 8 and 9) where bending

Table 4.1

Wavenumbers of FIR bands of various isolated H-bonded dimers in the gas phase

H-bonded dimer	Wavenumber (cm^{-1}) stretching Q_s	Wavenumber (cm^{-1}) bending Q_θ and Q_φ
F–H···NCH	155	70
F–H···OH_2	180	170, 145
F–H···$NCCH_3$	168	40
F–H···OMe_2	170	
Cl–H···$NCCH_3$	100	40
Cl–H···OMe_2	120	50
$(CH_3COOH)_2$	171 (B_u) 155 (A_g)	48.5(A_u), 56 (A_u) 99 (B_g), 120 (A_g)

Values for F–H and Cl–H complexes are from refs. (9) and (2). Only in the case of F–H···OH_2 can Q_θ and Q_φ be unambiguously distinguished. Values for acetic acid are from ref. (3). The C_{2v} symmetry of the cyclic dimers drawn in Figure 4.4 leads to modes that are symmetric (A) or antisymmetric (B) with respect to the twofold rotation C_2 around axis perpendicular to the plane of symmetry and symmetric (g) or antisymmetric (u) with respect to inversion (see ref. (3) for visualisation of these modes). IR bands are of type u, Raman bands of type g.

intermonomer vibrations, called librations in these cases, are found around 700–900 cm^{-1}, whereas the stretching intermonomer vibrations fall around 200 cm^{-1}. These exceptionally great values of wavenumbers of bending intermonomer modes are due to the especially small moments of inertia of the H_2O molecule, which corresponds to a particularly small effective mass of the H-atom.

Let us notc that, in opposition to force constants, which greatly differ when passing from an intramonomer vibration driven by covalent forces to an intermonomer vibration driven by forces due to an H-bond, the vibrational amplitudes are of the same order of magnitude for both these vibrations, as shown in eq. (5.A31) in the appendix of Ch. 5. The difference is that the amplitudes of intermonomer modes, with their Ω_s, Ω_θ or Ω_φ of the order of $kT/\hbar$ at room temperature (around 300 K), are sensitive to temperature changes. This is not the case of intramonomer modes: they have their ω always much greater than $kT/\hbar$ (or frequencies $\nu >> kT/h$), and it makes their corresponding bands hardly temperature sensitive.

Anharmonicities of intermonomer modes

So far, we have considered these intermonomer modes in the harmonic approximation, when the vibrational potential can be limited to terms in x^2, of order 2 in the displacements $x = Q_s$, Q_θ or Q_φ of atomic distances. This is often a good approximation that allowed us to characterize these bands by wavenumbers at which they appear in IR spectra and which are equal to $\Omega/2\pi c$. It may nevertheless appear insufficient in some cases. Thus, if one stretches the X$\cdots$Y distance, that is, if one makes Q_s bigger and bigger, the vibrational energy should tend towards the energy E_{HB} of the H-bond itself. In the harmonic approximation, it tends towards infinity. In order to correct this misrepresentation, we have to introduce supplementary terms of order greater than 2. Another more convenient way to introduce such anharmonic terms is to suppose that the vibrational potential takes on the form of a "Morse potential" for this stretching mode. The part of the vibrational potential energy (eq. (4.1)) that concerns Q_s is then written as

$$U(Q_s, Q_\theta, Q_\varphi) = \frac{\hbar\Omega_s}{\Delta}\{1 - e^{-\sqrt{M_s\Omega_s\Delta/2\hbar}(Q_s - Q_0)}\}^2 + \cdots \tag{4.6}$$

where dots refer to that part, which concerns Q_θ and Q_φ that we are not immediately interested in. This Morse potential osculates the harmonic potential for Q_s defined in eq. (4.1), that is, it has same tangent and same curvature for Q_s in the vicinity of Q_0. This may be seen by developing the exponential term of eq. (4.6) in powers of $Q_s - Q_0$. This development also shows that when the anharmonicity parameter Δ tends towards 0, the Morse potential then reduces to the harmonic one of eq. (4.1). A Morse potential is drawn in Figure 4.1, in the case of such a typical H-bonded complex as Cl–H$\cdots$O$(CH_3)_2$ for which experimental results exist in the literature. When $Q_s - Q_0$ is great, that is when the H-bond is greatly stretched, the Morse potential energy of this mode exponentially tends towards the dissociation energy E_{HB} of the H-bond. Eq. (4.6) allows relating this value to the anharmonicity parameter Δ:

$$\frac{\hbar\Omega_s}{\Delta} = E_{HB} \tag{4.7}$$

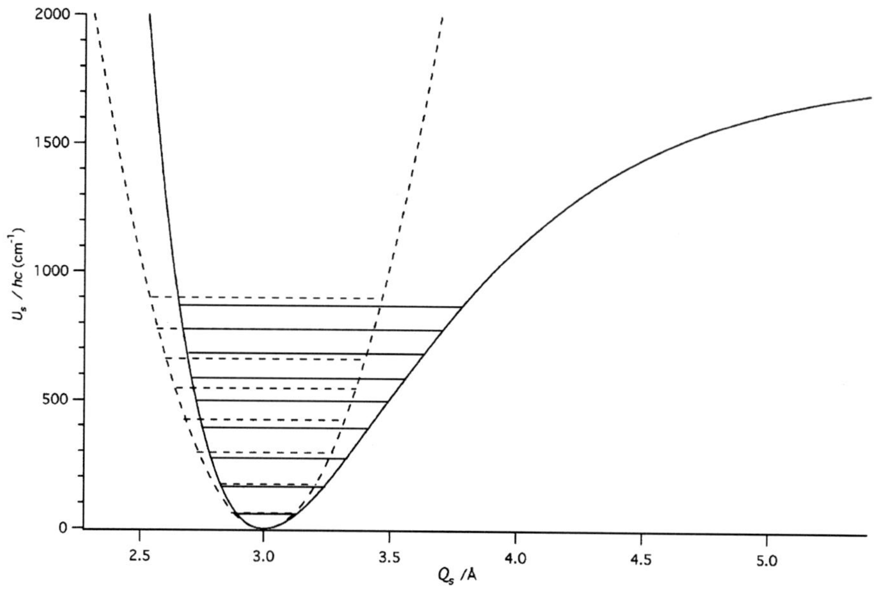

Figure 4.1 The vibrational potential U_s of the stretching intermonomer mode Q_s around its equilibrium position $Q_0 = 3$ Å, for such a complex as ClH···O$(CH_3)_2$ (2), in the form of a Morse potential (solid) or of the osculating harmonic potential (dashed). U_s is that part of the vibrational potential $U(Q_s, Q_\theta, Q_\varphi)$ of eq. (4.1), written in eq. (4.6), which depends only on Q_S. The lowest energy levels that are significantly thermally populated at 300 K, and correspond to smaller values of N_s (eqs. (4.4) and (4.8)), are drawn for both potentials. Ω_s (eq. (4.1) for the harmonic potential and eqs. (4.6) for the Morse potential) is such that $\Omega_s/2\pi c = 120\,\text{cm}^{-1}$ (c, speed of light in cm sec^{-1}). The anharmonicity parameter Δ is equal to 0.067 and corresponds to a dissociation energy E_{HB} (eq. (4.7)) of the H-bond equal to 22 kJ mol^{-1}. For Q_s tending towards infinity, U_s/hc tends towards $E_{HB}/hc \simeq 1800\,\text{cm}^{-1}$.

This Morse potential implies that the average X···Y distance increases with temperature, an effect well known as dilatation. This is because the average distance of Q_s in the N_sth excited state increases with N_s (Figure 4.1), and an increase of temperature populates levels with greater N_s. Such a dilatation is not accounted for in the harmonic approximation where the average distance remains equal to Q_0 in all states. The levels of the Hamiltonian with potential energy defined in eq. (4.6) are now

$$E_{N_s,N_\theta,N_\varphi} = \left(N_s + \frac{1}{2}\right)\hbar\Omega_s\left[1 - \frac{\Delta}{4}\left(N_s + \frac{1}{2}\right)\right] + \cdots \qquad (4.8)$$

Instead of being equally spaced, as in a harmonic oscillator (eq. (4.4) and Figure 4.1), their differences in energies decrease when N_s increases. The $N_s \to N_s + 1$ IR transition becomes smaller when N_s increases, which results in a shift towards lower wavenumbers when these excited levels become populated, for instance when temperature increases. As this shift is small with respect to the wavenumber of the centre of this band, 120 cm^{-1}, it is at the origin of a small broadening of this band.

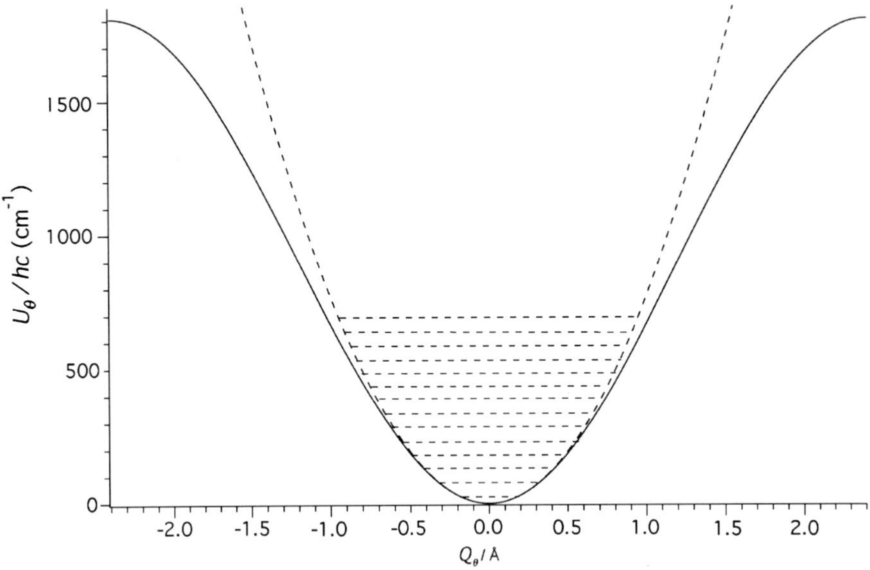

Figure 4.2 The vibrational potential U_θ of the bending intermonomer mode Q_θ, for such a complex as $ClH\cdots O(CH_3)_2$ (2), in the form of a periodic potential schematically drawn as a potential in $\sin^2\theta$ (solid line) and of the osculating harmonic potential (dashed line). U_θ is that part of the vibrational potential $U(Q_s, Q_\theta, Q_\varphi)$ of eq. (4.1) that depends only on Q_θ. The lowest energy levels that are significantly thermally populated at 300 K, and correspond to small values of N_θ (eq. (4.4)), are drawn for the harmonic potential. Ω_θ (eq. (4.1) for the harmonic osculating potential) is such that $\Omega_\theta/2\pi c = 50\,cm^{-1}$ (c, speed of light in cm sec^{-1}). The maximum of the periodic potential is $U_\theta/hc = E_{HB}/hc \simeq 1800\,cm^{-1}$ and corresponds to a dissociation energy of the H-bond, E_{HB} (eq. (4.7)), equal to 22 kJ mol^{-1}. This figure is valid for θ being replaced by φ.

The bending intermonomer modes also display similar 1D anharmonicity. By "1D" we imply that the anharmonicity concerns a single mode. It does not mix various modes, as a multidimensional anharmonicity does. This 1D anharmonicity of bending intermonomer modes arises from the 2π periodicity in θ or φ of the vibrational potential $U(Q_s, Q_\theta, Q_\varphi)$ in eq. (4.1). It is schematically represented in Figure 4.2 in the form of a potential in $\sin^2(Q_\theta/\Delta Q)$, together with the harmonic potential that osculates this periodic potential at Q_θ (or Q_φ) equal to 0. We may see that this periodic potential does not significantly differ from the harmonic one at not-too-big values of Q_θ. This type of anharmonicity has consequently no numerous implications, except perhaps in the case of the H_2O molecule in liquid water, which we have already seen to have such small moments of inertia that these modes have wavenumbers around 700 cm^{-1}, differing by more than one order of magnitude of the typical value, 50 cm^{-1}, adopted for $\Omega_\theta/2\pi c$ and $\Omega_\varphi/2\pi c$ in this figure, and which is typical of many H-bonded systems. This will make intermonomer bending vibrations (librations) of H_2O molecules in liquid water play a particularly important role that is usually devoted to stretching intermonomer vibrations in practically all other H-bonded systems.

Beyond these 1D anharmonicities, the intermonomer modes may also display a multidimensional anharmonicity that couples various modes. It means that the potential

$U(Q_s, Q_\theta, Q_\varphi)$ has terms of the form $Q_s^l Q_\theta^m$ with $l+m>2$. Their existence is strongly suggested by experiments where H-bonds are compressed. They then exhibit strong distortions and relatively small decreases of equilibrium X···Y distances Q_0. Such a behaviour is due to an anharmonic coupling of Q_s with both Q_θ and Q_φ. It is not easy to visualize. We represent part of it in Figure 4.3, where the equipotential curves for $U(Q_s, Q_\theta, Q_\varphi)$, or contour lines, are drawn in the plane Q_s, θ, with φ maintained at a fixed value. In the case of an isolated H-bond, which is the only one we consider here, similar equipotential curves are found in the Q_s, Q_φ plane with θ a constant. We see that for small amplitudes of Q_s and θ (or Q_θ), equipotential curves are ellipses. It is the domain of the harmonic approximation, where $U(Q_s, Q_\theta, Q_\varphi)$ of eq. (4.1) is a constant. With Q_φ also a constant, this equation in Q_s and Q_θ is that of an ellipse in the Q_s, Q_θ plane. At higher amplitudes of vibrations these equipotential curves depart from an ellipse and take on the shape of a cardioid, the transition between these two types of curves being one with a long path parallel to the θ axis. It means that when Q_s is negative, diminishing the value of this stretching coordinate requires a lot of energy. Compression then preferentially distorts the H-bond, instead of shortening it. In the cardioidic region, these torsional modes have limited amplitudes, but these amplitudes occur around two symmetric equilibrium positions that differ from $\theta = 0$. The presence of this cardioid shape explains the effect of pressure mentioned above. Mathematically, the effect of a constant applied pressure consists of adding to the potential $U(Q_s, Q_\theta, Q_\varphi)$ a term linear in Q_s. It somewhat modifies the equipotentials, displacing, for instance, the central elliptical characteristic of the harmonic approximation towards smaller values for Q_s. For greater values of Q_s and Q_θ, it also displaces equipotential curves that are not ellipses towards lower values of Q_s, but this displacement is accompanied by a (small) distortion. We may then see that pressure induces a shortening of the Q_s equilibrium position, which is not so much unexpected. What is more unexpected is that a strong distortion of these bonds accompanies this shortening, making the equilibrium positions for bending intermonomer vibrations different from 0. There is not, however, a single equilibrium position, but two that are symmetric in θ, that is they appear at θ_m and $-\theta_m$. This rupture of symmetry, which is well known in ordinary mechanics and is at the origin of buckling of materials, will have important consequences in the case of liquid water, where the number of H-bonds is exceptionally high. It will then be at the origin of many of its exceptional physical properties. This kind of anharmonic coupling is still qualitative, and no number has yet been put on it. This is because it does not give spectacular effects for most H-bonded systems, except for H_2O molecules in liquid water. Let us note that the potential in Q_s drawn in Figure 4.3 for $\theta = 0$ is of a Morse type, as drawn in Figure 4.1, and the potential in θ is periodic, as drawn in Figure 4.2 with, however, a flatter part when θ takes values between $\pi/2$ and π.

INTRAMONOMER VIBRATIONS IN THE MID-IR REGION

Intramonomer vibrations are found inside the X–H or Y molecules. They consequently exist even in the absence of an H-bond between these two molecules, in opposition to intermonomer bands we have seen above, which exist due to the formation of the H-bond. An H-bond has, however, marked effects on some intramonomer vibrations. It has a spectacular effect on the stretching vibration of the H-atom of X–H that establishes the H-bond.

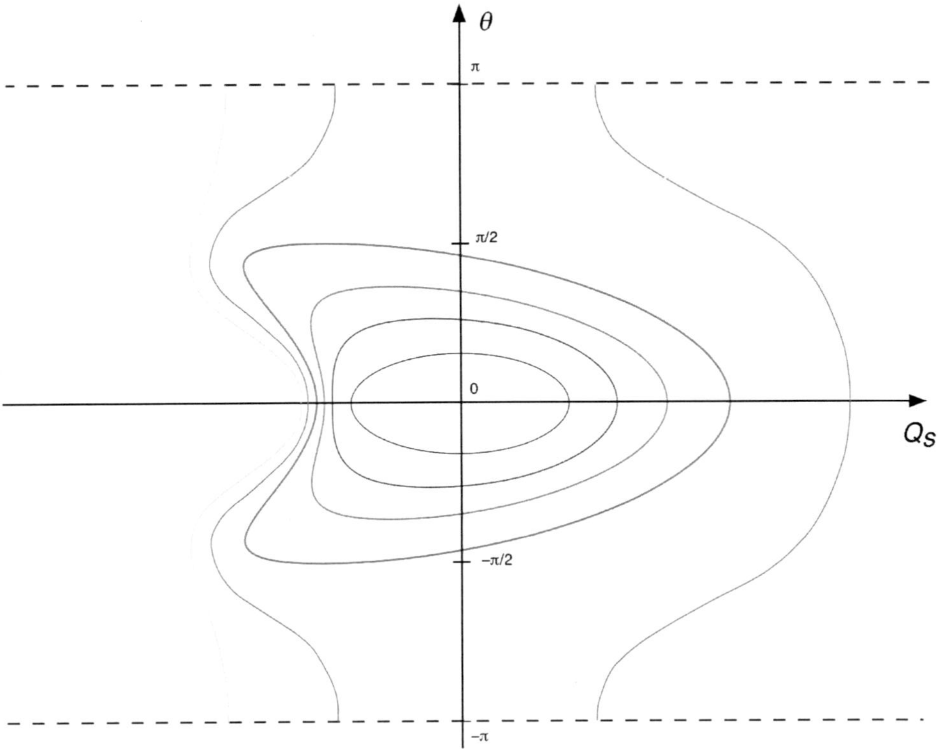

Figure 4.3 Schematics of equipotential curves in the Q_s, Q_θ (or θ) plane of the vibrational potential $U(Q_s, Q_\theta, Q_\varphi)$ that governs intermonomer modes of a single H-bond, Q_φ (or φ) keeping a constant value.

Stretching bands ν_s

Introduction

This band is usually labelled ν_s, or ν_s(X–H···) in the literature, which is somehow redundant, as ν usually labels stretching vibrations in vibrational spectroscopy. It is also sometimes labelled $\nu_{X-H\cdots}$. We use the shortest one, ν_s, but may occasionally use ν_s(X–H···) when we have to specify whether X–H is H-bonded or not. We illustrate the effect of the H-bond on ν_s in Figure 4.4, which shows the absorbance of acetic acid vapour in the 2000–3700 cm^{-1} region. Absorbance is defined in eq. (5.A8) of the appendix of Ch. 5. In this acetic acid vapour, monomers CH_3COOH are in equilibrium with cyclic dimers $(CH_3COOH)_2$. As a substitution of the H-atom in the COOH group by a D-atom proves (see Ch. 7) as also the changes of temperature or pressure, nearly all the bands that appear in this figure are ν_s(O–H···), the exception being features of very small intensities due to C–H stretching modes that we can disregard. The ν_s(O–H) band that originates from monomers, that is from O–H free groups that do not establish H-bonds, is that band that culminates around

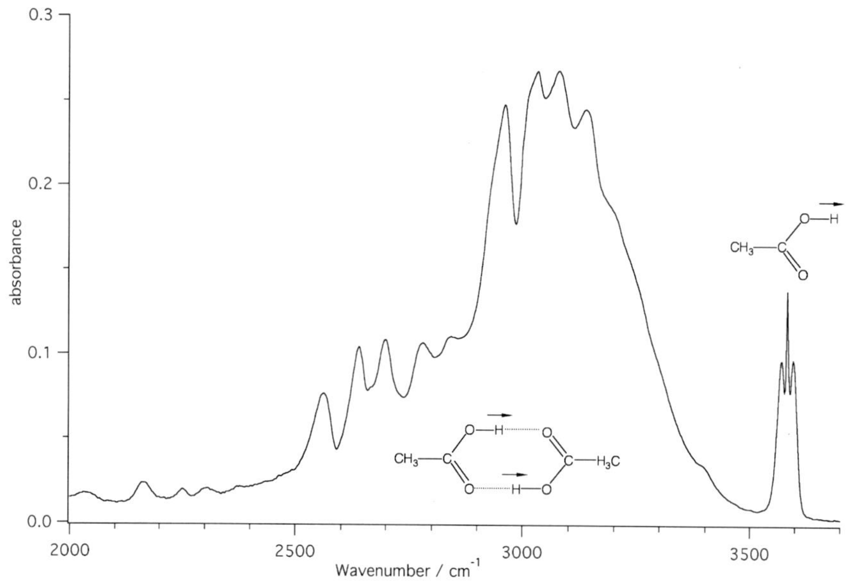

Figure 4.4 Absorbance spectrum in the mid-IR region of acetic acid vapour. The pressure and temperature are such that the number of monomers is approximately twice that of dimers.

$3600\,cm^{-1}$. It displays a typical rotational structure at the origin of its three peaks. All other bands due to this monomer display such a typical rotational substructure. They are not visible in this figure because they appear at wavenumbers lower than $2000\,cm^{-1}$. They are bands that are typical of harmonic vibrations of molecules in the gas phase. The appreciable width of ν_s(O–H) that is displayed in Figure 4.4 is almost entirely due to this rotational structure, that is due to transitions between various rotational levels that accompany the $0 \rightarrow 1$ transition in ν_s. The ν_s(O–H) band is thus quite a normal band that can be well described within the harmonic approximation for vibrations. We may note that this band in the vicinity of $3600\,cm^{-1}$ is typical of an O–H group that does not establish an H-bond: in addition to carboxylic acid vapours, it appears in such different compounds as water vapour, vapours of alcohols, vapours of other carboxylic acids, or of sulfuric acid, etc. It allows us to unambiguously characterize such an O–H group.

The rest of the spectrum displayed in Figure 4.4, that is, the *whole band that extends from 2000 to 3700 cm^{-1}* is due to H-stretching vibrations ν_s of the cyclic dimer, that is, ν_s(O–H···) vibrations of H-bonded O–H···O groups. When temperature increases, the intensity of this whole ν_s(O–H···O) band weakens, while that of the ν_s(O–H) band around $3600\,cm^{-1}$ increases. It means that H-bonds are disrupted when temperature increases, transforming the cyclic H-bonded dimers into monomers. The difference between these two bands, ν_s(O–H) and ν_s(O–H···O), is spectacular. The integrated intensity of ν_s(O–H···O) is greater than that of ν_s(O–H) by more than one order of magnitude, as may be seen in Figure 4.4 where pressure and temperature have been so chosen that the number of free O–H groups is of the same order of magnitude as that of H-bonded O–H··· groups. Its centre is shifted towards lower wavenumbers by about 500 cm^{-1} with respect to ν_s(O–H). Its width is increased also by more than one order of magnitude. It does not display any rotational structure. This is however not so much surprising, in view of the great enhancement of the moments of inertia of the dimer as compared to those of the monomer. ν_s(O–H···O) furthermore exhibits a well-marked substructure with numerous minima and maxima. The point worth noting is that these spectacular features appear in *all H-bonded systems*, be they found in the gas, liquid or solid states. In other words, these specific and greatly marked spectroscopic features are a *signature* of H-bonds. We examine them in more details in the following subsections.

Integrated intensities

The denomination "intensity" may be confusing as it is sometimes used for the value of the absorbance of a band at its maximum. This latter value, which is also called the "height" of the band, does not characterize it, but, as shown in the appendix of Ch. 5, the integrated intensity of a band equal to its moment M_0 of order zero does characterize it. The enhancement of the integrated intensity of the ν_s mode is the most spectacular effect of H-bonds. As shown in the next Ch. 5, it is due to the exceptionally great variation of the electric dipole moment of X–H···Y that accompanies a stretching of the X–H distance. It is not often invoked, however, because it requires measurements that are not easy to carry out and are most often imprecise. Thus, chemists rather use the shift of the centre of ν_s to characterize or extract information on H-bonds, and as we shall see in the next subsection it is much easier to measure quantity. Let us note, however, a few measurements of integrated intensities. Thus, the integrated intensities of ν_s(O–H···O) in carboxylic acids (Figure 4.4) has been found to be 25 times greater than that of ν_s(O–H) (4). In the case of liquid water-propanol mixtures, the integrated intensities of the ν_s(O–H···) bands of the propanol-water H-bonded complex are about 2 times more intense than ν_s(O–H···) of H_2O molecules in liquid water (5), itself about 17 times greater than that of H_2O molecules of water vapour (no H-bond), and about 0.7 that of ν_s(O–H···) of ice (6). The integrated intensity of ν_s(O–H···) in ice has also been estimated to be greater than that in water vapour by a factor equal to 20–30 (7), which is in agreement with preceding ones and gives an indication of the difficulty to precisely measure absolute integrated intensities.

Even if not frequently measured, this enhancement of intensity of ν_s makes IR spectroscopy a method that is *hypersensitive to H-bonds*. It allows a precise observation of H-bonds in the gas phase where they are often in small number. This is of a great interest, as it allows the study of isolated H-bonds, as shown by Legon and Millen (8–10) and others (11), which gives access to the intrinsic properties of H-bonds. Gases at weak pressures can be expanded at hypersonic speeds, a technique known as "jet-cooled expansion" that we have described in Ch. 3, as it has expertly been used in the case of microwave spectroscopy. As the dispersion of the speeds of H-bonded complexes is then extremely reduced, this is equivalent to studying theses gases at very low temperatures, in the vicinity of some K. The hypersensitivity of IR spectroscopy to H-bonds also allows the study of H-bonds in solid matrices of inert gases at low temperatures, that is, of H-bonded complexes that are trapped in solid Ar, N_2, etc. at cryogenic temperatures. As we shall see in Ch. 6, it gives valuable information on the transfers of protons in such simple, isolated H-bonded complexes as Cl–H···NH_3 where the transfer of protons that transforms it into its ionized form Cl^-···NH_4^+ could be studied and has been shown to be extremely sensitive to the presence of H-bonds in its vicinity (12–14). It also put into evidence the weak H-bonds N_2 molecules accept, a result that may have important consequences in atmospheric chemistry. We cannot sufficiently stress the interest of these results that have been collected using these methods, and which eventually gave us a precise understanding of the original properties of simple H-bonds, such as their spectroscopic properties we describe in this chapter, but also of their thermodynamic properties, and of the reactivity of H-bonds we describe in Ch. 6. This is crucial, because we shall see in aqueous systems the importance of the interactions of H-bonds, more precisely in the "H-bond networks" that appear with the presence of H_2O molecules, and which have collective properties we could have hardly understood and could not have defined if we had previously not got a precise knowledge of the properties of isolated H-bonds. It could be achieved thanks to the hypersensitivity of IR spectroscopy to H-bonds. Microwave spectroscopy, described in Ch. 3, is also a precise technique, in particular for what concerns the geometrical parameters of these simple H-bonded systems. The sensitivity of IR spectroscopy has more general impact, as IR spectroscopy is a method of a much wider application that may also be used with liquids and solids. Its sensitivity may then be used to study systems where the number of H-bonds is small.

This hypersensitivity may in a paradoxical way become an inconvenience in some circumstances. This is so when the number of H-bonds becomes too big. It may then lead to saturation of IR bands, typically when absorbance becomes greater than 3. From the definition of the absorbance in eq. (5.A8) of the appendix of Ch. 5, the intensity of the IR beam that comes out of the sample is then smaller than 10^{-3} times that of the incoming beam, causing the level of noise to become appreciable with respect to that of the signal. It ruins all the interest in IR spectra. We shall see in Ch. 9 that it happens for liquid water. In that particular case, IR can, however, be replaced by Raman spectroscopy which, in opposition to IR, displays a very low sensitivity to H-bonds. In Raman, H-bonds give spectra that have a shape similar to that found in IR spectra. They are not identical as selection rules may be different in Raman and IR. More generally, this drawback of IR occurs when H_2O molecules are present, as they are always at the origin of a very great number of H-bonds (Ch. 8), which makes IR absorbance easily become greater than 3 if measurements are made with no special care. We shall nevertheless see in Ch. 11 that methods exist that allow us to overcome this difficulty, with IR spectroscopy still being the most powerful method to observe H-bonds, even in the presence of numerous water molecules.

ν_s band centres

The shift $\Delta\bar{\nu}$ of the centre of ν_s(X–H···), in Figure 4.4 of about $-500\,\text{cm}^{-1}$, that is towards lower wavenumbers as compared to that of ν_s(X–H), is a general property of H-bonds that may be easily measured. Although less spectacular than the enhancement of the integrated intensity we have seen above, it provides a simple method to characterize H-bonds. It may be accurately measured by calculating the integral values of the first two moments M_0 and M_1 of ν_s (eq. (5.A14) of the appendix in Ch. 5). Its great interest, however, lies in the rapid estimation that can be visually performed on the spectrum itself, with an accuracy of about 30–50 cm^{-1} that is often sufficient, and remains much smaller than the width of ν_s. This shift always occurs towards lower wavenumbers, making $\Delta\bar{\nu}$ always negative, with the exception of the still controversial nonconventional C–H··· H-bonds mentioned in Ch. 1.

It implies that the force constant of ν_s(X–H···) is smaller than that of ν_s(X–H). This is an indication that the X–H covalent bond is weakened when this group establishes an H-bond, an effect well taken into account in the theoretical approach (15) that considers the formation of an H-bond as mainly due to a transfer of electronic density from the nonbonding orbital of Y to the excited antibonding σ^* orbital of X–H. This charge transfer is at the origin of this weakening of the H-atom stretching force constant that is in turn partly responsible for this shift of the centre of ν_s towards lower wavenumbers and entirely responsible for the lengthening of the X–H equilibrium distance with the establishment of an H-bond. It is also at the origin of the increase of the variation of the electric dipole moment we have seen to give the general great enhancement of the integrated intensity of ν_s. Let us note that this spectral shift towards lower wavenumbers is sometime called a "red shift". In IR spectroscopy this qualification is ambiguous, because, in opposition to the visible spectroscopical region from which it takes its origin, the region of red light is found at higher wavenumbers in IR. Consequently, we shall not use this qualification.

This shift of ν_s is often easily measured with the help of an IR spectrometer that is most common and easily operable in many laboratories. The existence of correlations of this shift with other characteristic quantities of H-bonds, such as their enthalpies ΔH or their X···Y equilibrium distances, allows a rapid estimate of the magnitudes of these quantities that would otherwise require more time and effort to directly measure them. This is the reason why IR spectroscopy is so often used in routine procedures. It should nevertheless be kept in mind that retrieving such quantities from an indirect IR measurement can only be done within the limits of such a procedure that are mentioned in the following subsections.

Shifts $\Delta\bar{\nu}$ of ν_s and enthalpies ΔH of H-bonds

The first relations between $\Delta\bar{\nu}$ and enthalpies ΔH of formation of H-bonds have been established by Badger and Bauer on a few H-bonded systems such as water, ethanol, cyanhydric and chlorhydric acids (16). The consequent "Badger–Bauer rule" states that within the same family of H-bonded complexes, the greater ΔH, the larger the absolute value of $\Delta\bar{\nu}$. IR allows more precise correlations to be established. The point however is, as we have already seen, that actual measurements of ΔH have been performed in a small number of systems only. They are therefore often replaced by *ab initio* calculations of the value of the enthalpy. It then becomes possible to define linear relationships between these two quantities. This is the case of systems made of various alcohols or phenols establishing H-bonds on various amines (17). It allows a rapid estimate of ΔH from a simple IR spectrum of the system. Let us repeat, however, that this is no general law and that even at the level of an estimate such a correlation should be handled with care. We may nevertheless give some values. Thus weak H-bonds such as water–benzene complexes that we have seen in Ch. 1 to have ΔH of about –2 kcal mol^{-1} (about –9 kJ mol^{-1}) have ν_s bands shifted (18) by about $-70\,\mathrm{cm}^{-1}$, a relatively easily measured quantity. Stronger bonds such as in phenol–acetone φ–O–H···O=C–$(CH_3)_2$ exhibit a shift of their ν_s centres of about $-260\,\mathrm{cm}^{-1}$ for a ΔH of about -4.7 kcal mol^{-1} (about -19.6 kJ mol^{-1}) (17).

Shifts $\Delta\bar{\nu}$ of ν_s and H-bond distances

The relation between $\Delta\bar{\nu}$ and average length $\bar{Q}$ of an H-bond (equal to Q_0 in the first approximation, see eq. (5.5) in Ch. 5 and eq. (5.A44) in the appendix of Ch. 5) is at the

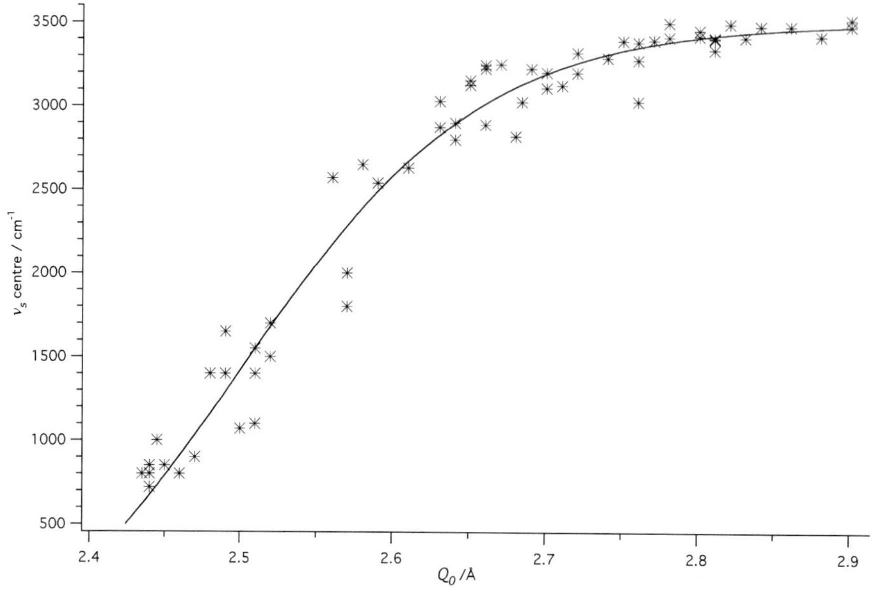

Figure 4.5 Novak's correlations between centres of ν_s bands and equilibrium distances Q_0 of a series of O–H···O bonds in crystals. Data are from Table 2 of Novak's article (19). The curve is the best-fit sigmoid for these data.

origin of more accurate correlations. It is illustrated in Figure 4.5 that reproduces, in a somewhat modified presentation, one of the correlations disclosed by Novak (19) between spectroscopic and geometrical quantities of H-bonds. In this figure, the wavenumbers of band centres of various O–H···O bonds of various crystals, most of them inorganic hydrates, are displayed along the y axis, while the corresponding O···O distances Q_0 are displayed along the x axis. It shows a marked correlation of these two quantities. The wavenumbers of band centres may not be exactly equal to the quantities $\overline{\nu}$ defined in the appendix of Ch. 5, as they were taken equal to the wavenumbers of peaks of ν_s bands. This is not exactly the same as the more rigorous mathematical definition given in this appendix of Ch. 5. These two quantities, however, do not differ much. It justifies making no distinction between the values Q_0 and $\overline{Q}$ of equilibrium distances, advocated above. The values of Q_0 displayed in Figure 4.5 were obtained by a compilation of crystallographic results in the literature. Such a correlation curve is most useful, as it allows us to deduce from an easily obtained IR spectrum a first value for the equilibrium O···O distance of an unknown H-bond. An example is given in Table 2.2 of Ch. 2, where values of Q_0 have been obtained using this method. Novak gives in the same article (19) other useful correlation curves, particularly for H-bonds of types other than O–H···O, such as N–H···N, or correlations of the wavenumbers of band centres with O–H equilibrium distances, which become greater when the strengths of the H-bonds are enhanced, that is when the O···O distances are shortened. Also, correlations between band centres of ν_s bands and O···O distances of O–D···O bonds have been published by other authors (20). These "Novak's curves" are therefore central for deducing properties of H-bonds other than spectroscopic properties from easily available IR spectra.

They should nevertheless be used with some care. First of all, the correlation drawn in Figure 4.5 is valid only for O–H···O bonds. Other curves should be used in the case of O–H···N, N–H···N, N–H···O, etc. bonds. Furthermore, even in the case of well-defined types of H-bonds, such as O–H···O bonds of Figure 4.5, an appreciable dispersion exists between the experimental points and the best-fit curve. This dispersion has several origins. One of them, maybe the most important one, is due to the fact that the centre $\overline{\nu}$ of ν_s depends on factors other than Q_0, even if this one is predominant. Thus the presence of other nearly lying H-bonds, for instance, modifies the value of the centre of ν_s, because non-negligible harmonic interactions between various H-bonds may then become effective. These interactions make centres of ν_s vary between gas and liquid or solid state, even when H-bonds are identical in these various states. It implies that points taken on isolated H-bonds should exhibit much smaller dispersions than those of Figure 4.5. Unfortunately, the number of studies presently available on such isolated H-bonds that have followed the pioneering works of Millen and coworkers (8, 9) is still limited for each category of H-bonds, and no such correlation that requires a minimum number of measurements could be established. Another possibility for dispersion we shall examine below in the next subsection is cooperative effects between H-bonds established with the same atom. All these considerations mean that deducing the equilibrium distance of an H-bond from the knowledge of the wavenumber of the centre of its ν_s band using such a correlation curve as shown in Figure 4.5 is, without any other knowledge of the H-bond, an estimate only. It thus does not allow us to conclude with another argument that an H-bond is weaker than another one because its centre of ν_s falls at a slightly greater wavenumber. In the next subsection, we nevertheless show that such a conclusion may be reached when more information is available. It may also be done when one can take the only points in Figure 4.5 that correspond to H-bonds with a similar environment, that includes similar interactions with neighbouring H-bonds, etc. This is what Mikenda did more recently in the case of hydrates (21), proposing a most useful correlation to interpret time-resolved nonlinear IR spectra of liquid water (22). Another point concerning Novak's correlation curves: they are only valid in the case of intermolecular H-bonds that do not suffer strong constraints from covalent bonds. In the case of an intramolecular H-bond, a small value of Q_0 may be found either in the case of a strong H-bond or in the case of a weak but strongly bent H-bond. Between these two extreme cases, the value of the wavenumber of ν_s may be completely different, another way to state that no correlation can be established between these quantities for intramolecular H-bonds.

As can be seen in Figure 4.5 the slope of the correlation curve monotonously increases when the equilibrium distance Q_0 decreases. It allows us to roughly classify H-bonds into three types (19): weak, medium-strength and strong H-bonds. Weak O–H···O bonds are those with $Q_0 > 2.7$ Å, where this curve displays a weak slope. Intermediate O–H···O bonds are those with 2.7 Å $> Q_0 >$ 2.6 Å, where this curve displays a greater slope that becomes really great for strong O–H···O bonds that have their Q_0 smaller than 2.6 Å. With this classification, weak O–H···O bonds have their band centres falling above 3200 cm^{-1}, medium-strength H-bonds have their falling between 3200 and 2600 cm^{-1}, and strong H-bonds have their falling below 2600 cm^{-1}. This classification retrieves that one we made in Ch. 1 on the basis of ΔH values of H-bonds. It is more precise.

Evidence of cooperativity between H-bonds

The sensitivity of the value of the centre $\overline{\nu}$ to small modifications of H-bonds allows us to put into evidence such small effects as for instance the existence of *cooperativity* between H-bonds. By cooperativity, we characterize the small increase of the energy of a single H-bond R–O–H···Y, for instance, when another H-bond is established on the O-atom of R–O–H. Thus the open trimer of methanol represented in Figure 4.6 has the D-bond established by its central CH_3OD molecule at the origin of a ν_s band centred at lower wavenumber (2490 cm^{-1}) than the other D-bond (2600 cm^{-1}). Let us note that in this case the H-atom of O–H is replaced by a D-atom, in order to unambiguously eliminate the small contribution of C–H stretch bands that appears in the same region as ν_s(O–H···). We shall see in Ch. 7 that energies or enthalpies of D-bonds hardly differ from those of H-bonds. This shift towards lower wavenumbers might be due to a harmonic coupling between the two ν_s(O–D···) vibrations of this trimer, and also possibly with the ν_s(O–D) of the end group of the trimer. There are several arguments against this possibility. First, ν_s(O–D···) at 2600 cm^{-1} of the first group should then have the wavenumber of its centre changed between dimer and trimer. This change is real and detectable but amounts only to a few wavenumbers, as may be seen by comparing the middle and bottom parts of Figure 4.6. Furthermore, no other band appears in tetramers, pentamers, etc. In these higher polymers, only the same three bands as those visible for the trimer are present. The band at 2490 cm^{-1}, that is, the band that appears at lower wavenumbers becomes more intense and broader when the *n*-mer becomes greater. The increase of its width is due to the harmonic interaction between neighbouring ν_s(O–D···) bands. It means that this harmonic interaction is of the order of this width, that is a few tens of wavenumber, too small to be at the origin of the separation between the two types of ν_s(O–D···) bands at 2600 and 2490 cm^{-1}. It implies that the band at 2600 cm^{-1} is that of the first O–D··· group that establishes one D-bond but accepts no D-bond. The band at 2490 cm^{-1} is then that of the various O–D··· groups that establishes one D-bond and accept another one, while the band at 2690 cm^{-1} is that of free O–D groups at the end of the *n*-mer. It shows that a O–D group that establishes one D-bond and at the same time accepts another one, has a ν_s(O–D···) band that appears at wavenumbers lower by about 110 cm^{-1} as compared to the first O–D group of the *n*-mer that establishes one D-bond but accepts none. We are now in a position to conclude that such an H(D)-bond is somewhat stronger. This is known as *cooperativity* of H-bonds. The magnitude of this cooperative effect is not negligible, as it currently amounts to an increase of ΔH of about 20–30%, which may reach a factor of 2 in polyamides (24) and might have yet unsuspected implications for our understanding of the structure of proteins. Another example of cooperativity of H-bonds may be found in the value of the static electric dipole moment of the water molecule, which we shall see in Ch. 8 to be in ice 1.7 times that of the isolated H_2O molecule. Cooperativity, a general property of H-bonds, is also a delicate property that requires to be discerned with a sensitive probe such as IR.

More recently, anticooperative effects between the two H-bonds established by H_2O molecules have also been displayed (25), by simply showing that an HDO molecule that establishes an H-bond with a pyridine or tetrahydrofuran molecule with its H-atom and none with its D-atom has a ν_s(O–H···) band centred at 3375 cm^{-1}. When the D-atom also establishes a D-bond with another pyridine molecule, the ν_s(O–H···) band centre shifts to 3420 cm^{-1}. This

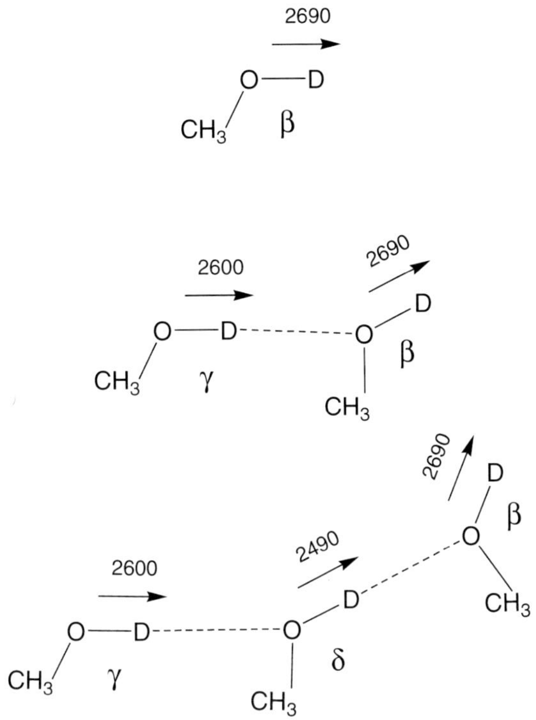

Figure 4.6 Wavenumbers of ν_s band centres of oligomers of deuterated methanol when dissolved in CCl_4 at various concentrations (23).

shift towards a higher wavenumber implies a small weakening of the O–H··· bond that is certainly also accompanied by a weakening of the O–D··· bond in the same HDO molecule. It shows that cooperativity or anticooperativity are effects that come from the electronic structure of the H-bond, as they are insensitive to the isotopic form of the H-atom.

Before closing this subsection on cooperativity, let us mention that another type of "non linear" interactions between H-bonds has also been shown by IR spectroscopy. It has also been labelled cooperativity, but it is different from the cooperativity described above, as it is then sensitive to the isotopic difference between H and D. It appears (26) in cyclic dimers, similar to those of carboxylic acids drawn in Figure 4.4. The distribution of H- and D-atoms is not completely random in these dimers, as it should be if the energies of the dimer were independent of the isotopic H/D substitution. The concentration of HD dimers is lower than expected and that of HH and DD dimers higher. The origin of this effect is not yet clear.

Conclusion regarding ν_s band centres

In conclusion of this subsection, we may stress that the shift of the centre of ν_s towards lower wavenumbers, which is found for all intermolecular H-bonds, provides an easy and rapid estimate of geometrical and thermodynamic properties of H-bonds. It may be performed for isolated H-bonds, where it then gives immediate results. It may also be performed when several H-bonds are present, be it in gaseous H-bonded clusters, in liquids or in solids. A more precise measurement of these quantities is also possible but it requires a more accurate analysis, as other effects we have seen above may also slightly affect the value of this shift.

Widths of ν_s bands

A spectacular enhancement of the width of ν_s accompanies the dramatic increase of its integrated intensity and the marked and highly useful shift towards smaller wavenumbers that follows the establishment of an H-bond. We illustrate this effect in Figure 4.7 that schematically reproduces the evolution of ν_s bands of free O–H groups and of weak, medium-strong and strong H-bonds, as defined above by the wavenumbers of their ν_s band centres. The ν_s(O–H) band of a free O–H group is, as already seen, a narrow band that nearly always falls in the vicinity of $3600\,cm^{-1}$. When this O–H group is that of an alcohol R–O–H that establishes a weak H-bond, the corresponding ν_s(O–H···) band appears at a lower wavenumber and starts having an appreciable width with features such as a shoulder that may appear, as drawn on its low wavenumber side. Its integrated intensity is greater than that of ν_s(O–H) by a factor that may reach a value of 10. When the H-bond established by such an O–H group becomes still stronger, as in the case of carboxylic acids, all these features become more marked. The integrated intensity of ν_s(O–H···) is more than 10 times greater than that of the corresponding ν_s(O–H) of the same molecule establishing no H-bond, its centre appears at lower wavenumbers, and its width is often greater than $500\,cm^{-1}$. In addition, the shoulder that appeared in the case of a weak H-bond now becomes a collection of several submaxima and subminima. When this H-bond becomes a strong H-bond, as in the case of carboxylate acid salts (28, 29), the integrated intensity of ν_s(O–H···) has still increased its value, its centre has been shifted towards still lower wavenumbers, and its width is now greater than $1000\,cm^{-1}$. The number of submaxima and subminima is often smaller than that found in the case of medium-strength H-bonds. The subminima are often called “Evan’s holes”, or “transmission windows”, because the absorbance is at a minimum there, and are due to Fermi resonances described in Ch. 5. They are often found as two Evan’s holes, due to resonance of ν_s with the two bending overtones of C–O–H, one in the plane of this group and the other one out of this plane (30, 31). As this ν_s(O–H···) band extends towards the region of wavenumbers smaller than $2000\,cm^{-1}$, it is overlapped by many other bands in its lower wavenumber part. Some of these other bands are schematically drawn in Figure 4.7. Their presence, often denser in reality than represented in Figure 4.7, makes a precise measurement of the first moments of ν_s(O–H···) hard to perform. It explains why no value of integrated intensities or widths of strong H-bonds have been performed. We know that they are greater than that for weaker H-bonds, and we also know that their centres also appear at lower wavenumbers. All this knowledge remains qualitative, however, due to this difficulty to isolate ν_s(O–H···), a problem that does not exist for weaker H-bonds. Let us finally note that the number of such bonds is limited, but there exist still stronger H-bonds, most of them, however, not being O–H···O bonds. The widths of the ν_s(X–H···) bands of the few of these extremely strong H-bonds that have been observed, are smaller than that of the strong H-bonds represented in Figure 4.7, and are more comparable to that of medium-strength or even

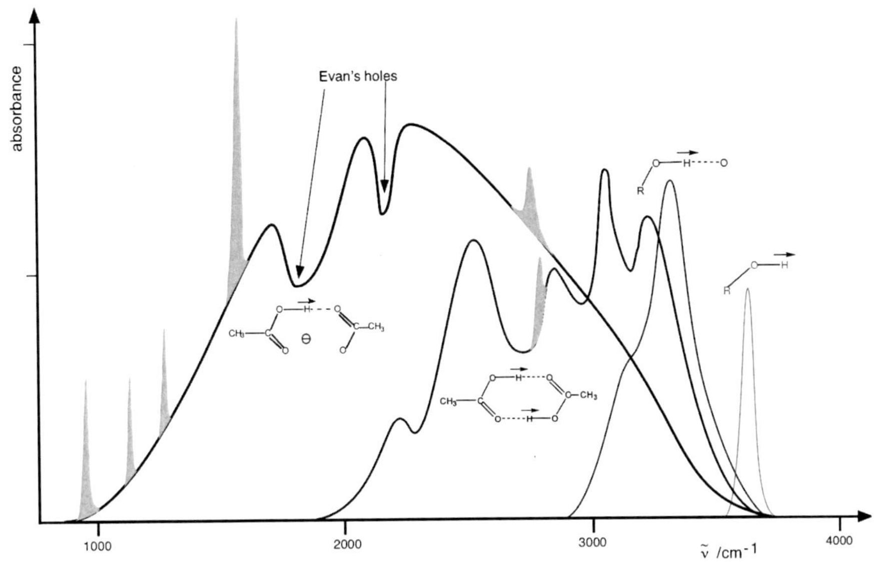

Figure 4.7 Schematic evolution of the ν_s bands of an O–H group, when it does not establish an H-bond as in a alcohol vapour R–O–H, or establishes a weak H-bond as in an alcohol R–O–H··· (27), or a medium-strong H-bond as in carboxylic acid cyclic dimers $(CH_3COOH)_2$ (the true experimental spectrum is displayed in Figure 4.4), or strong H-bond as in carboxylate acid salts (28, 29). Overlapping bands other than ν_s(O–H···) that appear in the experimental spectra are schematically represented filled in grey.

weak H-bonds. Almost nothing is known of their integrated intensities, while we know their centres are shifted towards very low wavenumbers.

We thus see that the width of ν_s is also a characteristic feature of an H-bond. From Figure 4.7 we see that as H-bonds become stronger, their ν_s(X–H···) bands exhibit a greater integrated intensity that is accompanied by a greater shift of their centres towards lower wavenumbers, *and* a greater increase of their widths. This is true for nearly all H-bonds, except for extremely strong H-bonds such as F–H···F^-. These specific properties of the first moments of ν_s(X–H···) are the spectral signature of an H-bond. The width σ of ν_s(X–H···), equal to the second centred moment and mathematically related to the moment of order 2 of this band (eqs. (5.A15) and (5.A16) of the appendix in Ch. 5), is nevertheless more difficult to measure than the wavenumber of its centre. This is because the value of the second moment M_2 of a band becomes sensitive to the limits of the band and consequently requires a more delicate definition of the limits of this band on its wings, where the factor $(\nu - \overline{\nu})^2$ takes on a great importance. σ is consequently a quantity that has not been often measured. As for the centre of ν_s(X–H···), an estimation of its value may be visually performed. It is, however, far from being as precise as the estimation of $\overline{\nu}$, and it is consequently a quantity that has not been much exploited. It might be worth doing it, as it is particularly sensitive to temperature changes around room temperature. This sensitivity enables it to be measured with an enhanced accuracy (32, 33) and is at the origin of original information on the properties of H-bonds that are described in Ch. 5.

The ν_s band of H-bonds: conclusion

The stretching band $\nu_s(X{-}H\cdots Y)$ is the most characteristic band of intermolecular H-bonds. Its huge integrated intensity, increased by 1 to 2 orders of magnitude with respect to that of $\nu_s(X{-}H)$ makes IR spectroscopy the most sensitive method to detect the presence of H-bonds, and also to study isolated H-bonds in gases or cryomatrices, where their numbers are very low. Systematic studies of such systems, initiated by D. J. Millen and coworkers, have conveyed invaluable information on the intrinsic properties of H-bonds. This hypersensitivity of IR spectroscopy to H-bonds furthermore provides a very precise mean to decide whether chemical groups establish H-bonds or not. An illustration is given in the case of liquid water (Ch. 9), where IR spectroscopy unambiguously shows that nearly all O–H groups of H_2O molecules establish H-bonds, a result that goes in opposition to the common thinking, and is in conflict with theoretical approaches that somewhat arbitrarily define an energetic cut-off for H-bonds. This hypersensitivity, however, conveys an inconvenience—the integrated intensity of the ν_s band of a sample containing identical H-bonds is equal, in a first approximation, to the integrated intensity for one H-bond multiplied by the number of H-bonds. As the first factor relative to one H-bond may vary in a great proportion when the strength of this H-bond is slightly modified, this integrated intensity of ν_s is difficult to use for measuring the total number of H-bonds in this sample. This will consequently be done most of the time using bands other than ν_s. Only when all possible sources of modification of the strength of these H-bonds, such as temperature, water vapour concentration, pressure, etc. are fully controlled, can ν_s be used for such a measurement. In other words, great care should be applied to calculate the number of H-bonds from the measurement of the integrated intensity of ν_s.

The important shift of ν_s towards lower wavenumbers, its not less important width, together with the presence of several submaxima, are also characteristic features of H-bonds. These shifts and widths also appear in Raman spectra of H-bonds. In opposition to IR spectra, however, the integrated intensity of ν_s in Raman spectra is lower than that of $\nu_s(X{-}H)$. Thus Raman spectra is scarcely used to study H-bonded systems, because this ν_s band may then be masked by more intense bands, such as $\nu(C{-}H)$, for instance. The exception is liquid water, where no C–H groups are present, and where the density of H-bonds is so high that the poor sensitivity of ν_s Raman bands to H-bonds is no problem whereas the great sensitivity of IR to H-bonds may be at the origin of difficulties. We shall see that point in more detail in Ch. 9 on liquid water and in Ch. 10 on water in macromolecules, particularly the ways that have been found to overcome these difficulties, thus keeping advantage of the exceptionally good signal-to-noise ratio displayed by IR spectroscopy in the study of these systems. The characteristic shape of $\nu_s(X{-}H\cdots Y)$ is now well understood, as described in next Ch. 5 and good simulations are proposed that put into light the degree of this understanding. It is due to a strong anharmonic modulation of the force constant of ν_s by slow intermonomer modes of X–H···Y, that is, relative vibrations of X–H and Y components. This strong modulation is due to the sensitivity of the electronic structure of the H-bond to the position of the H-atom. This sensitivity, together with the great sensitivity of the electric dipole moment to a change in the X–H distance that is responsible of the great integrated intensity of ν_s, is due to the special electronic structure of the H-bond, for a great part characterized by the transfer of electrons from the nonbonding orbital of Y towards the antibonding σ^* of X–H (15). This strong modulation is also at the origin of the important width of ν_s that furthermore allows the appearance of local transmission windows, called "Evan's holes", that are due to resonance interactions, called "Fermi resonances", described in more detail in Ch. 5 and that are also classified as anharmonic couplings. These features, particularly the shift towards lower wavenumbers, are easy to measure and are often used to quickly extract information on H-bonds. It is found in isolated H-bonds, and this calls for several "caveats", as arguments based on this shift may easily prove wrong when invoked without care.

The first caveat is that ν_s of single isolated H-bond being already broad the width of ν_s cannot be invoked to postulate the presence of a variety of H-bonds. In the case of identical H-bonds, this width of ν_s that we have seen to be due to a distribution of X···Y distances or equivalently an effective distribution of strengths of H-bonds, may be seen as reflecting the presence of H-bonds of various strengths. One should, however, keep in mind that this is a dynamic effect, as each H-bond rapidly oscillates from weaker to stronger at the frequency of intermonomer modes. Second caveat: as described in Ch. 5, the shift of the centre of ν_s is due for its greatest part to that strong coupling of ν_s(X–H···Y) with intermonomer modes. However, other mechanisms can also modify this shift. Even if this modification is small as compared to that due to this coupling, it often hinders concluding that such an H-bond is stronger than any other because the centre of its ν_s band appears at slightly lower wavenumbers. Further arguments should be given to reach such a conclusion, particularly the arguments concerning the width and shape of ν_s should be consistent with the previous one concerning its centre. This is what we have done above when putting into evidence cooperative or anticooperative effects of H-bonds. It means that correlations of the centres of ν_s bands with such quantities as equilibrium distances Q_0 or enthalpies of formation ΔH of H-bonds are only valid when one has eliminated other causes of shifts of ν_s, as in considering, for instance, similar H-bonds, or a restricted family of H-bonds.

The strong anharmonic modulation (or coupling) of ν_s that is detailed in Ch. 5, makes the analysis of ν_s somewhat laborious, and not so much instructive for what concerns chemical analyses. However, interesting information can be gained from the simple analysis of its first moments, particularly its centre and width. This may prove particularly useful when some external parameters may be varied, such as, for instance, temperature or deuteration. The centre and width of ν_s are sensitive to such variations, as shown in eq. (5.A49) of the appendix in Ch. 5. In the case of simple systems, made for instance of a single type of H-bonds, that may or may not interact, the analysis of the evolutions of these first moments allowed to measure the magnitudes of the modulation by slow intermonomer modes, of the harmonic interaction between closely lying H-bonds, and of Fermi resonances. In the case of liquid water, a critical analysis of the changes of these first moments when temperature increases will be seen in Ch. 9 to provide crucial information on the structure of its exceptional H-bond network.

All these characteristic features described above, are valid only for intermolecular H-bonds. Intramolecular H-bonds that are constrained by stronger covalent forces may display these features, when these constraints are not too big. They however display them in an attenuated way (34). Thus the integrated intensity of the ν_s band of an intramolecular H-bond with short X···Y distance may well be comparable to that of a free X–H when this short distance is imposed by covalent bonds and not a direct consequence of the existence of this H-bond. The established H-bond is then a weak one, with spectral features of a weak H-bond.

Other intramonomer bands

The establishment of an H-bond X–H···Y also induces changes in internal vibrations of X–H other than ν_s. These changes, which we examine in this section, are less spectacular than the changes on ν_s we have just discussed above, which means that in opposition to ν_s these vibrations remain harmonic and are not as strongly modulated as ν_s by slow intermonomer modes. They nevertheless may be as useful as those displayed in the ν_s region, in particular when the main goal is structural, chemical or biological analysis. The ν_s band is most useful in physics to study the dynamics of H-bonds and in chemistry to detect H-bonds or follow their formation. In biology, at molecular level, the spectacular properties of ν_s will be less interesting, mainly because bioreactions occur in aqueous media where the number of H-bonds is always extremely high and often different H-bonds are present, with their broad ν_s bands strongly overlapping and consequently difficult to separate. They consequently will be mostly used to check if H-bonds are established or broken, but the core of the analysis will most of the time be carried out on these other bands. Examples will be given in the following chapters, mainly those ones that concern water in macromolecules (Ch. 10) and the methods of observation of water molecules (Ch. 11).

Various categories of local vibrations may suffer changes upon the establishment of an H-bond and the IR bands that correspond to modes that have components on these vibrations are consequently changed. These vibrations are those that directly imply the H-atom that establishes the H-bond, such as the bending band $\delta_{C\text{–}O\text{–}H}$ of a C–O–H group, $\delta_{C\text{–}N\text{–}H}$ of a C–N–H group or $\delta_{H\text{–}O\text{–}H}$ of a H_2O molecule, all these being vibrations of C–O–H, C–N–H or H–O–H angles. The second type of vibrations that suffers changes upon the establishment of an H-bond are vibrations that occur in the vicinity of the H-bond, such as, for instance, the stretching $\nu_{C\text{–}OH}$, or $\nu_{C\text{–}NH}$ vibrations of C–OH or C–NH groups that may establish an H-bond. These vibrations are vibrations of the C–O or C–N distances in these groups. Finally, the third type consists of vibrations that imply an atom that accepts an H-bond. A typical example is $\nu_{C=O}$ stretching vibrations of carbonyl or carboxyl groups (vibration of C=O distances).

Various actions that affect the H-bond can be used to induce changes in the vibrations of these groups. These are, for instance, deuteration of X–H···Y, which becomes X–D···Y and which we have seen to have an important effect on ν_s, but may also have a well-marked one on these other bands. Another particularly simple action is changing the temperature, which leads to breaking or weakening of H-bonds. Also the addition of water molecule that we shall see has an exceptional ability to develop around it a particularly dense H-bond network, is another action that leads to modifying H-bonds, especially their numbers and then characterize their effects on particular vibrations. We may see how this kind of experiment works in Ch. 11 devoted to methods to observe water molecules. In the following subsections, we give examples of corresponding experiments before summarizing in Tables 4.2 and 4.3 typical changes that most encountered vibrations display after the establishment of an H-bond.

Table 4.2

Typical average wavenumbers $\bar{\nu}$ of centres of various stretching and bending bands of isolated free X–H groups when they do not establish H-bonds (second column) and their shifts $\delta\bar{\nu}$ after establishment of an H-bond (column 3)

Vibration	$\bar{\nu}$ (cm^{-1})	$\delta\bar{\nu}$ (cm^{-1})	Remarks
$\nu_{C\text{-}OH}$ (Alcohol I)	1010 to 1030	+20	CH_3OH slightly apart
$\nu_{C\text{-}OH}$ (Alcohol II)	1030 to 1130	+20	
$\nu_{C\text{-}OH}$ (Carboxylic acid)	1150 to 1200	+110 to +140	Refs. (4, 35)
$\nu_{C\text{-}NH}$ (amide III)	1300 to 1350	+15	A weak band in IR, relatively more intense in Raman
$\delta_{C\text{-}O\text{-}H}$ (alcohol)	1300 to 1450	>0, variable	
$\delta_{H\text{-}O\text{-}H}$ (H_2O)	1595	+10 to +60	Ref. (36)
$\delta_{C\text{-}N\text{-}H}$ (amide II)	1510	+50	Intense in both IR and Raman Ref (37)

A positive shift indicates a higher value for the wavenumber of the centre of the band in X–H···Y.

Table 4.3

Typical average wavenumbers $\bar{\nu}$ of centres of $\nu_{C=O}$ bands in various groups when they do not accept H-bonds (second column) and their shifts $\delta\bar{\nu}$ after acceptation of an H-bond (column 3)

Vibration	$\bar{\nu}$ (cm^{-1})	$\delta\bar{\nu}$ (cm^{-1})	Reference
$\nu_{C=O}$ carboxylic $(CH_3COOH)_2$	1790	−80	(4) and (38) (see text)
$\nu_{C=O}$ carboxylic $(HCOOH)_2$	1775	−65	(4) and (39) (see text)
$\nu_{C=O}$ carboxylate (COO^-)	1620*	−20*	(40)
$\nu_{C=O}$ carbonyl amide I in amide	1680–1710	−80 to −60	(37, 41)
$\nu_{C=O}$ carbonyl amide I in polypeptides and proteins	≈1700	−30 (−65 for C=O accepting two H-bonds)	(42)

A negative shift indicates a lower value for the wavenumber of the centre of the band in X–H···O=C. For carboxylate (*) the value $\bar{\nu}$ is that of the centre of the C=O··· band of COO^- in sodium hyaluronate that accepts one H-bond in its C=O group and $\delta\bar{\nu}$ is the shift after acceptation of a second H-bond in the same C=O group.

Stretching and bending vibrations of the donating X–H group

In order to illustrate the changes of $\delta_{C\text{-}O\text{-}H}$ and $\nu_{C\text{-}OH}$ bands upon the establishment of H-bonds, we show in Figure 4.8 the IR spectrum of cellulose I_α at room temperature and its variations after a temperature change. The structure of this typical biopolymer is displayed in Figure 2.3. The groups that presently interest us are COH of primary and secondary alcohols that establish H-bonds with neighbouring O-atoms. On increasing the temperature, some H-bonds are broken. In these spectra, we are especially interested in the bending $\delta_{C\text{-}O\text{-}H}$ band that appears at 1425 cm^{-1} in the spectrum of cellulose at room temperature (upper spectrum), in the stretching $\nu_{C\text{-}OH}$ band of some secondary alcohols that falls at 1060 cm^{-1} and in the stretching $\nu_{C\text{-}OH}$ band of primary alcohols that falls (43) at 1034 cm^{-1}. The lower spectrum is the spectrum equal to the upper spectrum minus that of the same sample at 115 °C. The variations of the $\nu_{C\text{-}OH}$ and $\delta_{C\text{-}O\text{-}H}$ bands are clearly visible in this difference spectrum as different type of bands that exhibit both a minimum and a maximum that correspond to a shift

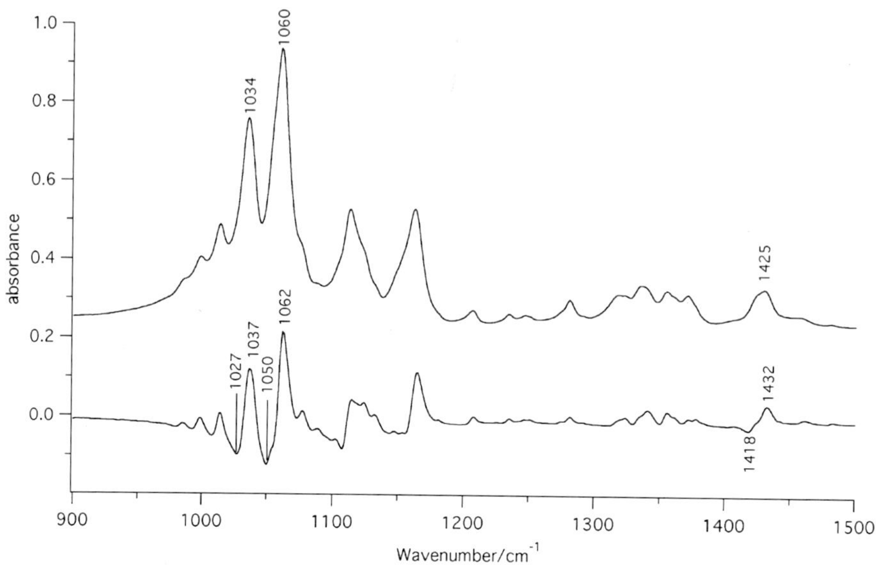

Figure 4.8 IR spectrum of cellulose I_α (Valonia) at room temperature (upper spectrum) and difference spectrum of sample at room temperature minus at 115 °C (lower spectrum). Spectra are offset for clarity.

of this band upon a temperature change. Thus the band at 1425 cm^{-1} consists of two bands: $\delta_{C\text{-}O\text{-}H\cdots}$ of H-bonded C–O–H··· groups and $\delta_{C\text{-}O\text{-}H}$ of free C–O–H groups. Increasing the temperature results in C–O–H··· groups being transformed into C–O–H groups. The corresponding band in the difference spectrum is consequently $\delta_{C\text{-}O\text{-}H\cdots}$–$\delta_{C\text{-}O\text{-}H}$. We may deduce from the positions of the maximum/minimum of this band in the lower spectrum that $\delta_{C\text{-}O\text{-}H\cdots}$ displays a peak around 1432 cm^{-1} whereas $\delta_{C\text{-}O\text{-}H}$ displays one around 1418 cm^{-1}. These would be exact values if the widths of these two bands in the upper spectrum were small compared to this difference 1432–1418 = 14 cm^{-1}. This is clearly not the case, the widths of these individual bands being comparable to this difference, or even greater. As a consequence the minima and maxima of the difference band in the lower spectrum do not exactly coincide with the maxima of $\delta_{C\text{-}O\text{-}H\cdots}$ and $\delta_{C\text{-}O\text{-}H}$.

In the same way, the $\nu_{C\text{-}OH\cdots}$ stretching band of secondary alcohols clearly appears around 1062 cm^{-1} while the corresponding $\nu_{C\text{-}OH}$ band appears around 1050 cm^{-1}. For primary alcohols, the $\nu_{C\text{-}OH\cdots}$ band appears around 1037 cm^{-1} while the corresponding $\nu_{C\text{-}OH}$ band appears around 1027 cm^{-1}. The same trend has been put into evidence in the case of simpler systems made of isolated methanol molecules in solid matrices in the presence of a variable number of H_2O molecules (44): the $\nu_{C\text{-}OH\cdots}$ band, which is relatively narrow in this case, is found at 1050 cm^{-1} and the $\nu_{C\text{-}OH}$ band at 1032 cm^{-1}. It could furthermore be established that this band shifts to 1028 cm^{-1} when the O-atom of this COH group, that does not establish an H-bond, accepts one in its lone pair orbitals. Most similar results have been found when the methanol molecule binds to a nanodroplet of water, instead of a single H_2O molecule (45). All these results are summarized in Table 4.2. Let us note that no variation has been put into evidence of the integrated intensities of all these bands that display shifts of their centres upon the establishment of an H-bond. Let us also note that both these vibrations find their average wavenumbers increasing with the strength of the H-bond. In that particular case of the H_2O molecule, the shift towards higher wavenumbers of $\delta_{H\text{-}O\text{-}H}$ could be precisely correlated to the shift towards lower wavenumbers of ν_s in various water-containing solids and liquids (36).

Carbonyl and carboxyl C=O groups

The acceptation of H-bonds by carbonyl or carboxyl C=O groups is always at the origin of a negative shift of the corresponding stretching $\nu_{C=O}$ band. These often-encountered bands are one of the most exploited bands in organic chemistry or biochemistry. The reason is that they are most intense bands that often display a very good signal-to-noise ratio. C=O groups are furthermore strong H-bond acceptors. The integrated intensities of the $\nu_{C=O}$ bands may be smaller than those of ν_s but, as they are narrower, their heights or intensities at maximum may be greater than those of ν_s. In opposition to ν_s, their intensities also remain great when they do not accept H-bonds. Consequently, they are bands that are immediately identified in the region 1600–1750 cm^{-1}. Their sensitivities to accepted H-bonds are illustrated in Figure 4.9 that represents the IR spectrum of a protein, Bovine Serum Albumine (BSA) at room temperature. This protein was left for several days in heavy water. The result is that nearly all N–H and O–H groups have been transformed into N–D and O–D groups, as shown by the absence of ν_s (X–H···) bands above 3000 cm^{-1} in the upper spectrum. It leads to elimination of the $\delta_{H\text{-}O\text{-}H}$ bands due to H_2O molecules that fall in the same region as $\nu_{C=O}$ bands, and also due to diminishing the coefficient of the bending vibration $\delta_{C\text{-}N\text{-}D}$ in the decomposition of the $\nu_{C=O}$ mode, called amide I in this case. It thus makes this amide I mode of deuterated amide or peptide groups to be composed of only $\nu_{C=O}$ vibrations, in a first approximation. The bands in the region 2800–3000 cm^{-1} are $\nu_{C\text{-}H}$ bands. Their appearance in the upper spectrum clearly shows that C–H groups have not been transformed into C–D groups. The band that extends from 2000 to 2700 cm^{-1} in the upper spectrum is the ν_s(N–D···) band of protein groups, also called amide a, that contains a smaller amount of ν_s(O–D···) groups that come from remaining D_2O molecules and O–D side groups. The amide I band is the intense band that culminates at 1651 cm^{-1}. The amide II band that appears at 1435 cm^{-1} has $\delta_{C\text{-}N\text{-}D}$ bending vibrations as main component. Its great integrated intensity indicates that it also has a nonnegligible component on $\nu_{C=O}$ vibrations of carbonyls. Finally, the band at 1577 cm^{-1} is an intense stretching band of COO^- groups of side groups of the protein (42). The sensitivity of the amide I and II bands to the presence of D-bonds established by N–D peptidic groups or accepted by C=O peptidic groups appears in the lower spectrum. The $\nu_{C=O\cdots}$ band (amide I) of C=O groups that accept a D-bond appears around 1632 cm^{-1}, whereas the corresponding one of free C=O groups appear around 1674 cm^{-1} (this difference spectrum is that of the sample at room temperature, where D-bonds are relatively numerous, minus that of the same sample at 116 °C, where D-bonds are less numerous). The corresponding shift $\delta\bar{\nu}$ is of about –40 cm^{-1}. It is opposite to that of the amide II band; its maximum appears at lower wavenumbers than its minimum, whereas the amide II that has its maximum corresponding to N–D··· groups appear at higher wavenumbers (1464 cm^{-1}) than the minimum (1416 cm^{-1}) that corresponds to free N–D groups. This is in agreement with results displayed in Table 4.2.

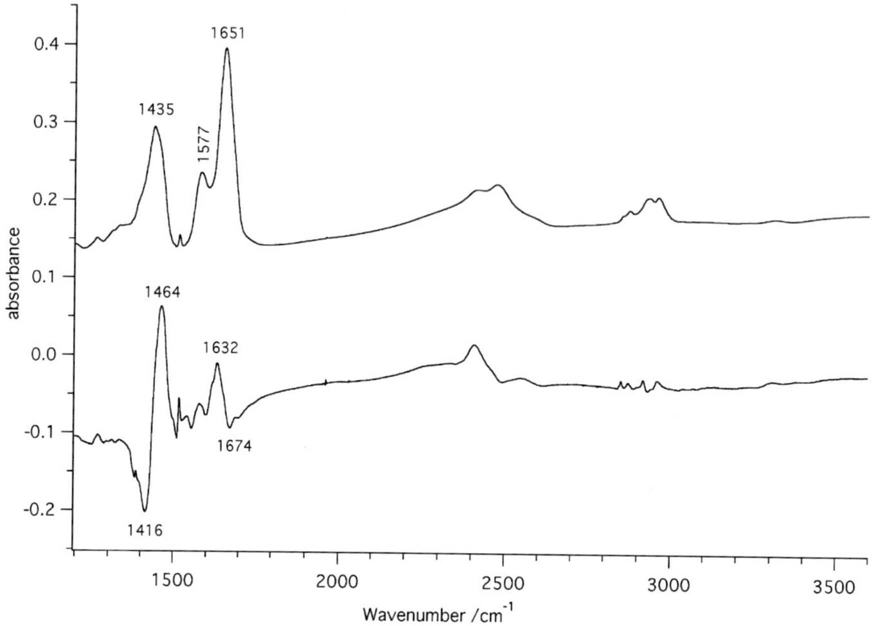

Figure 4.9 IR spectrum of deuterated Bovine Serum Albumine (BSA) at room temperature (upper spectrum). Same spectrum minus spectrum of the same sample at 116 °C (lower spectrum). Spectra are offset for clarity.

As these $\nu_{C=O}$ bands contain a lot of information (see Ch. 11), we display in Figure 4.10 the behaviour of C=O bands in isolated carboxylic acid monomers and cyclic dimers in the gas phase (their structures are displayed in Figure 4.4). The $\nu_{C=O}$ bands are the intense and well-defined bands just above 1700 cm^{-1}. The spectra of monomers and dimers are obtained from spectra at same temperature, here 81 °C, but at various pressures. This temperature around 80 °C is the optimal temperature to have a number of dimers comparable to that of monomers in that particular case of cyclic H-bonded dimers in the gas phase. The proportions of monomers and dimers being different at different pressures, one can easily combine two spectra recorded at various pressures to obtain the spectrum of either monomers or dimers. The shift of $\nu_{C=O}$ modes towards lower wavenumbers provoked by the establishment of an H-bond is directly visible when one compares the spectra of monomers and dimers. It clearly appears in the difference spectra of carboxyl groups, which is the representation that highlights it when these bands do not have the exceptionally good signal-to-noise ratio that these bands display in the case of Figure 4.10. The $\nu_{C=O}$ doublet of monomers has the same origin as the triplet of ν_s(O–H) in Figure 4.4: rotational structure. The measurement of the shift of $\nu_{C=O}$ upon the establishment of an H-bond is, however, somewhat more delicate than in the preceding case, because we have here two interacting $\nu_{C=O}$ vibrations, defined by coordinates t_1 and t_2. The effect of this interaction is to make the two $\nu_{C=O}$ corresponding modes symmetric $(t_1 + t_2)/\sqrt{2}$ and antisymmetric $(t_1 - t_2)/\sqrt{2}$ appear at $\overline{\nu}_{C=O\cdots H} + V_0$ for the symmetric one and at $\overline{\nu}_{C=O\cdots H} - V_0$ for the antisymmetric one. Here $\overline{\nu}_{C=O\cdots H}$ is the wavenumber of

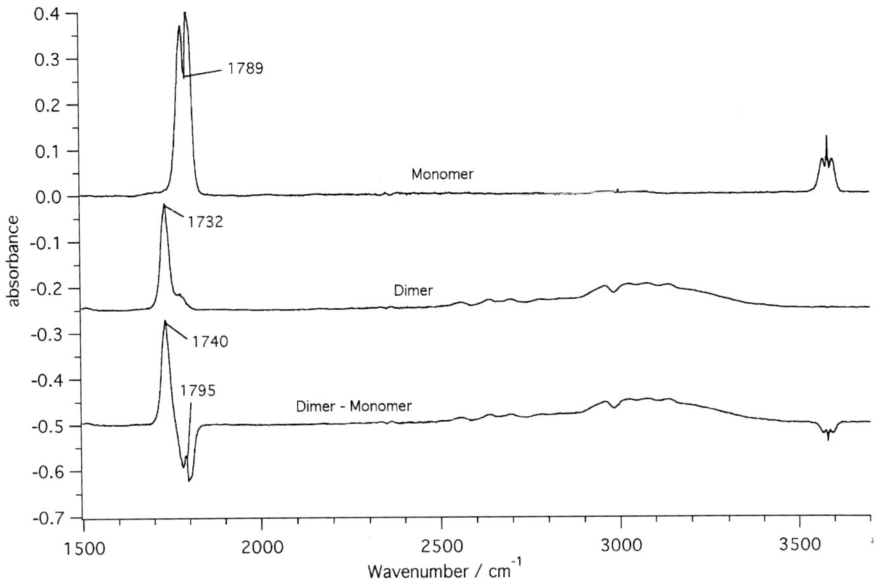

Figure 4.10 IR spectrum of monomers of acetic acid in the vapour phase at 81°C, and of cyclic dimers at same temperature. The bottom spectrum is the difference between these two spectra with a coefficient proportional to the ratio of the numbers of dimers to monomers (4). Spectra are offset for clarity.

the $\nu_{C=O}$ vibration in the absence of interaction, and this interaction is harmonic and proportional to $V_0 t_1 t_2$. As these two $\nu_{C=O}$ vibrations are parallel, the symmetric vibration does not appear in IR, as it does not change the electric dipole moment that remains equal to zero. Only the antisymmetric vibration appears in IR. The inverse is true for Raman displaying only the symmetric mode, which is the only one to induce a change of the polarizability of the dimer. The centre of $\nu_{C=O\cdots}$ of dimers appearing at 1732 cm^{-1} in IR (Figure 4.10) and at 1682 cm^{-1} in Raman (38) we deduce that $\bar{\nu}_{C=O\cdots H} = 1707\,\text{cm}^{-1}$. The shift of the centre of $\nu_{C=O}$ due to the establishment of H-bonds on the two carboxyl groups of these dimers is consequently quite marked, as it is equal to $1707 - 1789 = -82\,\text{cm}^{-1}$, with 1789 cm^{-1} being the wavenumber at which $\nu_{C=O}$ modes appear in monomers. We display in Table 4.3 the shifts induced by H-bonds on C=O groups of various types of molecules.

The integrated intensities of $\nu_{C=O}$ bands have been scarcely measured. Only in the case of these excellent models made of carboxylic acid cyclic dimers could it be shown that the integrated intensity of a $\nu_{C=O\cdots H}$ vibration of a group that has its C=O accept one H-bond is greater than that of the corresponding $\nu_{C=O}$ vibration of the same free group by about 30% (4). The exceptionally good signal-to-noise ratio that these $\nu_{C=O}$ bands display, together with their marked sensitivity to H-bonds described above, make these bands useful to measure such delicate quantities as equilibrium constants through especially precise measurements of their integrated intensities. We once more illustrate this point in the case of these cyclic dimers of carboxylic acids in the gas phase, where we have seen that these $\nu_{C=O}$ bands of dimers can be easily separated from the $\nu_{C=O}$ bands of monomers. By

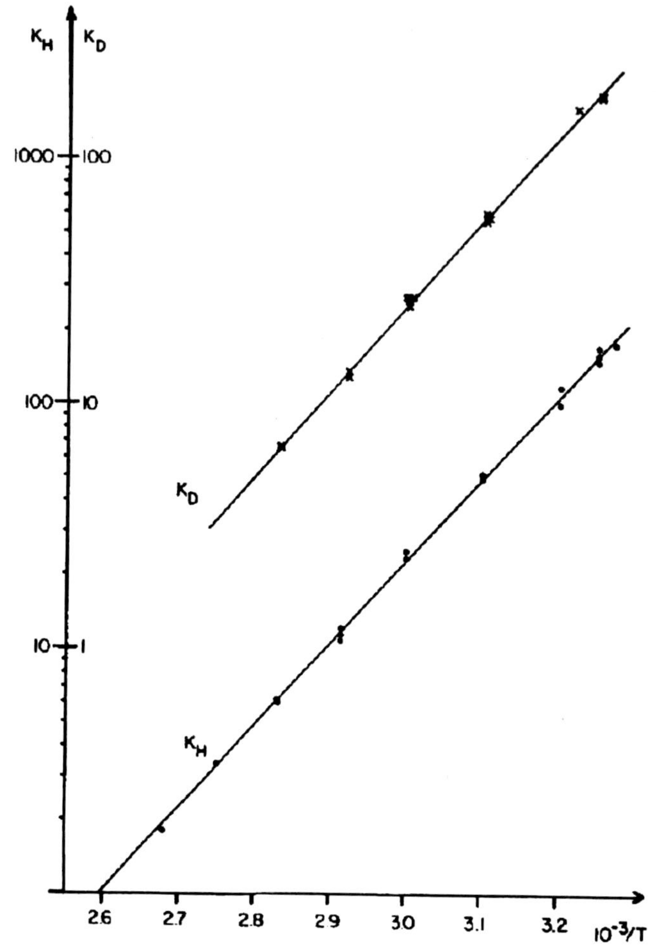

Figure 4.11 Ratios $M_d(T)/M_m^2(T) = K$ (logarithmic scale) of the integrated intensities of $\nu_{C=O\cdots H}$ (index *d*) and $\nu_{C=O}$ (index *m*) bands of acetic acid cyclic dimers in the gas phase as a function of T^{-1}. K_H is for CH_3COOH while K_D is for CH_3COOD. Reproduced with permission from Bournay and Maréchal (46). Copyright 2006, American Institute of Physics.

directly measuring the integrated intensities $M_m(T)$ and $M_d(T)$ of both these bands at various temperatures and pressures, we may thus deduce, with a good accuracy, the value of the enthalpy of formation ΔH of one H-bond. The ratio n_d/n_m^2 of the number of cyclic dimers to the number of monomers obeys the equation:

$$\ln \frac{n_d(T)}{n_m^2(T)} = \ln \frac{\mathcal{M}_d(T)}{\mathcal{M}_m^2(T)} + C = 2\frac{\Delta H}{kT} + C' \qquad (4.9)$$

where C and C' are temperature independent quantities. Applying this equation to the spectrum of acetic acid vapour gives (46) (Figure 4.11) $\Delta H = -7.5\,\text{kcal mol}^{-1} = -31.4\,\text{kJ mol}^{-1}$

for one H-bond, and also for one D-bond. The interest of this figure is to show the precision with which such a quantity is defined, thanks to the particularly high signal-to-noise ratio these $\nu_{C=O}$ bands display.

MULTIPHOTON VIBRATIONAL SPECTROSCOPIES: RAMAN AND NONLINEAR IR

Thus far, we have examined vibrational spectroscopy using IR absorption spectroscopy, what we called in Ch. 3 "one photon method", a general type that encompasses most experiments in spectroscopy. There exist, however, other types of spectroscopy to observe vibrations. These are for instance Raman spectroscopy, which is also of a current use in chemical physics and may be considered a routine method. Other less known methods are modern time-resolved IR spectroscopies. All these methods are "two-photon" or "multiphoton" spectroscopies. They do not involve a single photon, as in absorption, but the simultaneous absorption and emission of two photons, as in Raman and in other scattering experiments, or the successive absorption(s) and emission(s) of photons that are coherently delayed in time, as in time-resolved nonlinear spectroscopies. By "coherently", we assume the optical waves that carry these two photons keep a well-defined phase difference. In this latter type of spectroscopy, we include all modern set-ups that involve time-controlled laser spectroscopic techniques. We briefly sketch the interest of these various methods for the study of H-bonds in the following subsections.

Raman spectra

We have already encountered Raman spectroscopy in Ch. 3. We have seen that it gives information of the same type as that obtained by IR spectroscopy. For the study of H-bonds, Raman spectra are however much less useful than IR spectra. First of all, Raman is a more sophisticated technique that, even though routinely used now, displays a poorer signal-to-noise ratio. Furthermore, Raman bands are much less sensitive to H-bonds than IR bands. This is especially true for ν_s bands, which do not show this dramatic enhancement of intensities found in IR that we have already seen in this chapter. Also, ν_s(O–H) bands, and consequently ν_s(O–H···O) bands, display in Raman spectra intensities that are often weaker than those of $\nu_{C\text{–}H}$ bands. As these CH groups are in most organic materials more numerous than OH groups, the bands due to H-bonds are often masked, or at least strongly overlapped by intense CH bands. The consequence is that Raman spectra are scarcely used to study H-bonded materials. The one exception is liquid water, or aqueous materials that contain an appreciable quantity of water molecules. In these materials, the density of H-bonds is so high (Chs. 8 and 9) that IR bands may become saturated, especially those that are sensitive to H-bonds, destroying all interesting properties of IR spectroscopy. In these conditions where the numbers of H-bonds is huge and CH groups absent, Raman spectra may be useful. They nevertheless remain less informative on H-bonds, as can be seen with the spectra of supercritical water (47) or supercritical ethanol (48). In Ch. 11, we shall also see that these inconveniences displayed by IR spectroscopy for materials where the numbers of H_2O molecules is particularly high may be overcome in many systems, keeping IR the most powerful method to study H-bonds in these particularly important species, and leaving Raman spectroscopy to occasional use for studying H-bonds. We have already seen one such occasion—the study of H-bonds in the very far IR region below $100\,cm^{-1}$. It is, however, a region of a limited interest.

Time-resolved nonlinear IR spectroscopies

These are highly sophisticated modern techniques that consist of sending several coherent but delayed IR pulses on a sample. With respect to conventional IR spectroscopy described above, time-resolved nonlinear IR spectroscopy gathers supplementary information on the influence of the environment of the vibrations under study. In conventional IR spectroscopy, the vibrations are considered as during an infinite time. In nonlinear IR spectroscopy, their lifetimes are determined. The existence of finite lifetimes is mainly due to the environment of these vibrations. Its influence often appears in the form of "relaxation" times that convey, as in NMR spectroscopy described in Ch. 3, information of a statistical type on this environment and the system–environment interactions. This information is obtained using IR pulses instead of a continuous IR irradiation as in conventional IR spectroscopy. If irradiation sequences or irradiation pulses fall in NMR spectroscopy in the range of the ns (10^{-9} sec) that electronics handles quite well since a long time, pulses in IR spectroscopy require acting in a much more difficult range of about 10–100 fsec (1 fsec = 1 femtosecond = 10^{-15} sec; in the following we also consider 1 psec = 1 picosecond = 10^{-12} sec). The recent sophisticated technology used is consequently that of extremely short laser pulses. In view of setting things, the period of a vibration with wavenumber $3300\,cm^{-1}$ falls in the vicinity of 10 fsec. It means that during a 100-fsec pulse, such a vibration performs 10 oscillations. A vibration that displays a band at $330\,cm^{-1}$ performs a single oscillation during this same pulse. As in NMR, a series of short IR pulses are sent onto the sample with various delays and provide information on other dynamic aspects of vibrations than their only frequencies. Various optical set-ups may be used to do this. The simplest one consists of sending an intense and tunable "pump" IR pulse of a duration of about 100 fsec or less, and recording the spectrum some time τ after, by sending a weak IR pulse that is analyzed by a conventional dispersion (49, 50) optics. This second pulse probes the evolution of the sample after pumping as a function of τ. This set-up is known as a pump-probe set-up. Pump-probe experiments give information through correlation functions on vibrational relaxation or reorientational mechanisms that are poorly seen using conventional IR absorption spectroscopy, now often referred to as "linear IR" or "1D IR" spectroscopy (the unique variable is frequency, whereas with this new time-resolved nonlinear IR spectroscopy, we also have the evolution of the spectra with time, an additional dimension). With other set-ups, such as encountered in "photon echo" (51) or "2D IR" (52) experiments, the joint probability that an initial excitation at frequency ν_1 at time 0 is detected at frequency ν_3 at time τ is measured.

We do not here describe how these still highly specialized techniques work but evocate the kind of results they provide. Most of the bands that have been studied by this new technique are ν_s(X–H···) bands of H-bonds. We have seen earlier in this chapter that these bands are most intense in IR spectroscopy. In consequence, the first short intense pulse used in time-resolved nonlinear IR most rapidly populates the first excited states of ν_s, a favourable point to use these techniques. We have seen a strong modulation of this ν_s mode by low-frequency intramonomer modes reproduces quite precisely the shape of this band as obtained by 1D IR spectroscopy, in particular in the case of isolated H-bonds in gases. What can then time-resolved nonlinear IR convey in addition? It fully accounts for this modulation and moreover gives original information on the modulating intermonomer mode Q_s (eq. (5.2) in Ch. 5), which does not display in liquid water the behaviour of a regularly oscillating mode, but that of an "overdamped" oscillator (53, 54), a property that does not come directly out of the 1D IR spectra of liquid water. Overdamping means that when the O···O distance of an O–H···O bond is established to some value Q_s different from its equilibrium value Q_0, and after that left alone, it relaxes with no oscillation towards its equilibrium value Q_0. These slow intermonomer modes that modulate ν_s do not consequently exhibit an oscillatory behaviour. In these excellent models of H-bonds that are found in carboxylic acid dimers (Figure 4.4) diluted in CCl_4 the modulating intermonomer modes are, in opposition, underdamped, that is, they display marked oscillations during several psec (55) before reaching their equilibrium positions when they are found at a different position at time $t = 0$; a behaviour that was supposed to be true in the reconstitutions of IR spectra of H-bonds shown in Figure 5.2 of Ch. 5, and corresponds to potentials of Figure 5.1, also in Ch. 5. Nonlinear time-resolved IR also showed that exciting ν_s(O–H···) into its first excited state does not lead to a breaking of the H-bond (49). This is interesting, because the energy of an absorbed photon with wavenumber $\tilde{\nu}$ at 3000 cm^{-1} is $hc\tilde{\nu} = 8.6\,\text{kcal mol}^{-1} = 35.9\,\text{kJ mol}^{-1}$, a quantity that exceeds the energy of a single H-bond in these dimers we have previously seen equal to $7.5\,\text{kcal mol}^{-1} = 31.4\,\text{kJ mol}^{-1}$. This result ensures that irradiation by IR photons does not provoke disruptions of H-bonds. These disruptions occur, but are caused by other mechanisms. Nonlinear time-resolved IR spectroscopy may in some case convey information on these disruptions. It could thus show that after breaking of the D-bonds in liquid methanol CH_3OD (56), the photoresidues keep for sometime a memory of the former H-bonded species. It could also unambiguously show that nearly all OH groups of liquid water establish H-bonds (57), the very few of them that do not exhibit an H-bond relaxes very rapidly towards an H-bonded state. As seen in Ch. 9, this constitutes crucial information for understanding the structure of the H-bond network of liquid water.

Nonlinear time-resolved IR spectroscopy is furthermore an especially well-adapted method to measure the lifetimes of the first excited states of vibrations. The lifetime of the first excited state of ν_s(O–H···) in $(CD_3COOH)_2$ cyclic dimers, found in solutions of CCl_4, was thus found equal to 200 fsec, that of $\delta_{C\text{–}O\text{–}H}$ to 250 fsec (58). These vibrations furthermore display similarities in their relaxation paths, suggesting that these paths imply a multitude of modes. The analysis of the relaxation of $(CH_3COOD)_2$ cyclic dimers in solutions of inert CCl_4 that cannot establish or accept H-bonds, gave a lifetime of about 500 fsec for the first excited state of ν_s(O–D···). The lifetime of the first excited state of ν_s(O–H···) of H_2O molecules in liquid water was found to be of the order of 1 psec (53, 59, 60), that of ν_s(O–D···) equal to 1.45 psec (61) or 1.8 psec (60) when HDO was strongly diluted in ordinary liquid water. Comparable vibrational lifetimes of about 400 fsec have also been obtained for the bending band δ_{H-O-H} of H_2O molecules diluted in heavy water (62), which is appreciably shorter than that of δ_{H-O-H} of H_2O molecules of ordinary water, where it is then found equal to 1.4 psec (63). For stronger H-bonds the lifetime of the first ν_s(O–H···) excited state seems to be shorter, around about 200 fsec (64), which is ascribed to a more efficient relaxation channel by Fermi resonances. It also allowed to put into evidence in the ν_s(O–H···) band of HDO diluted in heavy water two components that have different relaxation times. One of these components exhibits great similarities with the spectrum of ice. It may be a valuable indication for understanding the dynamics of H_2O molecules in liquid water that we shall see in Ch. 9 is far from being established at a molecular level.

In addition to measurements of lifetime of these vibrational excited states, time-resolved nonlinear IR could also give precise information on the mechanisms of deexcitation of these states. It could thus be shown that relaxation of the first excited state of ν_s modes of water molecules in liquid water was mainly due to resonance interactions of these modes with excited bending modes (65). As a result of the analysis of 1D IR spectra shown above, Fermi resonance with bending modes allows the energy of the first excited state of ν_s to be transferred to the overtone of the bending band. It offers a fast relaxation path toward vibrational levels of a lower energy. Time-resolved nonlinear IR spectroscopy shows that this process is the main relaxation mechanism of ν_s and is at the origin of an unexpected increase of the relaxation time when temperature increases (66, 67).

These new data convey in addition original information on the structure of water molecules around, for instance, ions. They thus allow the distinguishing of the H_2O molecules that establish H-bonds with an anion X^- from H_2O molecules of the bulk. The lifetimes of their first ν_s excited state is much longer (2.6 psec in the vicinity of Cl^-) than those of HDO molecules dissolved in heavy water (68). It allows the collection of precise information on the acid–base reaction in water, putting into evidence at least three steps with different time constants (69) and enabling measurements of their lifetimes (70).

We thus see that these recent methods are promising. They are the modern aspects of IR spectroscopy, which allows mastering more parameters on which IR spectra depend than conventional (1D) IR spectroscopy. Up to now, they have been mainly aimed at studying the dynamics of H-bonds at times of the order of 100 fsec and longer. They provide a lot of original information on this time scale. There is no need to point that these newly recorded data are valuable. They are concerned with model systems where a single type of H-bonds is present. Will they be useful for the study of more complicated systems? The answer is affirmative for many systems encountered in chemistry. Hopefully, we may also be affirmative for biological systems, as they might well have the same fate as NMR spectroscopy that we have seen to be a pioneering method to propose sophisticated protocols to fully exploit time evolutions of spectra and has now become a central method to determine structures of macromolecules. Similar to NMR spectroscopy about 30 years ago, time-resolved nonlinear IR spectroscopy is still limited to chemically well-defined systems. Within a few years it might find its own ways for the full exploitation of its powerful properties so that it can be applied to observing H-bonds in much more complicated macromolecules. An example is given by the study of the amide I band of various dipeptides in trehalose/water glasses, taken as a model to mimic the lyoprotective effect of trehalose (71) described in Ch. 10.

Sum-frequency generation spectroscopy

This is a third-order nonlinear spectroscopic method that does not involve time delay. It consists of sending two coherent beams on a sample simultaneously; one in the visible-UV region and the other one in the IR region; and observing the photon that is emitted at the sum of their frequencies and is concomitant with the absorption of two photons, one in each of the two incident beams (72). The spectroscopic regions of the two incident beams are regions of transparency of the sample. The emitted photon requires absence of a centre of symmetry at molecular level to appear. It means that it practically does not appear in the bulk of a liquid, for instance, which is isotrope and consequently displays a centre of inversion in the average, but may appear on its surface, or at the interface between this liquid and another medium, where this centre of inversion disappears. It will consequently be most useful in the study of surfaces and interfaces, particularly the structures of the molecules thereon that can be deduced from the spectrum of these surfaces or interfaces (73). In many situations, it may be the unique tool to study liquid surfaces and interfaces and we shall see this in Ch. 9, which is devoted to liquid water-related examples.

This recent method has up to now been essentially used to study interfaces of liquid water (74). It allowed the measurement of the number of "dangling" O–H groups on its surface that is in contact with water vapour. This number represents about 30% of all O–H groups on this surface (73) and the ν_s(O–H) of these dangling O–H groups appear (75) at 3700 cm^{-1}, proving that, contrary to water vapour, the symmetric and antisymmetric O–H stretching vibrations of a single H_2O molecule are no longer normal modes in liquid water or on its surface. At the same time, it could be shown that H_2O molecules are more ordered on this surface than in the bulk of the liquid. This is quite an original result that seems to be general for liquids composed of molecules that establish H-bonds, and for mixtures of such liquids. At the same time, H-bonds between H_2O molecules on the surface of liquid water are weaker than those in the bulk (76). This is also true at the interface between liquid water and another nonmiscible liquid, and seems to be a characteristic of large contact areas. The opposite has been found for interfaces of small area, such appears for instance in cages developed inside liquid water to put into solution small molecules that display some hydrophobic character. As mentioned in Ch. 9, thermodynamics has shown that H-bonds between H_2O molecules positioned on the surface of these cages are stronger than in the bulk. This is quite in agreement with recent experiments using sum-frequency spectroscopy.

Beyond neat liquid water sum-frequency vibrational spectroscopy allowed the study of the alkyl chain conformation of surfactants such as dodecyl sulfate described in Ch. 9, that mediates the interfaces between water and CCl_4 when the head group or the cations of these amphiphile molecules varied (77, 78). It also gave experimental evidence that the water surface favours the presence of anions rather than cations (79). This effect is at the origin of the oxidative power of sea water, which is ascribed to Cl^- anions positioned on surfaces of liquid-water droplets. It has also been applied to other H-bonded liquid water, such as methanol, where it could be demonstrated that the CH_3 groups point away from the liquid at the interface with vapour (73). This is also the case for CH_3 groups of acetic acid (80). Also the surface of liquid water could be shown (81) to be disrupted by even such a small amount as 0.3% of acetic acid, which does not ionize, even at this low concentration. At a higher but still low concentration of 1.6% all characteristic spectroscopic features of surface H_2O molecules disappear.

We thus see that, as for the time-resolved nonlinear IR methods, sum-frequency vibrational spectroscopy is a valuable tool to convey information on H-bonds, particularly H-bonds at interfaces.

CONCLUSION

This whole chapter shows that IR spectroscopy has unique possibilities to observe H-bonds. These H-bonds first manifest themselves in the FIR region where specific bands due to the relative vibrations of the two parts X–H and Y of an H-bond X–H···Y appear. These bands disappear when these H-bonds disrupt. They are weak and are found in a somewhat difficult spectroscopic region. They are consequently not often considered. What makes IR a unique tool to observe H-bonds is their dramatic effects in the conventional mid-IR region, due to the huge changes their establishments induce on intramonomer vibrations. This is particularly true of the H-stretching band of X–H, ν_s(X–H···), that exhibits, upon formation of an *intermolecular* H-bond X–H···Y a huge enhancement of its integrated intensity, always accompanied by both a great shift of its centre of gravity towards lower wavenumbers and by an important width. These effects make IR spectroscopy an especially sensitive and powerful tool to study H-bonds, a tool that provides correlations between wavenumbers and distances in a somewhat unusual way for optical spectroscopy, as illustrated in Figure 4.5. The origins of these strong effects of H-bonds on IR spectra are now well understood, as discussed in Ch. 5. The huge increase of the intensity of ν_s is due to the relatively small displacement of electrons that accompanies an increase of the X–H distance in X–H···Y, thus creating an important electric dipole moment. The great modifications of the bandshape are due to a strong anharmonic modulation of the frequency of this stretching vibration by intermonomer modes. In addition to its own great changes this modulation induces on the bandshape, it furthermore allows secondary anharmonic effects such as resonance interactions, also called Fermi resonances, to appear. The understanding of this modulation gives a precise picture of the dynamics of the H-atom in H-bonds and is seen to provide arguments that will allow establishing the structure of the H-bond network of liquid water (Ch. 9). It forms the basis for interpretations of spectra obtained using modern nonlinear time-resolved IR spectroscopy. These quite recent techniques have shown that 1 psec is a good order of magnitude for lifetimes of excited ν_s vibrations, and that the lifetimes of H-bonds themselves is often considerably longer. In the near future, these methods will be valuable for studying the reactivity of H-bonds, that is proton transfers in acid/base systems, and certainly more important, H-atom transfers in aqueous media at pH 7, a mechanism that now appears to be the basic reaction mechanism in biomedia (Ch. 6).

Intramonomer bands other than ν_s are also sensitive to H-bonds, although in a much less spectacular way. This sensitivity proves to be most useful in chemistry and biology, and, as will be seen in the case of the H_2O molecule, allows precise measurements of the number of H-bonds from which much structural and dynamic information can be obtained. It will prove useful to follow the development of the H-bond network of a macromolecule during hydration (Ch. 10). In physics and chemistry, the hypersensitive ν_s band ensures the presence and formation of H-bonds and still conveys a lot of information when these H-bonds are of a single type. Extraction of information from ν_s becomes more difficult when various types of H-bonds are present, as in biomedia. Furthermore, the extreme sensitivity of the integrated intensity of this ν_s band to the strength of the H-bond makes it hard to use for the measurement of the number of H-bonds in a sample. These other intramonomer bands make it possible.

These properties make IR spectroscopy certainly the most precise and powerful tool to observe H-bonds not only in physics but also in chemistry and biology. In its 1D form, it requires a technique, conventional IR, which is a routine technique, easy to implement in many set-ups. In its recent version of time-resolved spectroscopy, it is not presently a routine technique, but it has proved to be most informative. The particularly important effects H-bonds have on IR spectra require, however, some care and method during the analysis of these spectra, which is thus scarcely a routine analysis. We shall see that with a minimum of care and methodology it can nevertheless be used with no difficulty.

REFERENCES

1. P. R. Griffiths and J. A. de Haseth, Fourier Transform Infrared Spectroscopy, Wiley, New York, 1986.
2. J. C. Lassègues and J. Lascombe, in J. R. Durig (Ed.), Vibrational Spectra and Structure, Vol. 11, Ch. 2, Elsevier, Amsterdam, 1982.
3. H. R. Zelsmann, Z. Mielke and Y. Maréchal, J. Mol. Struct., 237 (1990) 273.
4. Y. Maréchal, J. Chem. Phys., 87 (1987) 6344.
5. J.-J. Max, S. Daneault and C. Chapados, Can. J. Chem., 80 (2002) 113.
6. J. E. Bertie, A. M. Khalique and H. H. Eysel, J. Phys. Chem., 93 (1989) 2210.
7. Yu. E. Gorbaty and A. G. Kalinichev, J. Phys. Chem., 99 (1995) 5336.
8. A. C. Legon and D. J. Millen, Faraday Discuss. Chem. Soc., 73 (1982) 71.
9. A. C. Legon and D. J. Millen, Chem. Rev., 86 (1986) 635.
10. A. C. Legon and D. J. Millen, Chem. Soc. Rev., 16 (1987) 467.
11. M. Farnik, M. Weimann and M. A. Suhm, J. Chem. Phys., 118 (2003) 10120.
12. A. J. Barnes, T. R. Beech and Z. Mielke, J. Chem. Soc, Faraday Trans., 2, 80 (1984) 455.
13. A. J. Barnes, J. Nicholas, S. Kuzniarski and Z. Mielke, J. Chem. Soc, Faraday Trans., 2, 80 (1984) 465.
14. L. Andrews, X. Wang and Z. Mielke, J. Phys. Chem. A, 105 (2001) 6054.
15. P. Hobza, Phys. Chem. Chem. Phys., 3 (2001) 2555.
16. R. M. Badger and S. H. Bauer, J. Chem. Phys., 5 (1937) 839.
17. H. Ratajczak, W. J. Orville-Thomas and C. N. R. Rao, Chem. Phys., 17 (1976) 197.
18. M. Besnard, Y. Danten and T. Tassaing, J. Chem. Phys., 113 (2000) 3741.
19. A. Novak, in J. D. Dunitz, P. Hemmerich, R. H. Holm, J. A. Ibers, C. K. Jorgensen, J. B. Neilands and D. Reinen (Eds.), Structure and Bonding, Vol. 18, Springer-Verlag, Berlin, Heidelberg, New York, 1974.
20. B. Berglund, J. Lindgren and J. Tegenfeldt, J. Mol. Struct., 43 (1978) 179.
21. W. Mikenda, J. Mol. Struct., 147 (1986) 1.
22. S. Bratos, J. Cl. Leicknam, S. Pommeret and G. Gallot, J. Mol. Struct., 708 (2004) 197.
23. J. B. Asbury, T. Steinel, C. Stromberg, K. J. Gaffney, I. R. Piletic and M. D. Fayer, J. Chem. Phys., 119 (2003) 12981.

24. N. Kobko and J. J. Dannenberg, J. Phys. Chem. A, 107 (2003) 10389.
25. W. A. P. Luck, D. Klein and K. Rangsriwatananon, J. Mol. Struct., 416 (1997) 287.
26. H. T. Flakus, J. Mol. Struct., 646 (2003) 15.
27. L. England-Kretzer, M. Fritzshe and W. A. P. Luck, J. Mol. Struct., 175 (1988) 277.
28. A. Novak, J. Chim. Phys., 69 (1972) 1615.
29. B. Orel and D. Hadzi, Spectrochim. Acta., 32A (1976) 1731.
30. D. Hadzi, Chimia, 26 (1972) 7.
31. D. Hadzi, B. Orel and A. Novak, Spectrochim. Acta., 29A (1973) 1745.
32. G. Auvert and Y. Maréchal, J. Phys., (France), 40 (1979) 735.
33. P. Excoffon and Y. Maréchal, Chem. Phys., 53 (1980) 23.
34. V. Schreiber, A. Koll, A. Kulbida and I. Majerz, J. Mol. Struct., 348 (1995) 365.
35. J.-J. Max and C. Chapados, J. Phys. Chem. A, 108 (2004) 3324.
36. M. Falk, Spectrochim. Acta., 40A (1984) 43.
37. R. G. Snyder, L. X. Q. Chen, V. J. P. Srivatsavoy, H. L. Strauss and S. Kubota, J. Phys. Chem., 99 (1995) 2214.
38. J. E. Bertie and K. H. Michaelian, J. Chem. Phys., 77 (1982) 5267.
39. J. E. Bertie and K. H. Michaelian, J. Chem. Phys., 76 (1982) 886.
40. Y. Maréchal, M. Milas and M. Rinaudo, Biopolymers (Biospectroscopy), 72 (2003) 162.
41. K. Kwac and M. Cho, J. Chem. Phys., 119 (2003) 2247.
42. J. Grdadolnik and Y. Maréchal, Biopolymers (Biospectroscopy), 62 (2001) 40.
43. Y. Maréchal and H. Chanzy, J. Mol. Struct., 523 (2000) 183.
44. N. Bakkas, Y. Bouteiller, A. Loutellier, J. P. Perchard and S. Racine, Chem. Phys. Lett., 232 (1995) 90.
45. F. Huisken, S. Mohoammad-Pooran and O. Werhahn, Chem. Phys., 239 (1998) 11.
46. J. Bournay and Y. Maréchal, J. Chem. Phys., 59 (1973) 5077.
47. E. Vanardou, E. Garcia-Verdugo, S. J. Barlow, Y. E. Gorbaty and M. Poliakoff, Vibrational Spectrosc., 35 (2004) 103.
48. P. Lalanne, J. M. Andanson, J. C. Soetens, T. Tassaing, Y. Danten and M. Besnard, J. Phys. Chem. A, 108 (2004) 3902.
49. K. Heyne, N. Huse, J. Dreyer, E. T. J. Nibbering and T. Elsaesser, J. Chem. Phys., 121 (2004) 902.
50. Y. L. A. Rezus, D. Madsen and H. J. Bakker, J. Chem. Phys., 121 (2004) 10599.
51. J. Stenger, D. Madsen, P. Hamm, E. T. J. Niberring and T. Elsaesser, J. Phys. Chem. A, 106 (2002) 2341.
52. J. M. F. DeCamp, L DeFlores, J. M. McCracken, A. Tokmakoff, K. Kwac and M. Cho, J. Phys. Chem. B, 109 (2005) 11016.
53. S. Bratos, J-Cl. Leicknam, G. Gallot and H. Ratajczak, in T. Elsaesser (Ed.), Ultrafast Hydrogen Bonding Dynamics and Proton Transfer Processes in the Condensed Phase, Vol. 23, Ch. 2, Kluwer Academics Publishers, Dordrecht, 2002.
54. K. B. Møller, R. Rey and J. T. Hynes, J. Phys. Chem. A, 108 (2004) 1275.
55. K. Heyne, N. Huse, E. T. J. Nibbering and T. Elsässer, Chem. Phys. Lett., 369 (2003) 591.
56. J. B. Asbury, T. Steinel and M. D. Fayer, J. Phys. Chem. B, 108 (2004) 6544.
57. J. D. Eaves, J. J. Loparo, C. J. Fecko, S. T. Roberts, A. Tokmakoff and P. L. Geissler, Proc. Nat. Acad. Sci., 102 (2005) 13019.
58. K. Heyne, N. Huse, E. T. J. Nibbering and T. Elsaesser, Chem. Phys. Lett., 382 (2003) 19.
59. R. Laennen, C. Rauscher and A. Laubereau, J. Phys. Chem. B, 102 (1998) 9304.
60. R. Laenen, K. Simoneidis and A. Laubereau, J. Phys. Chem. B, 106 (2002) 408.
61. T. Steinel, J. B. Asbury, J. Zheng and M. D. Fayer, J. Phys. Chem. A, 108 (2004) 10957.
62. O. F. A. Larsen and S. Woutersen, J. Chem. Phys., 121 (2004) 12143.
63. A. Pakoulev, Z. Wang, Y. Pang and D. D. Dlott, Chem. Phys. Lett., 380 (2003) 404.

64. K. Heyne, E. T. J. Nibbering, T. Elsaesser, M. Petkovic and O. Kühn, J. Phys. Chem. A, 108 (2004) 6083.
65. J. Stenger, D. Madsen, J. Dreyer, E. T. J. Nibbering, P. Hamm and T. Elsaesser, J. Phys. Chem. A, 105 (2001) 2929.
66. A. J. Lock and H. J. Bakker, J. Chem. Phys., 117 (2002) 1708.
67. C. P. Lawrence and J. L. Skinner, J. Chem. Phys., 119 (2003) 3840.
68. M. F. Kropman and H. J. Bakker, J. Chem. Phys., 115 (2001) 8942.
69. M. Rini, D. Pines, B-Z. Magnes, E. Pines and E. T. J. Nibbering, J. Chem. Phys., 121 (2004) 9593.
70. O. F. Mohammed, D. Pines, J. Dreyer, E. Pines and E. T. J. Nibbering, Science, 310 (2005) 83.
71. C. H. Londergan, Y. S. Kim and R. H. Hochstrasser, Mol. Phys., 103 (2005) 1547.
72. M. R. Watry, M. G. Brown and G. L. Richmond, Appl. Spectr., 55 (2001) 321.
73. P. B. Miranda and Y. R. Shen, J. Phys. Chem. B, 103 (1999) 3292.
74. Q. Du, R. Superfine, E. Freysz and Y. R. Shen, Phys. Rev. Lett., 70 (1993) 2313.
75. E. A. Raymond, T. L. Tarbuck, M. G. Brown and G. L. Richmond, J. Phys. Chem. B, 107 (2003) 546.
76. L. F. Scatena and G. L. Richmond, J. Phys. Chem. B, 105 (2001) 11240.
77. J. C. Conboy, M. C. Messmer and G. L. Richmond, J. Phys. Chem., 100 (1996) 7617.
78. J. C. Conboy, M. C. Messmer and G. L. Richmond, J. Phys. Chem. B, 101 (1997) 6724.
79. M. J. Shultz, S. Baldelli, C. Schnitzer and D. Simonelli, J. Phys. Chem. B, 106 (2002) 5313.
80. E. Tyrode, C. M. Johnson, S. Baldelli, C. Leygraf and M. W. Rutland, J. Phys. Chem. B, 109 (2005) 329.
81. C. M. Johnson, E. Tyrode, S. Baldelli, M. W. Rutland and C. Leygraf, J. Phys. Chem. B, 109 (2005) 321.

– 5 –

Infrared Spectroscopy of H-Bonded Systems: Theoretical Descriptions

INTRODUCTION

The spectacular changes of the ν_s band of a X–H molecule upon establishment of an H-bond are now fairly well understood. In this chapter we give a rapid theoretical overview of these changes that provides an accurate view of the dynamics of H-bonds and is at the basis of interpretations of spectra of promising new methods of time-resolved nonlinear IR spectroscopy described in Ch. 4 and also of interpretations of conventional IR spectra that are especially aimed at interpreting ν_s bandshapes. Theory requires some basic formulation of quantum mechanics and of spectroscopy that are given in the appendix of this chapter. They are used only for what is necessary in this subsection, where, as much as possible, a "hands-on" interpretation accompanies equations. It may help to understand the physics that lies behind them. The understanding of this theoretical aspect is not necessary to interpret most IR spectra of H-bonds encountered in chemistry, biophysics and biochemistry as the knowledge of the characteristic features of these spectra presented in previous Ch. 4 is sufficient. The reader who is not interested in theoretical descriptions may consequently skip this whole chapter without being afraid of missing something necessary for the understanding of the following chapters.

INTEGRATED INTENSITIES OF ν_s BANDS

We start with the dramatic enhancement of the integrated intensity of ν_s after the establishment of an H-bond. This quantity is equal to the integral of the intensity over the whole ν_s band and, following eq. (5.A14) of the appendix of this chapter, is also equal to the moment M_0 of order 0 of the ν_s band of the spectrum in $\varepsilon''(\nu)$. Following eq. (5.A19), M_0 is equal to the Fourier transform quantity $S(t)$ of $\varepsilon''(\nu)$ at time $t = 0$ that is equal, following eq. (5.A4), to the correlation function of the dipole moment μ of the X–H$\cdots$Y complex. For the ν_s band, we need to consider only the dependence of μ on the coordinate q that defines the ν_s vibrational mode. Developing the value of μ in powers of this coordinate around its equilibrium position q_0 and considering only linear terms, gives then

$$M_0 = S(t=0) = \frac{2\pi}{\hbar V \varepsilon_0}\left\langle \mu^2(0) \right\rangle = \frac{2\pi}{\hbar V \varepsilon_0}\left(\frac{\partial \mu}{\partial q}\right)^2 \left|\left\langle 0 \left| q - q_0 \right| 1 \right\rangle\right|^2 \tag{5.1}$$

where $\langle 0 |q| 1\rangle$ is the matrix element (see eqs. (5.A1) and (5.A2) given in the appendix of this chapter for the definitions of the matrix element) that characterizes the transition between ground state 0 and first excited state 1 in ν_s. As $h\nu = hc\tilde{\nu}$, with $\tilde{\nu}$ in the range 3000–3600 cm^{-1} much greater than $KT/\hbar$ at 300 K, the $0 \rightarrow 1$ is the only transition in ν_s that has a significant intensity. $\partial\mu/\partial q$ is the first derivative of the electric dipole moment μ of X–H$\cdots$Y with respect to q. The fraction that appears in eq. (5.1) is a constant that we can disregard, as we are not interested in the absolute value of M_0, but in the order of magnitude of the ratio of the values of M_0 in X–H and X–H$\cdots$Y. It does not furthermore make difference considering the spectrum in $\varepsilon''(\nu)$, for which eq. (5.1) is strictly valid, instead of the spectrum in absorbance $A(\tilde{\nu})$ such as that drawn in Figure. 4.4, for instance, which is representative of most experimental IR spectra. When a more precise calculation is required, it may be performed using $A(\tilde{\nu})/\tilde{\nu}$ instead of $A(\tilde{\nu})$ for calculating moments (1), as this quantity is proportional to the spectrum in $k(\tilde{\nu})$ (eq. (5.A8) of the appendix of this chapter) and hence to $\varepsilon''(\tilde{\nu})$ if we neglect the often weak variation of $n(\tilde{\nu})$ with $\tilde{\nu}$ (eq. (5.A7)). As eq. (5.1) is also valid for X–H (no H-bond) and as the matrix element $|\langle 0 |q| 1\rangle|^2 = \hbar/2\, m\omega = \langle 0 |q^2| 1\rangle$ for a harmonic oscillator only slightly varies between ν_s(X–H$\cdots$) and ν_s(X–H), the great variation by more than one order of magnitude that is observed between the integrated intensities of ν_s(X–H$\cdots$) and ν_s(X–H) mainly comes from the variation with q of the dipole moment μ. It means that this variation, characterized by $\partial\mu/\partial q$, is greater by a factor falling in the range 3–10 when X–H establishes an H-bond than when it does not. In other words, *stretching the X–H covalent bond of a X–H group that establishes an H-bond causes the electrons to not follow the H-atom as much as when X–H does not establish an H-bond* (2). *This stronger decoupling of the average position of the electrons from the motion of the H-atom* is at the origin of an important change of the electric dipole moment of the whole complex, itself responsible for the great integrated intensities all ν_s(X–H$\cdots$) bands exhibit. The transfer of electrons between the nonbonding orbital of Y in X–H$\cdots$Y and the antibonding orbital of X–H (3) that we have seen in Ch. 1 to be an important feature of the electronic structure of H-bonds, is certainly for a great part responsible for the magnitude of $\partial\mu/\partial q$. Let us note that we have considered $\partial\mu/\partial q$ as a quantity that is independent of q, a first approximation that is most of the time a very good approximation in IR spectroscopy and is called the electrical harmonicity approximation.

ν_s BANDSHAPES OF ISOLATED H-BONDS: MODULATION BY INTERMONOMER MODES

The great width of ν_s(X–H$\cdots$), the presence of several submaxima and subminima, clearly show that this band has an anharmonic character. This was first noted by Stepanov (4) and later by Bratos *et al.* around 1955 (5). Anharmonicity, however, is a concept that needs more precision—as we have already seen, it is by definition present as soon as the development of the vibrational energy on vibrational coordinates requires including terms higher than an order of 2. It leaves many possibilities for ν_s bands that at the same time retain some harmonic character. Thus, the overtones of ν_s, that is $0 \to n$ transitions in q, remain very weak for n greater than 1. In order to take these features into account, one can then simply postulate (6) that ν_s is harmonic with its force constant depending on the geometry of the H-bond, that is on at least one of the intermonomer modes Q, with Q representing either the stretching mode Q_s, or one of the intermonomer bending mode Q_θ or Q_φ defined in Figure 2.1.

Modulation by intermonomer stretching modes

We first suppose Q is the stretching mode Q_s, which would be appropriate for nearly all H-bonded systems. In the isolated X–H···Y system where the X–H distance is q and the X···Y distance Q_s, with corresponding momenta $p = -i\hbar(\partial/\partial q)$ and $P_s = -i\hbar(\partial/\partial Q_s)$, the vibrational Hamiltonian may be written as

$$\begin{aligned} H(q,Q_s) &= \frac{p^2}{2m} + \frac{P_s^2}{2M_s} + V(q,Q_s) \\ &= \frac{p^2}{2m} + \frac{P_s^2}{2M_s} + \frac{1}{2}m\omega^2(Q_s)(q-q_0)^2 + \frac{1}{2}M_s\Omega_s^2(Q_s-Q_0)^2 \end{aligned} \tag{5.2}$$

where m is the mass of the H-atom, which is to be replaced by $2m$ in the case of a D-bond, and M_s the reduced mass of the intermonomer Q_s mode. In this equation constant terms in $V(q_0,Q_0)$ are omitted, which is allowed by the choice we have to define energy 0. These terms are irrelevant in vibrational spectroscopy but cannot be omitted when thermodynamics of H-bonds, such as considered in Chs. 1 and 7, are examined. The equilibrium distances q_0 and Q_0 are the distances q and Q_s for which the energy $V(q,Q_s)$ of the electronic ground state is minimum. With this form the potential for ν_s (coordinate q) may be seen as a harmonic potential that nevertheless exhibits an anharmonic coupling with the intermonomer mode defined by Q_s, which is otherwise also governed by harmonic terms. The origin of this anharmonic coupling between these two harmonic modes is to be found in the dependence of ω in Q_s that makes such terms as $q^2Q_s^n$ appear in eq. (5.2), with n a positive integer. Such terms are of order higher than 2, consequently anharmonic terms. This dependence of ω in Q_s looks natural if one refers to Novak's curve of Figure 4.5. One should, however, keep in mind that Novak's curve refers to experimental equilibrium positions Q_0, whereas in eq. (5.2) ω depends on the vibrational coordinate Q_s, not on its value Q_0 at equilibrium. This ω dependence in Q_s therefore describes a dynamic modulation of ν_s by intermonomer modes, whereas Novak's curve represents its average static effect.

Having in mind the dramatic effects the establishment of an H-bond has on the ν_s bandshape, we may anticipate that this anharmonic coupling is not small. It means that it cannot be handled by classical perturbation techniques. It may, however, be taken into account in the frame of the adiabatic separation (6) of rapid and slow motions. This adiabatic separation is already used to separate the motions of the electrons in the molecular complex from the vibrations of the atoms and is then called "Born–Oppenheimer" separation. In this approximation applied to the separation of ν_s from the intermonomer modes, the rapid vibration ν_s, which is ruled by $H(q,Q_s)$ of eq. (5.2) and displays characteristic wavenumbers around

3000 cm^{-1}, is thought to be sufficiently fast as to adapt itself immediately to any change of the coordinate Q_s of the slower intermonomer mode, which has characteristic wavenumbers smaller than 200 cm^{-1} (in other words a frequency at least some 15 times smaller). It means that the wavefunction that describes ν_s sees Q_s as a very slow motion and consequently depends on the value of Q_s, but not on its dynamics. Mathematically it is equivalent to supposing that the total wavefunction for ν_s (coordinate q) and intermonomer mode (Q_s), which is an eigenfunction of the Hamiltonian $H(q,Q_s)$ of eq. (5.2), has the form

$$\phi_n^N(q,Q_s) = \chi_n(q,Q_s)\alpha_n^N(Q_s) \tag{5.3}$$

where $\chi_n(q,Q_s)$ is the nth harmonic wavefunction in q. It parametrically depends on Q_s, and is shortly written $|n\rangle_q$ in eq. (5.A37) of the appendix of this chapter. It is a harmonic eigenfunction of that part of $H(q,Q_s)$ that depends on q only in eq. (5.2). It means it verifies

$$\left\{\frac{p^2}{2m} + \frac{1}{2}m\omega^2(Q_s)(q-q_0)^2\right\}\chi_n(q,Q_s) = \left(n+\frac{1}{2}\right)\hbar\omega(Q_s)\chi_n(q,Q_s) \tag{5.4}$$

With this form for the total wavefunction $\phi_n^N(q,Q_s)$ only the transition in q between levels n and $n \mp 1$ are allowed. As only the level $n = 0$ is populated at room temperature, it naturally incorporates the experimental finding that only $0 \rightarrow 1$ transitions in q have a significant intensity. The Hamiltonian H_n that governs the wavefunction $\alpha_n^N(Q_s)$ of the slow intermonomer mode Q_s when ν_s is in its nth state is then the value of the total Hamiltonian $H(q,Q_s)$ of eq. (5.2) averaged over the rapid mode q

$$\begin{aligned} H_n(Q_s) = \langle\chi_n(q,Q_s)|\,H(q,Q_s)\,|\chi_n(q,Q_s)\rangle_q &\simeq \frac{P_s^2}{2M} \\ + \frac{1}{2}M_s\Omega_s^2(Q_s - Q_0)^2 &+ \left(n+\frac{1}{2}\right)\hbar\omega(Q_s) \end{aligned} \tag{5.5}$$

It means that the slow motion only sees the average effect of the rapid one. In this equation we disregarded, as indicated by the sign $\simeq$, all terms of the form $\langle\chi_n(q,Q_s)\,|P_s|\,\chi_m(q,Q_s)\rangle_q$. It characterizes the adiabatic approximation. It means that the intermonomer mode Q_s is so slow that it cannot induce transitions between various levels of the rapid mode q. This rapid mode consequently adiabatically follows the slow intermonomer mode, adapting immediately to any change in the value of Q_s, but suffering no transition to another level that would be caused by this change. Once more, the same process that makes electrons adiabatically follow vibrations of nuclei, immediately adapting themselves, but with no transition to an excited electronic level. The Nth wavefunction $\alpha_n^N(Q_s)$ of the slow intermonomer mode when the rapid ν_s mode is in its nth state then verifies the equation

$$\left\{\frac{P_s^2}{2M_s} + \frac{1}{2}M_s\Omega_s^2(Q_s - Q_0)^2 + \left(n+\frac{1}{2}\right)\hbar\omega(Q_s)\right\}\alpha_n^N(Q_s) = E_n^N\alpha_n^N(Q_s) \tag{5.6}$$

where the E_n^N's are the total energies of the whole system composed of both modes q and Q_s, that is the eigenvalues of the initial Hamiltonian $H(q,Q_s)$ of eq. (5.2). When the dependence of ω on Q_s can be limited to a quadratic variation, $\alpha_n^N(Q_s)$ is still a harmonic wavefunction in Q_s. For all H-bonds that have been tested, even the linear dependence is sufficient. Adopting it means that we have to do with transitions that are displayed in Figure 5.1 and are most similar to those encountered in spectra of electronic transitions that fall in the UV-visible

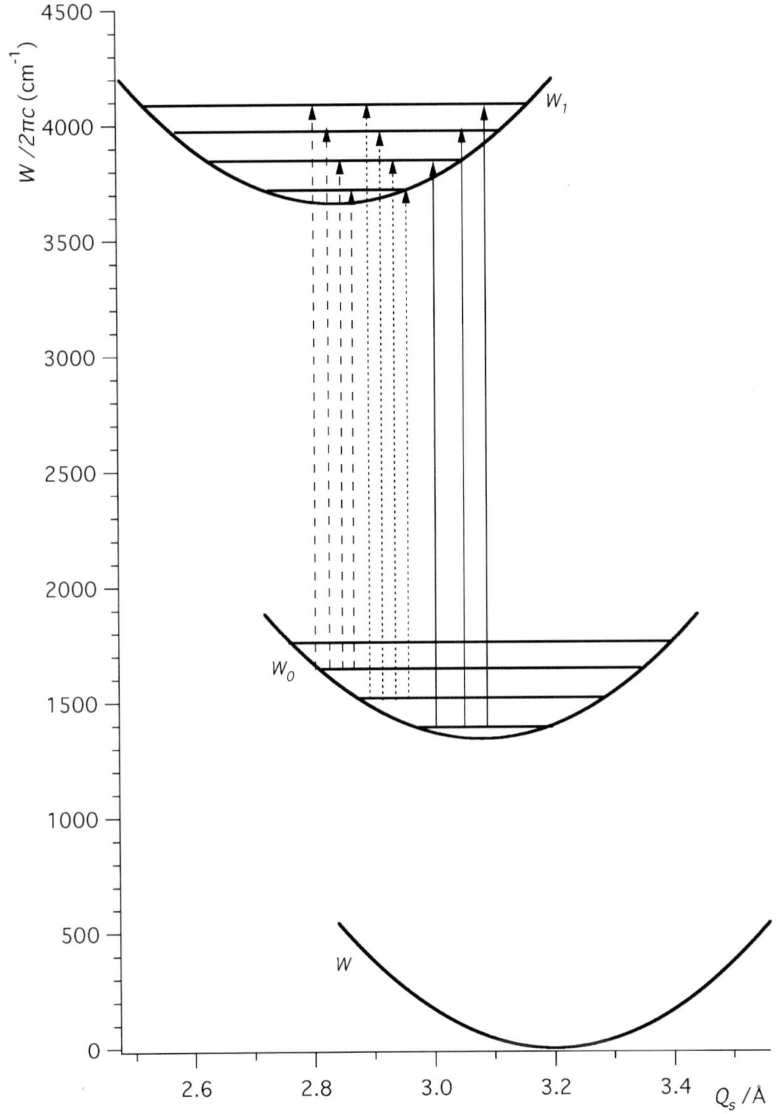

Figure 5.1 Vibrational potentials for the slow intermonomer Q_s mode of an isolated H-bond, typically Cl–H···O–$(CH_3)_2$ (7), with $\omega(Q_0)/2\pi c = 2500\,\text{cm}^{-1}$ and $\Omega_s/2\pi c = 120\,\text{cm}^{-1}$ (c, velocity of light in cm sec^{-1}). $W = \frac{1}{2}M\Omega_s^2(Q_s - Q_0)^2$. The potentials $W_n = W(Q_s) + (n+\frac{1}{2})\,\hbar\omega(Q_s)$ $(n = 0, 1)$ are the effective potentials that rule this slow-mode Q_s when the rapid mode is in its nth state. Vertical arrows indicate the various transitions that constitute the experimental ν_s band.

region and most often display a "Franck–Condon" vibrational structure characterized by a regular progression of bands that are displaced by a vibrational quantum of energy, presently Ω_s.

In Figure 5.1 the potential $W = \frac{1}{2} M_s \Omega_s^2 (Q_s - Q_0)^2$ is the harmonic potential for intermonomer mode Q_s that appears in eq. (5.2). Following eq. (5.6), the effective potential that rules Q_s is $W_0 = \frac{1}{2} M_s \Omega_s^2 (Q_s - Q_0)^2 + \frac{1}{2} \hbar\omega(Q_s)$, when the rapid ν_s mode in q is in its ground state $n = 0$, while it is $W_1 = \frac{1}{2} M_s \Omega_s^2 (Q_s - Q_0)^2 + \frac{3}{2} \hbar\omega(Q_s)$ when ν_s is in its first excited state with $n = 1$. The ν_s band corresponds to all transitions between the various levels E_0^N for Q_s of W_0 and the levels E_1^N of W_1. These various levels are represented as horizontal lines within the parabola W_0 and W_1 in Figure 5.1. Admitting that a linear dependence in Q_s of $\omega(Q_s)$ is a good approximation, one deduces from eq. (5.6) the values for these levels

$$E_n^N = \left(n + \frac{1}{2}\right)\hbar\omega(Q_0) + \left(N + \frac{1}{2}\right)\hbar\Omega_s - \left(n + \frac{1}{2}\right)^2 \frac{\hbar^2}{2M_s\Omega_s^2}\left(\frac{d\omega}{dQ_s}\right)^2 \tag{5.7}$$

The $0 \rightarrow 1$ transition in ν_s is accompanied by various transitions in the intermonomer mode Q_s that gives this band a structure, with possibly several minima and maxima, the same way a single electronic transition is accompanied by numerous vibrational transitions at the origin of its "vibrational or Franck–Condon structure". In Figure 5.1 we distinguish transitions that correspond to "cold bands" that are drawn as plain arrows. They all start from the ground vibrational state E_0^0, the only level populated at 0 K. The "hot bands" that correspond to transitions drawn as dotted or dashed arrows, appear at higher temperature, when states E_0^1, E_0^2, etc., start being populated, thus acting as possible initial states for transitions in ν_s. The schematic shape of such a simple spectrum with a regular progression is shown in the upper drawings of Figures 5.2 and 5.3, examined later in more detail.

Such an approach works quite well in the case of isolated H-bonds. Thus, the predicted Franck–Condon type progressions are clearly visible in the experimental IR spectra of $Cl{-}H\cdots O{-}(CH_3)_2$ in gas (7) or $F{-}H\cdots B$ in gas, with B being various H-bond acceptors (8). All these systems initially studied by D. J. Millen and his group display maxima of absorption at wavenumbers $E_n^N/\hbar$ that can be deduced from eq. (5.7). They also clearly appear in the IR spectra of *N*-methyl-acetamide (9). The predictions of this theory sketched above has for the first time been directly compared (6, 10) with experimental spectra of gaseous carboxylic acid dimers, which are represented in Figure 4.4. These spectra are much easier to record than spectra of isolated H-bonds such as those found in the $Cl{-}H\cdots O{-}(CH_3)_2$ family. They consequently display a much better signal-to-noise ratio, with a well-defined baseline. Also, the geometry of these cyclic dimers is more precisely known. A complication, however, is that in these isolated systems, we have two interacting H-bonds, 1 and 2, instead of a single H-bond. This can nevertheless be precisely taken into account by adding harmonic coupling terms between the two ν_s vibrations of the form $V_0 q_1 q_2$ between the two H-bonds 1 and 2 in the vibrational potential. These terms are called "Davydov" coupling terms in this case. The preceding theory and the simple scheme of transitions drawn in Figure 5.1 leads to a satisfactory simulation of the ν_s bandshape of these simple models of H-bonded systems, as shown in Figure 5.2. The capacity of this theory to predict and reproduce the spectrum of the corresponding cyclic dimer when the H-atom that establishes an H-bond is changed into a D-atom has been a decisive criterium of its validity. For this change only the

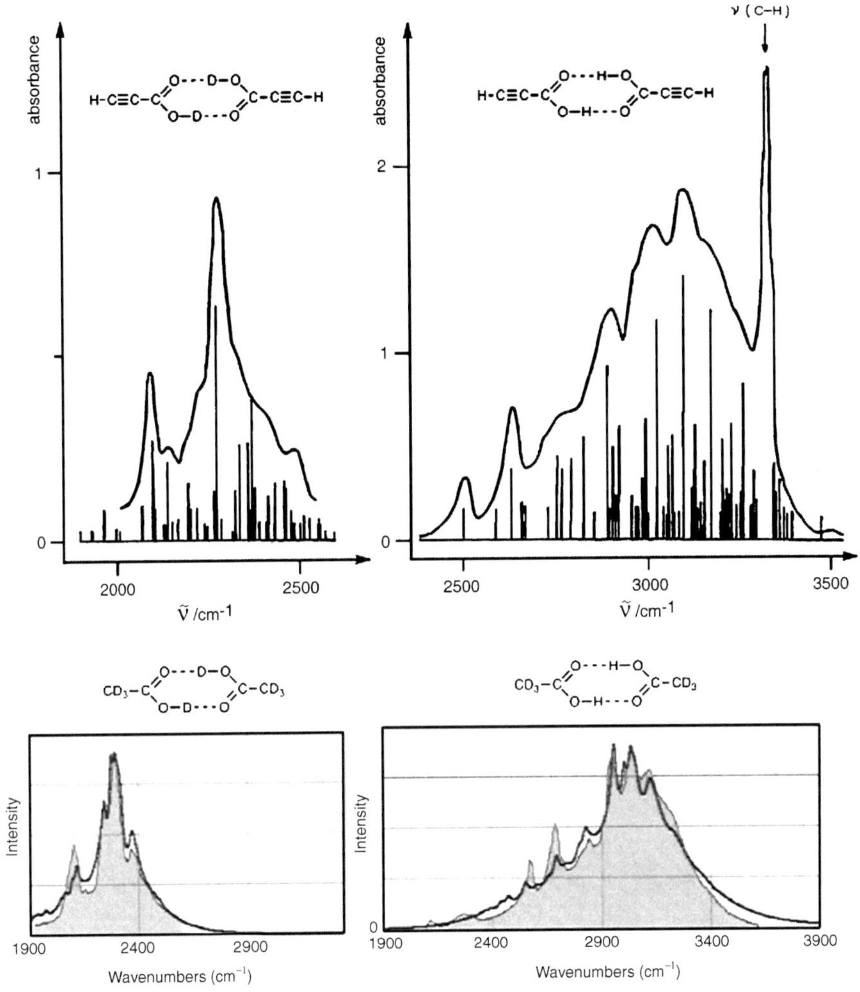

Figure 5.2 Simulation of experimental ν_s bands of carboxylic acid dimers. Propynoic and propynoic-D acids simulated on the basis of the first simplest theory (6) (upper spectra). Experimental spectra are drawn with full lines and predicted transitions as vertical bars. More recent simulations of IR spectra of acetic acid dimers are shown in the lower spectra, where greyed portion is experiment and thick line is theoretical reconstitution. Reproduced with permission from J. Bournay and Y. Maréchal, J. Chem. Phys., 55 (1971) 1230 and from ref. (13). Copyright 2006, American Institute of Physics.

mass m of the H-atom in eqs. (5.2) and (5.4) becomes $2m$. As may be seen in Figure 5.2, it has dramatic effects on ν_s that are well reproduced by this theory. In this figure, propynoic acid has been chosen because other anharmonic couplings, such as those due to Fermi resonances we shall see in a forthcoming subsection, are small. In cyclic dimers of many other carboxylic acids, Fermi resonances are present, leading to a less satisfactory

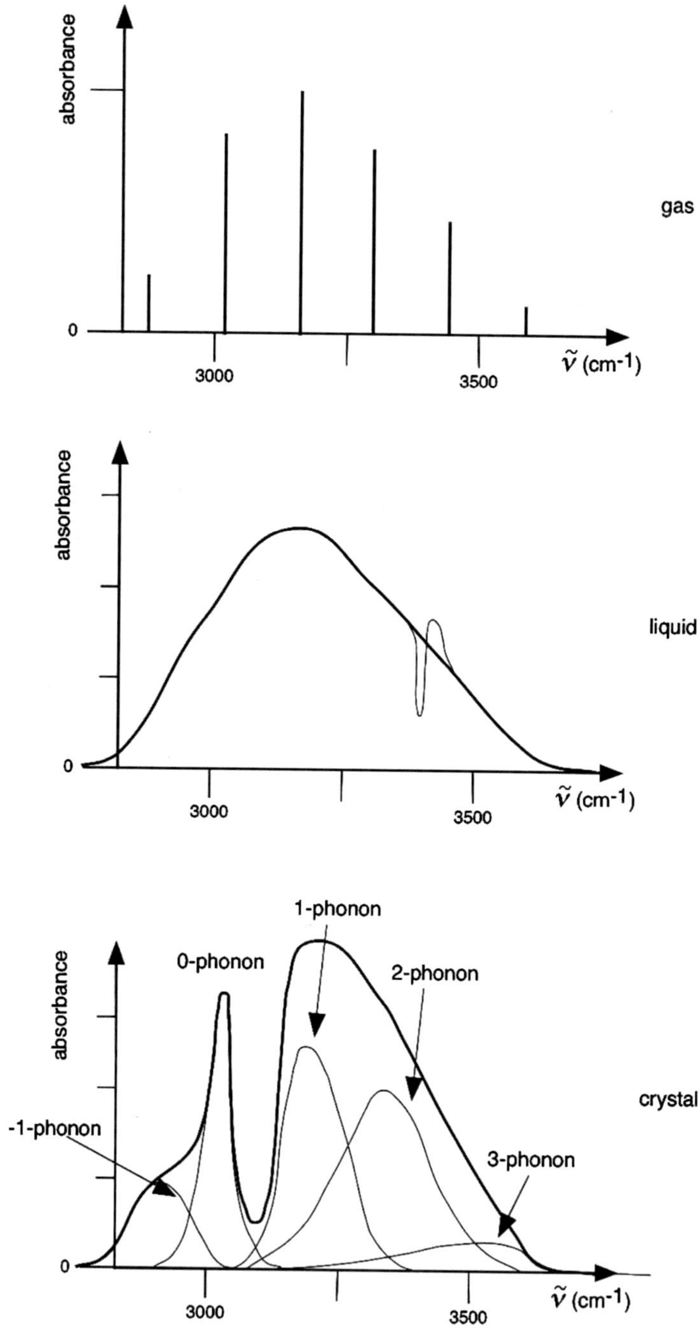

Figure 5.3 ν_s bands of an isolated H-bond in a gas, a liquid and a crystal. An "Evan's hole" due to a Fermi resonance is added in the middle spectrum as a thin line. In the bottom spectrum, the "n-phonon" components are drawn as thin lines.

simulation. Inclusion in recent theories (11, 12) of these extra anharmonic couplings, together with consideration of bandshapes of individual transitions of Figure 5.1 due to quantum dumping of both q and Q modes, leads to good reconstitutions of these spectra (bottom spectra of Figure 5.2) (13). In Figure 5.2, Fermi resonances that, as will be seen below, have only a local influence on bandshapes are neglected in order to limit the number of parameters.

Modulation by intermonomer bending modes

This modulation of ν_s by intermonomer modes of the H-bond is thus a fundamental mechanism inherent to the dynamics of the H-bond. For nearly all H-bonds the slow modulating intermonomer mode is the stretching mode Q_s considered above. An important exception, however, exists: liquid water, described in Ch. 9 where modulation by the bending intermonomer mode is predominant. More generally, H-bonds established by H_2O molecules are likely to enter this category of H-bonds where the modulation by intermonomer bending modes is important. The reason of this peculiarity is to be found in the very small value of the moments of inertia of the H_2O molecule, which is at the origin of rotations of this molecule of great amplitudes. Most developments that have been described above can then be reproduced with no difficulty for that case. The result is, however, slightly different, a consequence of the dependence of $\omega(\theta)$ on θ (or φ, the second intermonomer bending mode; see Figure 2.1) that is different from that on Q_s for most H-bonds that appears in eq. (5.2). We have seen this Q_s dependence resembles that illustrated in Novak's curve of Figure 4.5 and can most of the time be reduced to a linear variation with $(d\omega/dQ_s) > 0$ with, in some few cases, addition of a quadratic dependence with $(d^2\omega/dQ_s^2) < 0$. In the case of a modulation by a bending intermonomer mode, the linear dependence of $\omega(\theta)$ on θ is zero for symmetry reasons, and its quadratic dependence is such that $(d^2\omega/dQ_\theta^2) > 0$, at least for θ not approaching π. The absence of linear terms in $d\omega/dQ_\theta$ does not suppress the appearance in ν_s of a progression analogous to that in Ω_s as appears in eq. (5.7). It is, however, less marked. A greater difference is indeed found in the correlation of the shift of the centre of ν_s with its width: increasing this modulation by both stretching Q_s or bending θ (or φ) has the effect of making ν_s broader but, because $d\omega/dQ_s$ is positive and $d^2\omega/dQ_\theta^2$ is also positive, its centre shifts towards lower wavenumbers when the modulation is due to stretching intermonomer modes (eq. (5.A43) in the appendix of this chapter), but towards higher wavenumbers when performed predominantly by bending intermonomer modes (eq. (5.A49)). This correlation in an opposite direction will provide precious information on this modulation in the case of liquid water examined in Ch. 9, putting into evidence a specificity of this exceptional liquid.

ν_s BANDSHAPES OF NONISOLATED H-BONDS

This strong modulation of ν_s by intermonomer slow modes is a central and ubiquitous feature of all H-bonds. Most of the time it does not, however, give the regular "Franck–Condon" type progression defined by eq. (5.7) that corresponds to transitions drawn in Figure 5.1 and that clearly appear in the relatively few isolated H-bonds in gases. This is because most of the time H-bonds are not isolated as in these simple systems but are part of a bigger system that influences them. Thus, even in these simple gaseous cyclic dimers of carboxylic acids, this progression is no longer regular (Figures 4.4 and 5.2). We have seen it is for a great part due to the presence of two interacting H-bonds, at the origin of a harmonic interaction of the form $V_0 q_1 q_2$ between the two H-bonds that already disrupts this regular progression. Such a term has been shown to be indeed important in all cyclic H-bonded dimers, beyond those of carboxylic acids (14, 15). They may be thought to be general terms that describe vibrational interactions between relatively close H-bonds. Such terms have an effect neither on the integrated intensity of ν_s, nor on its width, but change its centre following eq. (5.A51) of the appendix of this chapter.

Furthermore, the great majority of H-bonds are found in liquids or solids. It modifies the detailed lineshape of ν_s in a way that is schematically represented in Figure 5.3. When this H-bond is diluted in a solvent that contains no other H-bond, the frequencies of the slow intermonomer modes extend over a wide range, in opposition to the above-considered case of isolated gaseous H-bonds where they have a well defined value Ω_s. As shown by Bratos (16) and later by Robertson and Yarwood (17) and Yarwood *et al.* (18), the predominant factor is now the rather broad statistical distribution of Q_s. The effects of the dynamics of the individual Q_s vibrators, characterized above by a single value Ω_s at the origin of a single frequency for these modes, is masked by this broad distribution. The consequence is the replacement of the Franck–Condon progression by a smooth distribution as illustrated in the middle spectrum of Figure 5.3.

When this H-bond that does not interact with other H-bonds is part of a crystal, ν_s then adopts another shape, illustrated in the lower drawing of Figure 5.3. The intermonomer modes of such a crystal are phonons, a set of modes that have frequencies extending in a definite region, called a phonon band defined by a momentum k that takes on discrete values. The result is that ν_s is now composed of several bands called "*n*-phonon bands" (19–21) that are drawn at the bottom of Figure 5.3. They correspond to transitions characterized by the number of quanta of the intermonomer modes that accompany the $0 \rightarrow 1$ transition in ν_s. Thus, the "0-phonon band" is that narrow band situated around 3000 cm^{-1} in the bottom spectrum of Figure 5.3 that corresponds to a ν_s transition accompanied by no phonon, that is no transition of intermonomer modes. The equivalent of this band for isolated H-bonds is the set of $E_0^N \rightarrow E_1^N$ transitions in eq. (5.7) and Figure 5.1, with N taking various values. The "*l*-phonon band" corresponds to a $E_0^N \rightarrow E_1^{N+l}$ transition for various values of N. The "–1-phonon band" is a "hot band", that is, as defined above, a band that starts from states E_0^N with N greater than 0 (the corresponding E_1^{N-1} final state should have $N-1>0$). The 0-phonon band is sharp. The 1-phonon band is broader, because the energies of the phonons are spread over a band of finite width. It makes the 2-phonon band broader, and the following 3-phonon bands even more so. The centres and total widths of these bands are still given by eqs. (5.A43) of the appendix of this chapter, or eq. (5.A51) if several interacting H-bonds are present in the crystal.

ν_s BANDSHAPES OF H-BONDS: FERMI RESONANCES

Another mechanism also often appears in the ν_s bands of H-bonded systems. It has been put into evidence in dimers of acetic acids (22), where the wavenumbers of various submaxima of ν_s were different in the spectra of $(CH_3COOH)_2$ and $(CD_3COOH)_2$. These spectra display shapes that are otherwise most similar. The difference of wavenumbers of submaxima could not be explained supposing the presence of only harmonic interactions between the two ν_s vibrations of the two H-bonds of the dimer and that of the strong anharmonic coupling discussed above, even refining it by transforming for instance the harmonic potential for intermonomer modes (eqs. (5.2) and (5.6)) into more realistic potentials such as a Morse potential (23). It is indeed due to another anharmonic interaction (5, 24), called resonance interaction, or Fermi resonances. We describe in more details this mechanism that is not specific to H-bonds, as is the strong anharmonic coupling described above, but is also encountered in spectra of species with no H-bonds.

The nature of these Fermi resonances is illustrated in Figure 5.4, taking as an example a C–O–H group. We are interested in two vibrations of this group: one, defined by coordinate q, with wavenumber falling around $3000\,cm^{-1}$ and the other harmonic one being the bending vibration defined by coordinate q_δ with wavenumber falling around $1500\,cm^{-1}$. We momentarily neglect the strong anharmonic coupling of $\nu_s(q)$ with slow intermonomer

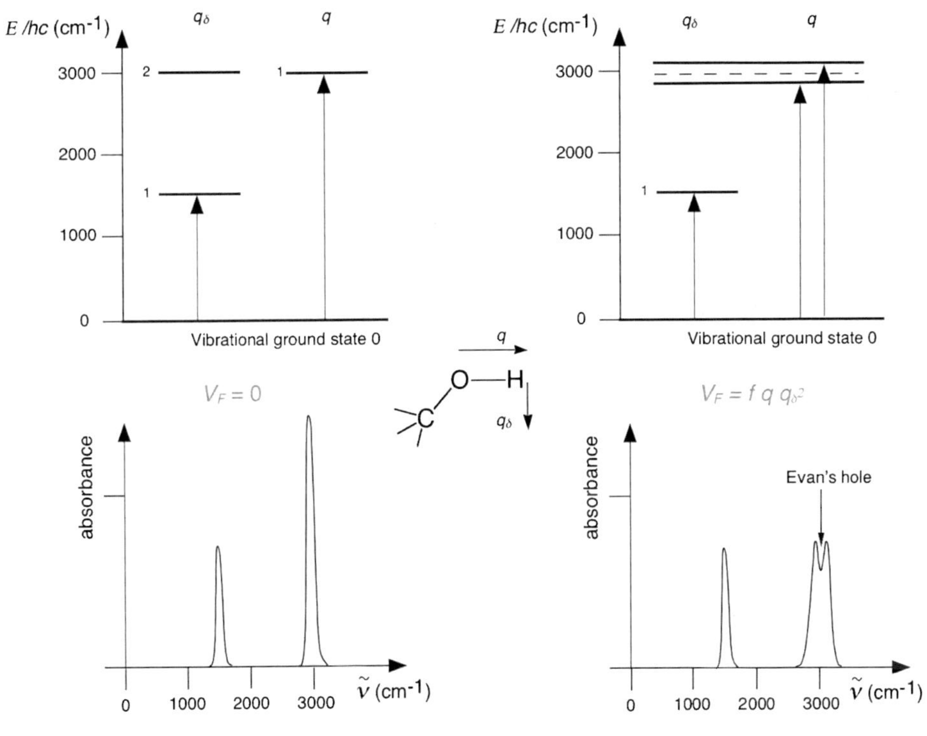

Figure 5.4 Fermi resonances: energies E of the first vibrational levels of a mode q and of a mode q_δ in a C–O–H group (upper diagrams) and corresponding IR spectra (lower diagrams). The values E/hc (c, velocity of light in cm sec^{-1}) of these two modes are represented when no interaction is present (left upper diagram) and when an anharmonic interaction of the form fqq_δ^2 is present (right upper drawing; f is a constant). Arrows indicate transitions that appear in vibrational spectroscopy (IR or Raman).

modes, which is justified by the argument that Fermi resonances are encountered in other systems than H-bonds where this strong coupling is absent. The total vibrational energy E of this C–O–H group is zero when both q and q_δ are in their ground states. E/hc is equal to 1500 cm^{-1} when q_δ is in its first excited state 1. When no anharmonic interaction between q and q_δ is present, E/hc is equal to 3000 cm^{-1} when q_δ is in its second excited state 2 (upper left drawing). A band consequently appears at 1500 cm^{-1} in the IR spectrum of such a group, as shown in the lower left spectrum. It corresponds to the $0 \rightarrow 1$ transition in q_δ, as the $1 \rightarrow 2$ transition, which is also allowed in the harmonic approximation, has a very weak intensity due to the very weak population of this first excited state in q_δ at room temperature ($E \gg kT$). The intensity of the "overtone" $0 \rightarrow 2$ transition in q_δ is zero in the harmonic approximation. The band that appears in this IR spectrum at 3000 cm^{-1} is the $0 \rightarrow 1$ transition in q, which has nothing to do with q_δ.

When an anharmonic interaction between q and q_δ of the form $V_F = fqq_\delta^2$ is present, with f a constant coefficient, it immediately appears in the IR spectrum, even when small. The band at 1500 cm^{-1} is hardly changed, but that at 3000 cm^{-1} takes now the form of a doublet (upper right drawing). This is because this interaction mixes the first excited state in q with the overtone in q_δ (the second excited state in q_δ). This mixing is important in that particular case of a resonance between the first excited state in q and the first overtone in q_δ, that is when both these states appear at same energy or wavenumber, here 3000 cm^{-1}. It rapidly becomes negligible when they depart from resonance. The corresponding linear combinations of these states, 1 in q and 2 in q_δ (upper left diagram), are the eigenfunctions of the total vibrational Hamiltonian. Their energies, or corresponding eigenvalues, differ by a quantity of the order of V_F in the case of a full resonance. In that case both these eigenstates have an equal component (50%) in the first excited state in q. The intensities of the two doublets are then equal. The integrated intensity of this doublet is equal to that of the singlet due to the $0 \rightarrow 1$ transition that appears at 3000 cm^{-1} when $V_F = 0$. In other words, Fermi resonance does not modify the integrated intensity of the $0 \rightarrow 1$ transition in q, and it can be shown that it neither modifies the wavenumber of its centre. The appearance of a doublet instead of a singlet shows that it modifies its width. The presence of a Fermi resonance starts having an effect only on the moments of order higher than 1, that is on the width and shape of the band in q, but it has no effect on its first two moments.

From the above considerations we could deduce that Fermi resonances have a very low probability to occur, because the resonance condition is severe and departing from it makes the doubling of the band in q rapidly vanish off. This is not true for the ν_s band of H-bonded systems, where the ubiquitous strong anharmonic coupling with low frequency intermonomer modes makes this band have an appreciable width that clearly appears in the two lower diagrams of Figure 5.3. It makes the resonance condition easy to fulfil, as the first excited state in ν_s is now broad. Fermi resonances then become visible in many vibrational spectra of H-bonded systems. In these spectra the "Evan's hole" of the simple model, drawn in the right lower spectrum of Figure 5.4, becomes the more complex one, drawn in the middle spectrum of Figure 5.3 (16). When more than one resonance occurs with overtones of various intramonomer vibrations, several Evan's holes appear with minima at wavenumbers close to that of the overtone of the implied vibrations. Also, "combination bands" can be at the origin of Fermi resonances. The overtone level of q_δ at 3000 cm^{-1} in Figure 5.4

is now replaced by a level that is a double excitation of two different vibrations δ and δ''. If the sum of their energies divided by hc is equal to $3000\,\text{cm}^{-1}$ when each of these modes is in its first excited state 1, resonance with the first excited state of ν_s appears and the corresponding ν_s band may be a doublet if an interaction of the form $f' q_\delta q_{\delta'}$ is present.

The appearance of these Evan's holes slightly modifies the shape of ν_s. These transmission holes can be identified when one can for instance modify the mode δ that is at the origin of these resonances. It happens when one passes from the $(CH_3COOH)_2$ cyclic dimers in the gas phase to $(CD_3COOH)_2$. The $\delta_{C\text{–}O\text{–}H}$ bending vibration drawn in Figure 5.4 is not a normal mode in these dimers. The corresponding normal mode that has this $\delta_{C\text{–}O\text{–}H}$ vibration as main component also has another non-negligible component, the bending $\delta_{C\text{–}C\text{–}H}$ vibrations of the CH_3 group. Changing this group into a CD_3 group consequently changes both the energy and the composition of the normal bending modes, making Fermi resonances different between these two dimers. We have seen it constituted the first strong evidence of the presence of Fermi resonances. Evan's holes are not always so easily identified, and it is not always evident to distinguish an Evan's hole from, for instance, the transmission dip that separates the 0-phonon band from the n-phonon band in the lower spectrum of Figure 5.3. The impact of these resonance interactions on ν_s is, however, limited in the general case. The creation of Evan's holes thus modifies neither the integrated intensity of ν_s nor its centre (25, 26). It starts having an effect on its width σ', which may be shown to have two quadratically additive components, one σ due to the strong anharmonic coupling with the slow intermonomer modes (eq. (5.A42) of the appendix of this chapter) and the other one σ_F due to Fermi resonances

$$\sigma'^2 = \sigma^2 + \sigma_F^2 \tag{5.8}$$

Fermi resonances are thus responsible for local effects that superimpose on the consequent broadening of ν_s due to the fundamental modulation of ν_s by slow intermonomer modes. Because they may occasionally sufficiently modify the shape of ν_s to be clearly visible, they have been given in the years 1970 an importance that is now strongly reduced. They may be directly taken into account (27, 28) or eliminated calculating the corresponding "peeled-off" spectra (29), the spectra that would appear in their absence. In the case of liquid water and other simple H-bonded systems, the resonance interaction between ν_s and the overtone of the bending band $\delta_{H\text{–}O\text{–}H}$ of H_2O has been shown to provide an efficient relaxation mechanism for the first excited state of ν_s (30–33), providing this excited state a very short lifetime. This finding is one of the original information on H-bonds that is conveyed by these modern promising "nonlinear time-resolved IR spectroscopy" techniques described in preceding Ch. 4. A surprising consequence of this relaxation path is the increase of the relaxation time when temperature increases (34). This is really an exception, as relaxation times usually decrease when temperature increases, making the relaxation faster. It is characteristic of this particular path from ν_s towards the bending overtone $2\delta_{H\text{–}O\text{–}H}$ of H_2O that becomes less rapid when temperature increases because the shift of ν_s towards higher wavenumbers when temperature increases makes the resonance condition efficient for a smaller number of intermonomer distances.

CONCLUSION ON ν_s BANDS

We may conclude from this theoretical point of view on the dynamics of the H-atom stretching motion in H-bonds that this ν_s vibration is particularly complicated. This is not reality. It is a nearly harmonic vibration that induces a relatively great decoupling from it of the average positions of the electrons. It results in an exceptionally great integrated intensity. It is furthermore strongly modulated by slow intermonomer modes. This strong modulation is due to the sensitivity of the electronic structure of the H-bond to the position of the H-atom. This sensitivity, together with the great sensitivity of the dipole moment to a change in the X–H distance that is responsible for the great integrated intensity of ν_s, is due to the special electronic structure of the H-bond, for a great part characterized by the transfer of electrons from the nonbonding orbital of Y towards the antibonding σ^* of X–H (3). This strong anharmonic modulation makes furthermore ν_s shift towards lower wavenumbers and gives it a particular shape characterized by an unusually great width that allows transmission windows (Evan's holes) to appear around some definite wavenumbers. The above equations show that all these anharmonic features can now be precisely taken into account within a rather simple theory that can be easily described in terms of physical intuitive effects, as described above. The exception is the case of strong H-bonds, for which quantization of the slow intermonomer modes is necessary to have a reliable basis, but has not yet been performed, despite interesting attempts (35). Thus, an appreciable amount of work still remains to be done before having a fair description of the dynamics of H-atoms in these strong H-bonds. The best argument in favour of the relative simplicity of the above description is the precision that can be achieved in simulations of such a complicated spectrum as that of cyclic dimers of carboxylic acids (Figure 4.4). Such a simulation that includes these various anharmonic couplings, together with the harmonic one between the two ν_s vibrations of the dimer, called Davydov interaction, is drawn in Figure 5.2 (13). It demonstrates the reality of these exceptional mechanisms that govern the dynamics of this ν_s vibration, and also the understanding we have of them. Such a precise understanding is necessary for the interpretation of time-resolved nonlinear IR spectroscopy that appears to be a new method able to convey a lot of original information on the dynamics of H-bonds, as mentioned in preceding Ch. 4.

APPENDIX: IR SPECTROSCOPY

In this appendix we briefly sketch, without any demonstration and with the minimum of mathematical formula, some basic aspects of IR spectroscopy that are invoked all along this book. We first define some mathematical but nevertheless simple notations that are currently used in quantum physics because they avoid heavier writing with integrals. Thus, an integral quantity that implies two vibrational wavefunctions Ψ_n and Ψ_m and "an operator" that may be a function of a coordinate q, or a differential operator such as its conjugated momentum p, defined as

$$p = -\mathrm{i}\hbar\frac{\mathrm{d}}{\mathrm{d}q} \tag{5.A1}$$

with $\hbar = h/2\pi$ (h = Planck's constant) is written as

$$\langle m|p|n\rangle = -i\hbar \int_{-\infty}^{\infty} \Psi_m^*(q) \frac{d\Psi_n(q)}{dq} dq \tag{5.A2}$$

where * means complex conjugate (i changed into $-i$, with $i^2 = -1$). In this "bracket" notation the operator that appears between the two vertical bars (p in first member of eq. (5.A2)) acts on the wave function on its right. If this operator is a coordinate such as q, or any power q^n of it, the "operation" reduces to a simple multiplication. As we are mainly dealing with IR spectroscopy, most wavefunctions will be those that describe vibrations of molecules in their electronic ground state. It means that in such an equation as eq. (5.A2) n and m most often represent the nth and mth wavefunction for the vibration defined by coordinate q, and most often these will be harmonic wavefunctions. The harmonic approximation is defined in section "Normal modes in the harmonic approximation".

The following first section of this appendix describes quantities that are measured when registering spectra obtained using various experimental set-ups and their relations with molecular quantities. These relations form the basis of the interpretations of molecular spectra. The second section describes some general properties of a distribution that are used in various chapters of this book when this distribution is the band of a spectrum. The third section deals with such concepts as normal modes in the harmonic approximation, while the fourth section deals with force constants, reduced masses, etc., and offers comparisons of these various quantities. The last section provides a more specific calculation of the first and second moment of a band such as ν_s, which corresponds to a normal mode characterized by a strong anharmonic coupling with a much slower mode.

Experimental spectroscopy: measured quantities and set-ups

The word "spectrum" encompasses different entities that we have to precise and to relate between them and with the various experimental quantities that are measured. Spectroscopy itself consists of characterizing the response of a medium to irradiating waves that oscillates at various frequencies. This response depends on the set-up used to obtain it. In this section we limit ourselves to absorption spectroscopy, which is used in the great majority of spectroscopic experiments in physical chemistry. This type of spectroscopy has been called in Ch. 3 "one-photon spectroscopy", in opposition to multiphoton spectroscopy that encompasses a greater number of more elaborated experimental types, such as Raman, time-resolved nonlinear IR, etc. It consists of the absorption of a single photon of the electromagnetic wave by the irradiated sample. It is treated within the frame of the linear response of a medium to the irradiation by an electromagnetic field that materializes with the appearance, at each position of the medium, of an electric dipole moment that is proportional to the intensity of the electric field of the electromagnetic wave at this same position. Even in this restricted class of experiments, several set-ups may be encountered that have their own advantages and disadvantages. It is thus necessary to relate the corresponding measured quantities to characteristic spectral quantities of the medium itself. Spectroscopy has furthermore a somewhat complicated history, coming from various fields of physics such as optics or electromagnetism, which have been unified not so long ago. As a consequence, various characteristic spectral quantities have been defined. They are not independent and relations between the most common ones, the optical and dielectric constants of a medium are first established. Also, the failure of classical physics to quantify energy levels of atoms or molecules has been at the origin of quantum mechanics where light is considered as an assembly of photons, that is quanta of energy that can be absorbed or emitted by a molecule and the number of which is not fixed. Relations between measured spectroscopic quantities and molecular quantities are therefore also given within the frame of quantum mechanics.

Following the approach of electromagnetism, we characterize the linear response of the medium to the electromagnetic field by a frequency-dependent dielectric constant ε. The induced electric dipole moment is equal to this dielectric constant multiplied by the value of the electric field of the irradiating electromagnetic wave. ε is a complex quantity, having the form, with $i^2 = -1$

$$\varepsilon(\nu) = \varepsilon'(\nu) + \mathrm{i}\varepsilon''(\nu) \tag{5.A3}$$

The real part ε' of this dielectric constant characterizes that part of the dipole moment that is in phase with the electric field $\vec{E}_0 \mathrm{e}^{2\mathrm{i}\pi\nu t}$ at a fixed position of the sample and therefore does not cause any transfer of energy between electromagnetic field and medium. Its imaginary part ε'', with its coefficient i, characterizes that part of the dipole moment that displays a phase difference of $\pi/2$ with the electromagnetic field. It is then responsible for the energy exchange between medium and electromagnetic wave. Time-dependent quantum perturbation theory allows defining the quantum of energy exchanged, which is the energy of a photon $h\nu$. It also relates the value of ε'' to the Fourier transform of the correlation function of the dipole moment of the isolated medium (36)

$$\varepsilon''(\nu) = \frac{2\pi}{\hbar V \varepsilon_0}[1 - \mathrm{e}^{-h\nu/kT}] \int_{-\infty}^{\infty} \mathrm{e}^{-2\mathrm{i}\pi\nu t} \left\langle \overline{\mu(t)\mu(0)} \right\rangle \mathrm{d}t \tag{5.A4}$$

where $\mu(t)$ is the component of the electric dipole moment along the electric field at time t. In quantum mechanics, this is an "operator" that acts on a wavefunction. V is the molar volume of the medium, ε_0 the susceptibility of the vacuum and the horizontal bar within the bracket $\langle\ \rangle$ with nonspecified wavefunctions means quantum thermal average. This equation allows us to calculate in the frame of quantum mechanics spectra in ε'', through the calculation of dynamics of dipole moments of molecular systems. Quantum mechanics does not allow to directly calculate the real part ε' of the dielectric constant as it is not at the origin of exchange of energy between the medium and the electromagnetic wave. It is, however, a quantity that is not independent of the imaginary part ε''. The quite general causality principle, which states that the effect of any action cannot precede the cause of this action (music cannot be heard before the pianist hits the keys!), leads to general relations between real and imaginary parts. These are known as "Kramers–Kronig" integral relations, written in eq. (5.A9) in the case of real and imaginary optical constants. In the following, the molecule is described by these constants. We presently retain that these Kramers–Kronig relations allow the real part of the dielectric or optical constant to be indirectly calculated from quantum mechanics of molecular systems through calculations of the imaginary parts of these constants.

In optical spectroscopy, the properties of the medium are not defined by its complex dielectric constant $\varepsilon(\nu)$ (eq. (5.A3)) but by its optical complex constant

$$\hat{n}(\nu) = n(\nu) - \mathrm{i}k(\nu) \tag{5.A5}$$

The real part n of this optical constant is the well-known refractive index of the medium, while its imaginary part k is its absorption coefficient. The propagation of an optical wave inside the medium along axis z, for instance, is then defined by the propagation of its electric vector $\vec{E}$ as

$$\vec{E} = \vec{E}_0 \mathrm{e}^{2\mathrm{i}\pi\nu(t-\hat{n}z/c)} = \vec{E}_0 \mathrm{e}^{2\mathrm{i}\pi\nu(t-nz/c)} \mathrm{e}^{-2\pi\nu kz/c} \tag{5.A6}$$

where c is the velocity of light. In the last equation the first exponential describes the phase of the propagating wave, as its argument is purely imaginary, whereas the second exponential, with its purely real argument, defines its (decreasing) amplitude. The phase factor is constant for $t - nz/c$ remaining a constant. It is thus the equation of a wave that propagates at a velocity equal to c/n. These optical constants are related to the above-defined dielectric constants of the same medium by the equations

$$\begin{aligned} \varepsilon'(\nu) &= n^2(\nu) - k^2(\nu) \\ \varepsilon''(\nu) &= 2n(\nu)k(\nu) \end{aligned} \tag{5.A7}$$

The energy of the optical wave that propagates along axis z is proportional to the square of the amplitude of its electric vector. It is equal to I_0 at the entrance of the medium defined by $z = 0$, and to I at the output of the medium defined by $z = l$. In a classical absorption set-up, also called the transmission set-up, which is used in the great majority of experiments and is displayed in Figure 5.A1, the measured quantity is the *absorbance* A equal to the decimal logarithm of the ratio of these two energies (also called intensities). With the help of eq. (5.A6), we may write

$$A(\tilde{\nu}) = \log\left(\frac{I_0(\tilde{\nu})}{I(\tilde{\nu})}\right) = \frac{4\pi}{2.3}\tilde{\nu}k(\tilde{\nu})l \tag{5.A8}$$

where 1/2.3 is the value of the decimal logarithm of e. In eq. (5.A8), we have replaced frequencies ν by wavenumbers $\tilde{\nu} = \nu/c$, which are the effectively measured quantities. From this equation the absorption coefficient k of a sample can be calculated from the measurement of its absorbance. It should, however, be noted that eq. (5.A8) refers only to the absorption of the sample, whereas in most absorption set-ups (Figure 5.A1), the quantity $\log(I_0(\tilde{\nu})/I(\tilde{\nu}))$ includes in addition to this absorption the effects due to reflections of the optical beam on the interfaces of the sample with its surrounding. In the great majority of experiments, these reflections are at the origin of constant terms, independent of $\tilde{\nu}$ (they are always independent of l). They can consequently be discarded with no serious consequence. This is, however, not always the case, as this simplification implies that the refractive index n of the medium, on which the intensity of this reflection depends, displays but a small variation with wavenumbers. In the case of such a medium as liquid water, with strong absorption coefficient k, and consequently appreciable dependence of n on $\tilde{\nu}$ (see eq. (5.A9)), this

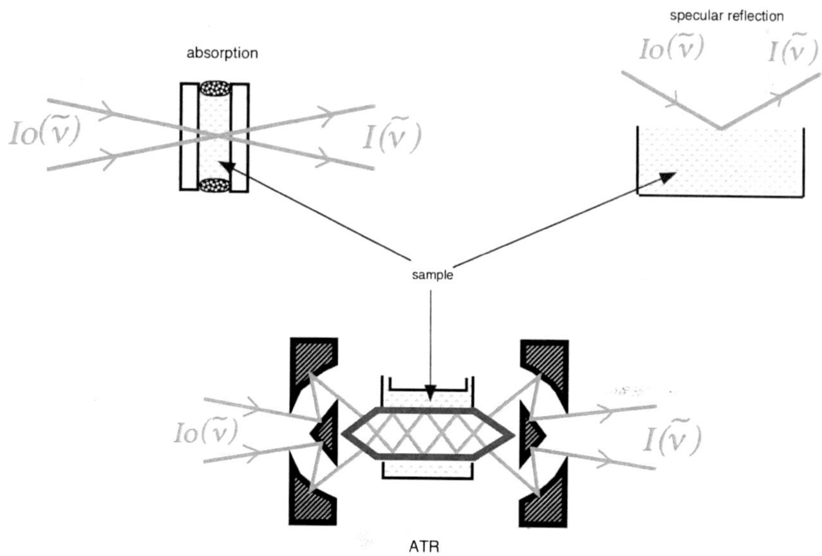

Figure 5. (A1) Three optical set-ups used in molecular spectroscopy: absorption (also called transmission), specular reflection and ATR. The sample is drawn as a liquid.

is no longer true, and reflections at these interfaces should be carefully subtracted to obtain a true value for k.

As announced previously, n and k are related by Kramers–Kronig relations, which write (37)

$$n(\nu)-n(\infty)=\frac{2}{\pi}\int_0^\infty \frac{\nu' k(\nu')}{\nu'^2-\nu^2}\,\mathrm{d}\nu'$$
$$k(\nu)=\frac{-2\nu}{\pi}\int_0^\infty \frac{n(\nu')}{\nu'^2-\nu^2}\,\mathrm{d}\nu' \tag{5.A9}$$

and may be equivalently expressed in terms of double Fourier transforms, which might be more practical equations to implement in a program

$$n(\nu)-n(\infty)=4\int_0^\infty \mathrm{d}\tau\,\cos(2\pi\nu\tau)\int_0^\infty k(\nu')\sin(2\pi\nu'\tau)\,\mathrm{d}\nu'$$
$$k(\nu)=4\int_0^\infty \mathrm{d}\tau\,\sin(2\pi\nu\tau)\int_0^\infty n(\nu')\cos(2\pi\nu'\tau)\,\mathrm{d}\nu' \tag{5.A10}$$

These equations are used to subtract the contribution of reflections on the interfaces of such a strongly absorbing medium as liquid water, when one wishes to calculate the optical constant k and at the same time n. It is also used (37) in the calculation of these optical constants n and k when the measured value $\log(I_0(\tilde{\nu})/I(\tilde{\nu}))$ is that obtained using another set-up, for instance an Attenuated Total Reflection set-up, commonly denominated ATR, or a specular reflection set-up, both displayed in Figure 5.A1. In the case of liquid water, the ATR set-up gives spectra that display a much better signal-to-noise ratio than the commonly used absorption set-up. The principle of ATR is to "dilute" the IR absorption, by making it operate only on the evanescent wave that gets out of the crystal inside which the IR beam performs total reflections. This evanescent wave, which propagates parallel to the interface between crystal and sample, represents but a small proportion of the whole propagating wave. The same value for $\log(I_0(\tilde{\nu})/I(\tilde{\nu}))$ is thus obtained with a thickness l equal to 1 μm for liquid water in an ordinary absorption set-up, but with a crystal some centimetres long in an ATR set-up. This "dilution" of the absorption is thus of four orders of magnitude. The same is true in some cases with a specular reflection set-up, also drawn in Figure 5.A1, and which is interesting in the study of some strongly absorbing H-bonded liquids.

All the $\tilde{\nu}$ dependent quantities we have evocated in this appendix may be represented in the form of spectra. It means that speaking of spectra requires defining which quantity we are exactly considering. As the object of this appendix is to establish relations between these various quantities, we display some of them in Figure 5.A2. We may see that although all quantities except n show similar spectral features, or similar bandshapes, the ratios of the intensities of the various bands considerably differ between these quantities (38). Finally, let us add another quantity, often used in chemistry, which is the molar extinction coefficient $\in$. It appears in the Beer–Lambert law that writes

$$I(\tilde{\nu}) = I_0(\tilde{\nu}) \times 10^{-Cl\in} \tag{5.A11}$$

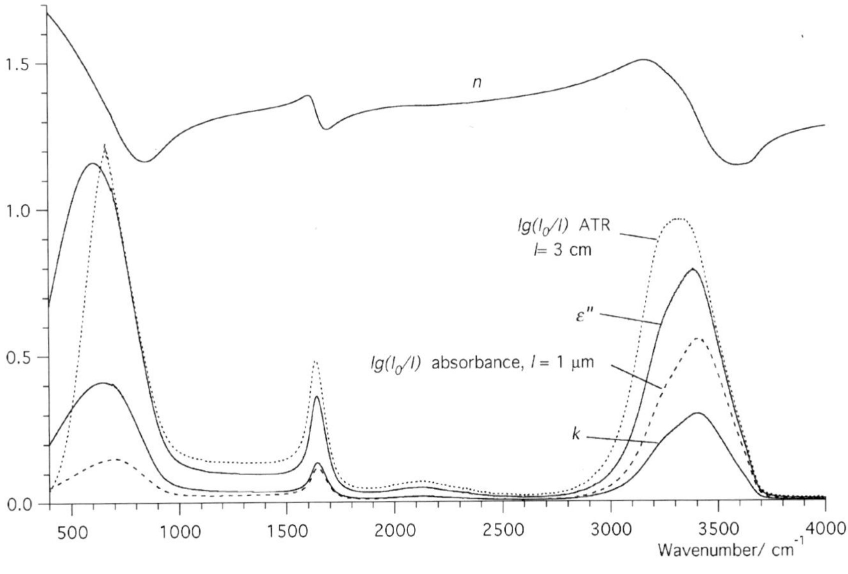

Figure 5. (A2) Various spectral quantities related to liquid water in the mid-IR region. Experimental values of $\log(I_0(\tilde{\nu})/I(\tilde{\nu}))$ are shown in the case of an absorption set-up with a 1 μm thick sample and of an ATR cell with an immersed portion of the crystal about 3 cm long (41). The optical constants n and k are also displayed together with the imaginary dielectric constants ε''.

where C is the concentration of a molecular species within a solvent. Comparing this equation with eq. (5.A8), we deduce

$$\in(\tilde{\nu}) = \frac{4\pi}{2.3}\frac{\tilde{\nu}k(\tilde{\nu})}{C} \tag{5.A12}$$

We do not use these quantities in this book, as they are essentially useful in the case of dilutions, and are mostly used in the case of UV-visible spectroscopy that concern electronic transitions. We use the more general physical quantities we have defined above. Using both of them would convey the threat to confuse c, the velocity of light with C, the concentration and ϵ, the molar extinction coefficient with ϵ, the dielectric constant.

First moments of a distribution or of a spectral band

In only very particular cases can a spectral band $s(\nu)$ be represented by a well-defined mathematical function, such as a Lorentzian or a Gaussian distribution. Most often the shape of a band cannot be reduced to such a simple mathematical form. We may nevertheless characterize it by its moments. We show below that the knowledge of all moments of a band is equivalent to knowing its exact shape. In practice, only the moments of order 0, 1 and 2 can be more or less easily measured, and their knowledge may thus be considered as a first characterization of the band. The moment M_n of order n of $s(\nu)$, is equal to

$$M_n = \int_{-\infty}^{+\infty} \nu^n s(\nu)\,\mathrm{d}\nu \tag{5.A13}$$

The moment M_0 is equal to the integrated intensity of $s(\nu)$, while its moment M_1 of order 1 is related to the frequency of the centre of the band $\bar{\nu}$

$$M_0 = \int_{-\infty}^{+\infty} s(\nu)\,\mathrm{d}\nu$$
$$\bar{\nu} = \frac{M_1}{M_0} \tag{5.A14}$$

The square of the width σ of $s(\nu)$ is the "centred moment" of order 2, also called variance in the case of a statistical distribution. It is defined as

$$\sigma^2 = \frac{1}{M_0}\int_{-\infty}^{+\infty} (\nu-\bar{\nu})^2 s(\nu)\,\mathrm{d}\nu \tag{5.A15}$$

and related to M_2 and M_1 by the equation

$$\sigma^2 = \frac{M_2}{M_0} - \left(\frac{M_1}{M_0}\right)^2 \tag{5.A16}$$

In order to demonstrate that the knowledge of the whole series of moments is equivalent to knowing the exact shape of this band, let us first consider the Fourier transform $S(t)$ of this band. If $s(\nu)$ and $S(t)$ are convergent, that is if they fall sufficiently rapidly to 0 at high values of ν and t, a condition which we consider as always fulfilled, they are related by the equations

$$S(t) = \int_{-\infty}^{+\infty} s(\nu)\mathrm{e}^{2\mathrm{i}\pi\nu t}\,\mathrm{d}\nu$$
$$s(\nu) = \int_{-\infty}^{+\infty} S(t)\mathrm{e}^{-2\mathrm{i}\pi\nu t}\,\mathrm{d}t \qquad (5.\mathrm{A}17)$$

$s(\nu)$ is the "inverse Fourier transform" of $S(t)$. The "conjugated" quantities ν and t are frequency and time. They might also be wavenumber and distance, as in the computation of a spectrum from an interferogram (see the introductory part of Ch. 4), defined as

$$\tilde{\nu} = \frac{1}{\lambda} = \frac{\nu}{c}, \quad d = ct \qquad (5.\mathrm{A}18)$$

with c being the velocity of light, or many other quantities, in the case of a band which is not a spectral band. Developing the exponential which appears in the integral of eq. (5.A17) for $S(t)$ introduces the moments in these equations:

$$S(t) = \int_{-\infty}^{+\infty} s(\nu)\sum_{n=0}^{\infty}\frac{(2\mathrm{i}\pi\nu t)^n}{n!}\,\mathrm{d}\nu = M_0 + \sum_{n=1}^{\infty}\frac{(2\mathrm{i}\pi t)^n}{n!}M_n \qquad (5.\mathrm{A}19)$$

which we may write in the form of a "cumulant" expansion (39) as

$$S(t) = M_0\mathrm{e}^{2\mathrm{i}\pi\bar{\nu}t}\mathrm{e}^{-(4\pi^2t^2/2)\sigma^2}\ldots \qquad (5.\mathrm{A}20)$$

where "..." means that the following multiplicative terms are exponentials of a series of terms in $(2\mathrm{i}\pi t)^n/n!$ with $n > 2$. Eq. (5.A20) shows that the centre $\bar{\nu}$ and width σ of a band appear as the coefficients of $(2\mathrm{i}\pi t)^n/n!$ in the cumulant expansion of the Fourier transform of $s(\nu)$ with $n = 1$ and 2, respectively. With eq. (5.A19) it also shows that the knowledge of all moments of a band defines the Fourier transform of this band and consequently the band itself (eq. (5.A17)).

Let us note that these considerations do not apply to a Lorentzian band of the form

$$l(\nu) = \frac{M_0}{1+\left(\dfrac{\nu-\bar{\nu}}{\Delta\nu}\right)^2} \qquad (5.\mathrm{A}21)$$

because all even moments of order greater than 1 diverge (they are equal to infinity), even if the Fourier transform $L(t)$ of $l(\nu)$ is well defined

$$L(t) = M_0\pi\Delta\nu\mathrm{e}^{2\mathrm{i}\pi\bar{\nu}t}\mathrm{e}^{-2\pi\Delta\nu|t|} \qquad (5.\mathrm{A}22)$$

$L(t)$ exhibits a discontinuity of its odd derivatives at $t = 0$, due to the presence of the absolute value of t in the real exponential. This discontinuity is at the origin of the divergence of its even moments. In opposition, these considerations fully apply to more rapidly convergent forms, such as a Gaussian, or a band of the form (40) $1/\cosh(2\pi\nu/\sigma)$ with cosh being the hyperbolic cosine function, or any experimental band which is considered to be equal to 0 for ν greater than some threshold ν_{max}. For a Gaussian band, we thus have

$$\begin{aligned} g(\nu) &= \frac{M_0}{\sqrt{2\pi}\sigma} e^{-(\nu-\bar{\nu})^2/2\sigma^2} \\ G(t) &= M_0 e^{2i\pi\bar{\nu}t} e^{-(4\pi^2 t^2/2)\sigma^2} \end{aligned} \tag{5.A23}$$

We see that $G(t)$ has the form defined in eq. (5.A20), with no terms in t^n with $n > 2$. In other words, stopping the cumulant expansion of a band to the term of order 2 is equivalent to defining the best Gaussian approximate of this band.

Normal modes in the harmonic approximation

In this section we define "normal modes", a concept that is central in the harmonic approximation. We start defining this approximation in the case of a molecule, remembering that this harmonic approximation is much more general and is indeed encountered in any mechanical system, macroscopic systems comprised. A molecule is made of many atoms that are held by covalent bonds or H-bonds that precisely determine their relative configurations at equilibrium. These atoms may nevertheless display vibrations around their equilibrium positions. Each interatomic vibration j may be characterized by a vibrational coordinate x_j that describes its departure from its equilibrium value. The potential energy of the molecule $V(x_1, x_2, \ldots)$ depends on all these x_j's, as moving the atoms away from their equilibrium positions requires providing energy. This energy is called vibrational energy because it does not imply electrons that remain in their ground state, as transitions between electronic states, require much bigger energies than those necessary to have atoms vibrate. When these vibrations keep a small amplitude as compared to their interatomic distances at equilibrium, which is most often true, we may develop V in powers of the x_j's. Because equilibrium corresponds to all x_j's equal to 0, V is minimum when all x_j's are equal to 0 and its first derivative with respect to any x_j is also 0. It implies that no linear term appears in the development of V in powers of the x_j's and that consequently the first terms are quadratic, that is of order 2. This is the *harmonic approximation*, when all higher order terms are neglected. It is an approximation that is sufficient to describe most vibrations.

The x_j's are variations from equilibrium position of interatomic distances or angles. In this harmonic approximation we have terms in x_j^2, but also crossed terms in $x_j x_{j'}$ that represent harmonic interactions between two different vibrations j and j' that are often related to nearly lying atoms. The presence of such terms means that one cannot excite the vibration defined by x_j without also exciting that defined by $x_{j'}$. One can, however, eliminate such cross terms in the harmonic approximation by performing a linear transform on the x_j's, which depends on the various coefficients found in the harmonic development of V, which are called force constants. The newly defined coordinates are those of *normal modes* of the molecule. They are linear combinations of the x_j's, which implies that the x_j's may also be expressed in the form of the reciprocal linear combination of these normal modes. The kinetic energy, expressed as quadratic terms in momenta (eq. (5.A1) for the definition of the conjugated momentum of a coordinate in quantum mechanics) of these normal mode can also be shown to have no cross terms. The consequence of the absence of cross quadratic terms, both in the potential and in the kinetic energy part, make these normal modes completely independent. It means that one can independently excite them. These normal modes are the vibrations that are at the origin of bands in IR or Raman spectroscopy, and to each of them corresponds one band. They belong to a symmetry group of the molecule. In other words, they may be classified according to the symmetry group of the molecule. The two O–H stretching normal modes of a H_2O molecule, for instance, are the symmetric mode defined by coordinate $(x_1 + x_2)/\sqrt{2}$ and antisymmetric mode defined by coordinate $(x_1 - x_2)/\sqrt{2}$, where x_1 and x_2 are the two O–H distances in H_2O (the factor $\sqrt{2}$ is irrelevant for the definition of normal coordinates, as any mode with coordinate proportional to that of a normal mode is still a normal mode; it, however, keeps the metrics of the coordinates). The symmetric mode keeps the plane of symmetry of H_2O, while the antisymmetric one does not keep it. It means that when the symmetric mode has a nonzero amplitude, the H_2O molecule is still symmetric with respect to this plane. This is not true of the antisymmetric mode. One of these modes can also be excited independently of the other one. The third normal coordinate of H_2O, its bending mode $\delta_{\text{H–O–H}}$, is defined by vibrations of the angle H–O–H around its equilibrium position. It is a symmetric mode, as it keeps the plane of symmetry of H_2O.

Reduced masses, force constants and vibrational amplitudes

Let us consider an H-bonded complex, X–H···Y, made of two molecules, X–H with mass $M_{\text{X–H}}$ and Y with mass M_{Y}. These two molecules perform relative vibrations. The mass that appears in the kinetic part of the vibrational Hamiltonian of the complex and is the conjugate momentum of the relative coordinate $Q_s - Q_0$ (Figure 2.1. Q_0 is the value of Q_s at equilibrium) of the two parts of the complex, is the "reduced mass" $M_{\text{X–H}\cdots\text{Y}}$ of the complex, which is defined as

$$\frac{1}{M_{\text{X–H}\cdots\text{Y}}} = \frac{1}{M_{\text{X}}} + \frac{1}{M_{\text{Y}}} \tag{5.A24}$$

The other vibrational coordinates of X–H···Y are those related to the other two intermonomer vibrations, those related to internal vibrations in X–H and Y, and those of the centre of gravity of the whole system, which separates from all other coordinates. When this complex is isolated, these coordinates of the centre of gravity do not appear in the potential energy. They can consequently be discarded as they are independent of the other ones. The coordinates of internal vibrations are driven by force constants due to covalent bonds within molecules X–H and Y. They are, as seen in the following, much greater than the force constants due to H-bonds that drive the intermonomer vibrations. These much faster intramonomer vibrations consequently hardly mix with intermonomer vibrations, even if cross terms between these two kinds of coordinate appear in the potential energy: they are well out of resonance, that is each of them displays vibration frequencies that are different, and the effect of these possible cross terms remains small in all cases. We are then left with two kinds of normal modes of the complex: those that are mainly composed of intramonomer vibrations on one hand, that is their corresponding normal coordinates are a linear combination of mainly coordinates of intramonomer vibrations, and those that are mainly composed of intermonomer vibrations on the other hand.

We compare the force constants of these two types of modes, taking as an example X–H being HCl, with $M_{\text{Cl-H}} = 36m_{\text{H}}$ (m_{H} is the mass of an H-atom) and Y being CH_3–O–CH_3, with $M_{CH_3-O-CH_3} = 46m_{\text{H}}$. Eq. (5.A24) shows that the reduced mass $M_{\text{Cl-H}\cdots\text{O-(CH}_3)_2}$ of the corresponding H-bonded complex Cl–H···$O(CH_3)_2$ falls in the vicinity of $20m_{\text{H}}$. The force constant of a vibration or a normal mode defined by coordinate x is the coefficient that multiplies $x^2/2$ in the vibrational potential energy. This term $\frac{1}{2}kx^2$ means that a force equal to $-kx$ tends to drive the corresponding vibration towards its equilibrium position defined by $x = 0$. In the case of molecules, force constants themselves are not easy to measure. The related quantity Ω, that appears in eq. (4.1), for instance, and is equal to

$$K = M\Omega^2 \tag{5.A25}$$

is much easier to measure quantity, in IR or Raman spectra, for instance, as it is equal to the frequency of the normal modes multiplied by 2π (eq. (4.5)). The quantity M is then the reduced mass of these two atoms or molecules. The ratio of the force constant $K_{\text{Cl-H}\cdots\text{O(CH}_3)_2}$ of the intermonomer stretching vibration of the H-bonded complex Cl–H···$O(CH_3)_2$ we have defined above, to the force constant $k_{\text{Cl-H}}$ of the intramonomer stretching vibration of Cl–H is then equal to

$$\frac{k_{\text{Cl-H}}}{K_{\text{Cl-H}\cdots\text{O(CH}_3)_2}} = \frac{m_{\text{ClH}}}{M_{\text{Cl-H}\cdots\text{O-(CH}_3)_2}}\left(\frac{\omega_{\text{Cl-H}}}{\Omega_{\text{Cl-H}\cdots\text{O-(CH}_3)_2}}\right)^2 \tag{5.A26}$$

where the reduced mass m_{ClH} of the stretching Cl–H vibration in HCl is equal to $m_{\text{ClH}} = 0.97m_{\text{H}}$ (eq. (5.A24)). The reduced mass $M_{\text{Cl-H}\cdots\text{O-(CH}_3)_2}$ has been seen equal to $20m_{\text{H}}$. The wavenumber of the stretching intermonomer vibration of Cl–H···O–$(CH_3)_2$ appears in the far infrared (FIR) region at $120\,\text{cm}^{-1}$ (Table 4.1 of Ch. 4) and that $\omega_{\text{Cl-H}}$ of the stretching $\nu_{\text{Cl-H}}$ vibration in HCl (7, 42) appears in the mid-IR region around $2500\,\text{cm}^{-1}$. We then have

$$\frac{k_{\text{Cl-H}}}{K_{\text{Cl-H}\cdots\text{O(CH}_3)_2}} = \frac{1}{20}\left(\frac{2500}{120}\right)^2 \simeq 22 \tag{5.A27}$$

The force constant due to an H-bond is thus weaker than that due to a covalent bond by more than one order of magnitude.

Let us now consider vibrational amplitudes. For a harmonic oscillator of (reduced mass) m and frequency $\nu = \omega/2\pi$, defined by coordinate $q - q_0$, the square amplitude of vibration in the nth state, of energy $E_n = (\text{n} + (1/2))\hbar\omega$, is

$$\left\langle n\left|(q-q_0)^2\right|n\right\rangle = \left(n+\frac{1}{2}\right)\frac{\hbar}{m\omega} \tag{5.A28}$$

so that the average thermal quadratic amplitude of vibration is

$$\overline{(q-q_0)^2} = \frac{1}{Z}\frac{\hbar}{m\omega}\sum_{n=0}^{\infty} e^{-(n+\frac{1}{2})(\hbar\omega/kT)}\left(n+\frac{1}{2}\right)$$
$$Z = \sum_{n=0}^{\infty} e^{-(n+\frac{1}{2})(\hbar\omega/kT)} \tag{5.A29}$$

with $k = 1.38 \times 10^{-23}$ J K^{-1} the Boltzmann constant and $\hbar = 1.05 \times 10^{-34}$ J sec the Planck constant h divided by 2π. The summations that enter eqs. (5.A29) can be exactly calculated. It gives the value of the average quadratic amplitude for such an oscillator

$$\sqrt{\overline{(q-q_0)^2}} = \sqrt{\frac{\hbar}{2m\omega}}\sqrt{\frac{1+e^{-\hbar\omega/kT}}{1-e^{-\hbar\omega/kT}}} \tag{5.A30}$$

The quantity $\sqrt{(1+e^{-\hbar\omega/kT})/(1-e^{-\hbar\omega/kT})}$ is equal to 1 for intramonomer modes that are such that $\hbar\omega \gg kT$ and equal to 1.89 for the intermonomer mode of such an H-bonded complex as Cl–H···O$(CH_3)_2$ at 300 K, with $\Omega_{\text{Cl–H}\cdots\text{O–(CH}_3)_2}/2\pi c = 120$ cm^{-1} (c = velocity of light). We find, in the case of the above-considered complex Cl–H···O$(CH_3)_2$ at 300 K, the ratio of vibrational amplitudes for the intramonomer Cl–H··· stretching mode and the intermonomer Cl–H···O$(CH_3)_2$ stretching mode

$$\frac{\sqrt{\overline{(q_{\text{Cl–H}\cdots}-q_0)^2}}}{\sqrt{\overline{(Q_s-Q_0)^2}}} = \sqrt{\frac{M_s\Omega_s}{m_{\text{ClH}}\omega_{\text{Cl–H}}}}\frac{1}{1.89} = \sqrt{\frac{20\times 120}{2500}}\frac{1}{1.89} \simeq 0.52 \tag{5.A31}$$

It means that the vibrational amplitudes are comparable for intramonomer and intermonomer modes. At very low temperatures it falls in the vicinity of 1 and decreases when temperature is raised because intermonomer modes see their amplitude increased when kT becomes comparable to $\hbar\Omega_s$, whereas faster intramonomer modes are temperature independent. Let us calculate the absolute value of this amplitude. Following eq. (5.A28), the mean square vibrational amplitude of the stretching mode of the H-atom in this complex is, with $m_H = 1.67 \times 10^{-27}$ kg and the velocity of light $c = 3 \times 10^{10}$ cm sec^{-1}

$$\sqrt{\overline{(q_{\text{Cl–H}\cdots}-q_0)^2}} = \sqrt{\frac{1.05\times 10^{-34}}{2\times 1.67\times 10^{-27}\times 2\pi\times 2500\times 3\times 10^{10}}} \tag{5.A32}$$
$$\simeq 0.810^{-11}\ \text{m} \simeq 0.08\ \text{Å}$$

This is the order of magnitude for the amplitudes of stretching vibrations of H-atoms. For the stretching intermonomer mode Cl–H···O$(CH_3)_2$ at 300 K, this amplitude is divided by 0.52 (eq. (5.A30)) and consequently falls around 0.16 Å. These amplitudes are to be compared with distances of atoms in molecules, of the order of 1 Å.

Centre and width of ν_s

In this section, we calculate the first moments of the IR band that corresponds to a rapid mode (ν_s), defined by coordinate q, which is strongly coupled to a slow mode by anharmonic terms we have defined in this chapter in eq. (5.2) and the following. We start the calculation in the simplified case of an isolated H-bond, and indicate at the end of this section how the instructive and intuitive results (eq. (5.A42)) can be generalized to more complex systems of interacting H-bonds. The Hamiltonian $H(q,Q)$ that governs these two modes of an isolated H-bond is that of eq. (5.2), where Q stands here for the coordinate of the three slow intermonomer vibrations defined in Figure 2.1, with corresponding reduced mass M

$$H(q,Q) = \frac{p^2}{2m} + \frac{1}{2} m\omega^2(Q)(q-q_0)^2 + \frac{P^2}{2M} + \frac{1}{2} M\Omega^2(Q-Q_0)^2 \qquad (5.A33)$$

m is the mass of the H-atom. The Fourier transform $C(t)$ of the spectrum $s(\omega) = (1/\omega)\,A(\omega)$, with $\omega = 2\pi\nu$, ν the frequency of the vibration, and $A(\omega)$ the experimental absorbance spectrum (eq. (5.A8)) is proportional to

$$C(t) = \int_{-\infty}^{+\infty} s(\omega)e^{i\omega t}\, d\omega \qquad (5.A34)$$

It is proportional (25, 43) to the average autocorrelation function of the electric dipole moment μ of the molecule

$$C(t) = \overline{\langle \mu(t)\mu(0)\rangle}_{q,Q} \qquad (5.A35)$$

where indices q and Q indicate that quantum average values of the quantities within brackets should be first evaluated by integrating over q and Q, and the bar over these quantities indicate that a thermodynamic average with Boltzmann statistics should then be calculated. In eq. (5.A35), the quantity of the second member should be multiplied by a constant factor. As this factor plays no role in the following, we omit it for simplicity, in the same way we have omitted the factor ½π which should appear before the integral in eq. (5.A34) to be consistent with the definition of the Fourier transform given in eq. (5.A17). The value of the electric dipole operator $\mu(t)$ at time t is equal to

$$\mu(t) = e^{-i(Ht/\hbar)}\,\mu(0)e^{i(Ht/\hbar)} \qquad (5.A36)$$

where H is the q, Q-dependent Hamiltonian defined in eq. (5.A33), and $\hbar$ the Planck constant h divided by 2π. The adiabatic separation between the rapid q vibration and the slow Q vibration allows us to write, with $|n\rangle_q$ representing the nth harmonic vibrational level in q

$$e^{i(Ht/\hbar)}\,|n\rangle_q = |n\rangle_q\langle n|\,e^{i(Ht/\hbar)}\,|n\rangle_q = |n\rangle_q e^{i(H_n t/\hbar)} \qquad (5.A37)$$

with (see eq. (5.5))

$$H_n(Q) = \langle n|H(q,Q)|n\rangle_q = \frac{P^2}{2M} + \frac{1}{2}M\Omega^2(Q-Q_0)^2 + \left(n+\frac{1}{2}\right)\hbar\omega(Q) \qquad (5.A38)$$

Eq. (5.A37) is equivalent to supposing that the motion in Q is sufficiently slow that its kinetic energy term P cannot induce transitions between various levels $|n\rangle_q$ of the rapid motion. Mathematically it means that all matrix elements of P of the form $\langle m|P|n\rangle_q$ are zero, unless $m = n$, in which case it is then equal to P (it still acts on a wavefunction of the slow motion on its right hand). This is the basis of the Born–Oppenheimer separation between electrons and nuclei in molecules. We may then write, supposing that the electric dipole moment $\mu(0)$ displays a linear dependence on normal mode q, an approximation called electrical harmonicity that reveals excellent for most molecular systems

$$\begin{aligned} C(t) &= \left(\frac{\mathrm{d}\mu(0)}{\mathrm{d}q}\right)^2 \overline{\langle \mathrm{e}^{-\mathrm{i}(\mathrm{H}_0 t/\hbar)} \langle 0|q|1\rangle_q \, \mathrm{e}^{\mathrm{i}(\mathrm{H}_1 t/\hbar)} \langle 1|q|0\rangle_q \rangle_Q} \\ &\simeq \left(\frac{\mathrm{d}\mu(0)}{\mathrm{d}q}\right)^2 \frac{\hbar}{2m\omega(Q_0)} \overline{\left\langle \mathrm{e}^{-\mathrm{i}(\mathrm{H}_0 t/\hbar)} \mathrm{e}^{\mathrm{i}(\mathrm{H}_1 t/\hbar)} \right\rangle_Q} \end{aligned} \quad (5.A39)$$

where only the vibrational coordinate q is considered, because we limit our interest to a transition in the ν_s mode, and we have neglected the weak Q dependence of $\langle 1|\mathrm{q}|0\rangle_q = \sqrt{\hbar/2m\omega(Q)}$ for a harmonic oscillator.

We have seen in eq. (5.A19) that the nth moment of a spectral band is the coefficient of $(2i\pi t)^n/n!$ when one develops $C(t)$ in powers of t. We consequently write eq. (5.A39) as

$$\begin{aligned} C(t) = \left(\frac{\mathrm{d}\mu(0)}{\mathrm{d}q}\right)^2 \frac{\hbar}{2m\omega(Q_0)} \Bigg[1 + \frac{\mathrm{i}t}{\hbar} \overline{\langle H_1 - H_0 \rangle_Q} \\ - \frac{t^2}{2\hbar^2} \overline{\left\langle (H_1 - H_0)^2 + [H_1, H_0] \right\rangle_Q} + \cdots \Bigg] \end{aligned} \quad (5.A40)$$

with $[H_1, H_0] = H_1 H_0 - H_0, H_1$. As $H_1 - H_0 = \hbar\omega(Q)$ (eq. (5.A38)), we find

$$\begin{aligned} C(t) &= \left(\frac{\mathrm{d}\mu(0)}{\mathrm{d}q}\right)^2 \frac{\hbar}{2m\omega(Q_0)} \left\{ 1 + \mathrm{i}t \overline{\langle \omega(Q) \rangle_Q} - \frac{t^2}{2} \overline{\langle \omega^2(Q) \rangle_Q} + \cdots \right\} \\ &= \left(\frac{\mathrm{d}\mu(0)}{\mathrm{d}q}\right)^2 \frac{\hbar}{2m\omega(Q_0)} \mathrm{e}^{\mathrm{i}t \overline{\langle \omega(Q) \rangle_Q}} \mathrm{e}^{-(t^2/2) \overline{\langle (\omega(Q) - \langle \omega(Q) \rangle_Q)^2 \rangle_Q}} \cdots \end{aligned} \quad (5.A41)$$

From eq. (5.A20) we deduce

$$\bar{\nu} = \frac{\overline{\langle \omega(Q) \rangle_Q}}{2\pi}, \quad \sigma = \frac{\sqrt{\overline{\langle (\omega(Q) - \bar{\omega})^2 \rangle_Q}}}{2\pi} \quad (5.A42)$$

The centre of the band is thus equal to the average value of ω divided by 2π, while its width is the average quadratic value of its fluctuations (due to Q) around this value also divided by 2π, an intuitive result.

It is interesting to calculate these values in simple cases. The first one, common to most H-bonds, is the one where Q is the stretching intermonomer vibration Q_s of the H-bond

X–H···Y, that is the X···Y distance. In that case, the variation of ω with Q_s is mostly linear, with coefficient $d\omega/dQ_s$ positive (Figure 4.5). With the help of eq. (5.A38), we find

$$2\pi\bar{\nu} = \omega(\bar{Q}) = \omega(Q_0) - \frac{\hbar\left(\dfrac{d\omega}{dQ_s}\right)^2}{2M\Omega_s^2}, \qquad 4\pi^2\sigma^2 = \left(\frac{d\omega}{dQ_s}\right)^2 \overline{\left\langle (Q_s-\bar{Q})^2\right\rangle_Q} \tag{5.A43}$$

with H_0 (eq. (5.A38)) being equal, up to a constant irrelevant term to

$$H_0(Q_s) = \frac{P_s^2}{2M_s} + \frac{1}{2}M_s\Omega_s^2(Q_s-\bar{Q})^2, \qquad \bar{Q} = Q_0 - \frac{\hbar}{2M_s\Omega_s^2}\frac{d\omega}{dQ_s} \tag{5.A44}$$

which means that the centre of ν_s is that for the equilibrium distance $\bar{Q}$ for X···Y, and its width is proportional to the average means square amplitude of vibration around this distance. Let us note that as $\bar{Q}$ is shorter than Q_0 (eq. (5.A44) with $(d\omega/dQ_s) > 0$), the centre of ν_s is shifted towards lower wavenumbers, a consequence of the q, Q_s anharmonic coupling. Let us also note that the mean square amplitude of vibration $\left\langle (Q_s-\bar{Q})^2\right\rangle_Q$, as given in eq. (5.A43) and which can be easily calculated in the case of a harmonic vibration governed by the Hamiltonian $H_0(Q_s)$ of eq. (5.A44), increases with temperature T. Adding corrective quadratic terms in Q_s to the development of ω in powers of Q_s, that is writing

$$\omega(Q_s) = \omega(Q_0) + (Q_s - Q_0)\frac{d\omega}{dQ_s} + \frac{(Q_s-Q_0)^2}{2}\frac{d^2\omega}{dQ_s^2} \tag{5.A45}$$

with $(d^2\omega/dQ_s^2) < 0$ (Figure 4.5) only slightly modifies preceding results: eq. (5.A44) is still valid if one replaces Ω by Ω', with

$$\Omega_s'^2 = \Omega_s^2\left(1 + \frac{\hbar\dfrac{d^2\omega}{dQ_s^2}}{2M_s\Omega_s^2}\right) \tag{5.A46}$$

and $\bar{\nu}$ and σ take the form, neglecting terms of order higher than $(d\omega/dQ_s)^2$ and $d^2\omega/dQ_s^2$

$$\begin{aligned} 2\pi\bar{\nu} &= \omega(Q_0) - \frac{\hbar\left(\dfrac{d\omega}{dQ_s}\right)^2}{2M_s\Omega_s^2} + \frac{1}{2}\frac{d^2\omega}{dQ_s^2}\left[\overline{\left\langle (Q_s-\bar{Q})^2\right\rangle_Q} + \left(\frac{\hbar\dfrac{d\omega}{dQ_s}}{2M_s\Omega_s^2}\right)^2\right] \\ &= \omega(\bar{Q}) + \frac{1}{2}\frac{d^2\omega}{dQ_s^2}\overline{\left\langle (Q_s-\bar{Q})^2\right\rangle_Q} \\ \sigma^2 &= \left(\frac{d\omega}{dQ_s}\right)^2\overline{\left\langle (Q_s-\bar{Q})^2\right\rangle_Q} + \cdots \end{aligned} \tag{5.A47}$$

This effect of quadratic terms in Q_s has scarcely been invoked, as it leads to small corrections, and shows that most of the effect of the anharmonic coupling between ν_s and intermonomer modes Q_s can be accounted for by considering linear variations of ω with Q_s only. The importance of quadratic terms emerges, however, from the analysis of the IR spectrum of liquid water (Ch. 9). In that case they do not only involve the stretching intermonomer vibration Q_s of an H-bond, but also one or both of the bending intermonomer vibration(s) Q_θ and/or Q_φ. For simplicity we represent both θ and φ by a single index δ for bending intermonomer vibrations. The dependence of ω on intermonomer modes then takes on the form

$$\omega(Q_s,Q_\delta)=\omega(Q_0,0)+(Q_s-Q_0)\frac{\partial\omega}{\partial Q_s}+\frac{Q_\delta^2}{2}\frac{\partial^2\omega}{\partial Q_\delta^2} \tag{5.A48}$$

The equilibrium position for such an H-bond is found at $Q_\delta=0$, and is quadratic for symmetry reason (Q_δ and $-Q_\delta$ have the same effect). As the aim of this appendix is to illustrate calculations of first moments, we have supposed for simplicity that only one bending vibration is present. Generalization to the case of several such vibrations is straightforward. From eq. (5.A42), we find then

$$2\pi\bar{\nu}=\overline{\langle\omega(Q_s,Q_\delta)\rangle_{Q,Q_\delta}}=\omega(Q_0)-\frac{\hbar\left(\frac{\partial\omega}{\partial Q_s}\right)^2}{2M\Omega^2}+\frac{\partial^2\omega}{\partial Q_\delta^2}\frac{\overline{\langle Q_\delta^2\rangle_{Q_\delta}}}{2}=\omega\,(\bar{Q},\overline{\langle(Q_\delta^2\rangle_{Q_\delta}})$$

$$4\pi^2\sigma^2=\overline{\langle(\omega(Q_s,Q_\delta)-\bar{\omega})^2\rangle_{Q,Q_\delta}}=\left(\frac{\partial\omega}{\partial Q_s}\right)^2\overline{\langle(Q-\bar{Q})^2\rangle_Q}+\frac{\left(\frac{\partial^2\omega}{\partial Q_\delta^2}\right)^2}{4}\overline{\langle(Q_\delta-\langle Q_\delta^2\rangle)^2\rangle_{Q_\delta}} \tag{5.A49}$$

where $\bar{Q}$ is again given by eq. (5.A44), with $d\omega/dQ_s$ replaced by $\partial\omega/\partial Q_s$. The great difference that occurs between quadratic terms in Q_s and Q_δ is that the coefficient $\partial^2\omega/\partial Q_\delta^2$ is positive for bending terms (Figure 4.2) around their equilibrium position, in opposition to the coefficient $\partial^2\omega/\partial^2 Q_s^2$ (eq. (5.A45)) that is negative (Figure 4.5). We shall see in Ch. 9 devoted to liquid water the importance of this point. Another difference that may be used to differentiate modulations by stretching and bending intermonomer modes is that the bending mode, with its term in $\overline{\langle Q_\delta^2\rangle_{Q_\delta}}$ (eq. (5.A49)) make the centre $\bar{\nu}$ of ν_s shift towards higher wave numbers when temperature increases, whereas the stretching intermonomer mode displays no such effect.

The introduction of Fermi resonances, with a supplementary term in the starting Hamiltonian (eq. (5.A33)), of the form $(q-q_0)q_\delta q_\delta'$, leads to an unchanged value for the centre of the band. It changes its width σ (eq. (5.A42)) into σ' with

$$\sigma'^2=\sigma^2+\sigma_F^2 \tag{5.A50}$$

where σ_F is the term due to Fermi resonances only. In other words, Fermi resonances have no effect on the integrated intensity and position of the band and start having an effect on its width, which is quadratically additive to the effects due to the q/Q anharmonic couplings. These results, including the effects of Fermi resonances, can be generalized with no difficulty to an assembly of identical interacting H-bonds (25, 26) in a crystal, or in the simplest case of an H-bonded dimer. In that case, we find that $\overline{\nu}$ becomes $\overline{\nu}'$ with

$$\overline{\nu}' = \overline{\nu} - \frac{\breve{V}_0 \omega(Q_0)}{2\pi} \tag{5.A51}$$

where $\overline{\nu}$ is that quantity defined in eq. (5.A42), and $\breve{V}_0$ defines the harmonic interaction between neighbour H-bonds n and $n+1$ in a crystal or 1 and 2 in such a dimer as that shown in Figure 4.4. In this case the starting Hamiltonian is the sum of the two Hamiltonians of eq. (5.2) for each H-bond 1 and 2 and furthermore displays a supplementary term of the form $m\omega^2(Q_0)\breve{V}_0 q_1 q_2$, with q_1 and q_2 the coordinates of the two stretching vibrations of the H-atoms that establish H-bonds in these dimers. The width is unchanged and defined by σ' in eq. (5.A50).

All these equations enable to deduce, from simple measurements of the 1st and 2nd moments of ν_s and their evolutions with temperature changes, the values of the various parameters that enter the potential that governs this strongly anharmonic vibration.

REFERENCES

1. J. Bournay and Y. Maréchal, J. Chem. Phys., 59 (1973) 5077.
2. D. Berckmans, H. P. Figeys, Y. Maréchal and P. Geerlings, J. Phys. Chem., 92 (1988) 66.
3. P. Hobza, Phys. Chem. Chem. Phys., 3 (2001) 2555.
4. B. I. Stepanov, Nature, 137 (1946) 808.
5. S. Bratos, D. Hadzi and N. Sheppard, Spectrochim. Acta., 8 (1956) 249.
6. Y. Maréchal and A. Witkowski, J. Chem. Phys., 48 (1968) 3697.
7. J. E. Bertie and D. J. Millen, J. Chem. Soc., (1965) 497.
8. M. Couzi, J. Le Calvé, P. Van Huong and J. Lascombe, J. Mol. Struct., 5 (1970) 363.
9. W. A. Herrebout, K. Clou and H. O. Desseyn, J. Phys. Chem. A, 105 (2001) 4865.
10. Y. Maréchal, in H. Ratajczak and W. J. Orville-Thomas (Eds.), Molecular Interactions, Vol. 1, Ch. 8, Wiley, Chichester, 1980.
11. O. Henri-Rousseau and P. Blaise, in I. Prigogine (Ed.), Advances in Chemical Physics, Vol. 103, John Wiley & Sons, 1998.
12. A. M. Yaremko, H. Ratajczak, J. Baran, A. J. Barnes, E. V. Mozdor and B. Silvi, Chem. Phys., 306 (2004) 57.
13. P. Blaise, M. J. Wojcik and O. Henri-Rousseau, J. Chem. Phys., 122 (2005) 64306.
14. H. T. Flakus and A. Machelska, J. Mol. Struct., 447 (1998) 97.
15. H. T. Flakus, A. Miros and P. G. Jones, J. Mol. Struct., 604 (2002) 29.
16. S. Bratos, J. Chem. Phys., 63 (1975) 3499.
17. G. N. Robertson and J. Yarwood, Chem. Phys., 32 (1978) 267.
18. J. Yarwood, R. Ackroyd and G. N. Robertson, Chem. Phys., 32 (1978) 283.
19. S. F. Fischer, G. L. Hofacker and M. A. Ratner, J. Chem. Phys., 52 (1970) 1934.
20. N. Rösch and M. A. Ratner, J. Chem. Phys., 61 (1974) 3344.
21. G. L. Hofacker, Y. Maréchal and M. A. Ratner, in P. Schuster, G. Zundel and C. Sandorfy (Eds.), The Hydrogen Bond, Recent Theory and Experiments, Vol. I, Ch. 6, North-Holland, Amsterdam, 1976.

22. M. Haurie and A. Novak, J. Chim. Phys., 62 (1965) 146.
23. C. A. Coulson and G. N. Robertson, Proc. R. Soc. Lond., A342 (1975) 289.
24. S. Bratos, J. Lascombe and A. Novak, in H. Ratajczak and W. J. Orville-Thomas (Eds.), Molecular Interactions, Vol. 1, Ch. 10, Wiley, New York, 1980.
25. G. Auvert and Y. Maréchal, J. Phys., (France), 40 (1979) 735.
26. P. Excoffon and Y. Maréchal, Chem. Phys., 53 (1980) 23.
27. A. Witkowski and M. Wojcik, Chem. Phys., 1 (1973) 9.
28. M. J. Wojcik, Mol. Phys., 36 (1978) 1757.
29. Y. Maréchal, Chem. Phys., 79 (1983) 85.
30. K. Heyne, E. T. J. Nibbering, T. Elsaesser, M. Petkovic and O. Kühn, J. Phys. Chem. A, 108 (2004) 6083.
31. D. Cringus, S. Yeremenko, M. S. Pshenichnikov and D. A. Wiersma, J. Phys. Chem. B, 108 (2004) 10376.
32. C. P. Lawrence and J. L. Skinner, J. Chem. Phys., 117 (2002) 5827.
33. C. P. Lawrence and J. L. Skinner, J. Chem. Phys., 119 (2003) 3840.
34. A. J. Lock and H. J. Bakker, J. Chem. Phys., 117 (2002) 1708.
35. S. Bratos and H. Ratajczak, J. Chem. Phys., 76 (1982) 77.
36. R. G. Gordon, J. Chem. Phys., 43 (1965) 1307.
37. J. E. Bertie and H. H. Eysel, Appl. Spectrosc., 39 (1985) 392.
38. Y. Maréchal, J. Mol. Struct., 648 (2003) 27.
39. R. Kubo, in D. ter Haar (Ed.), Fluctuation, Relaxation and Resonance in Magnetic Systems, Oliver and Boyd, Edinburgh, 1962.
40. H. R. Zelsmann and Y. Maréchal, Chem. Phys., 5 (1974) 367.
41. Y. Maréchal, J. Chem. Phys., 95 (1991) 5565.
42. A. C. Legon and D. J. Millen, Chem. Rev., 86 (1986) 635.
43. D. Chamma and O. Henri-Rousseau, Chem. Phys., 248 (1999) 53.

– 6 –

Reactivity of Hydrogen Bonds: Transfers of Protons and of H-Atoms

In the preceding chapters, we have seen that H-bonds display two fundamental properties: they are directional and have energies equal to a few kT's at room temperature, here "few" means in the range of 3–10. These properties are for a good part at the origin of their importance, as they allow for the appearance of stable but flexible molecular structures, which are essential in biology. In this chapter, we establish a third property of H-bonds, which will turn out to be at least as crucial as the two former ones but are only emerging in our vision of H-bonds, which means that we strongly suspect their impact is presently largely underestimated: under certain circumstances, which are not yet clearly established and still convey many pending questions, H-bonds form a very efficient path to transfer protons or H-atoms between two molecules. These transfers are the basic chemical mechanisms at the origin of reactivity of these particular media where the density of H-bonds is especially great, aqueous media. The reactivity of these media is well described in classical chemistry, including biochemistry, where such reactions as hydrolyses, for instance, are well depicted in terms of ordinary chemical equations. The fact that the common molecular origin of most of these reactions that occur in these media is to be found in H-bonds is certainly much less known. Also the fact that two different types of transfers are active in chemistry and in biology, where conditions are different, is still largely ignored, but in the near future is likely to become important upon an examination of these mechanisms.

GREAT AMPLITUDE MOTIONS IN ISOLATED H-BONDS

How can we approach the molecular mechanisms of such reactions? We have already seen in Chs. 4 and 5, devoted to the description and analysis of the specific properties H-bonds exhibit in IR spectra, that the stretching vibrations ν_s of the H-atoms that establish H-bonds are strongly anharmonic, which is due to ν_s having unusually large amplitudes. The analysis of IR spectra has shown that the coupling between ν_s, described by coordinate q, with the relative stretching vibration Q_s of the two molecular parts X–H and Y when they are H-bonded and form a molecular complex X–H···Y, is at the origin of most of the exceptional properties H-bonds display in IR spectroscopy. We have seen (in eqs. (5.2) and

(5.A33)) that the vibrational potential for these two modes defined by q and Q_s takes on the form:

$$V(q, Q_s) = 1/2m\omega^2(Q_s)[q - q_0]^2 + 1/2M\Omega_s^2(Q_s - Q_0)^2 \tag{6.1}$$

The Q_s dependence of the force constant $k(Q_s) = m\omega^2(Q_s)$ creates an anharmonic coupling between both modes q and Q_s. It is at the origin of the exceptional width of ν_s bands of H-bonds, and of their shifts towards lower wavenumbers when compared to bands of the same X–H molecules, when they do not establish H-bonds. These most intense ν_s bands are $0 \rightarrow 1$ transitions, or transitions between the ground vibrational state of q and its first excited state. The other possible transitions, $0 \rightarrow n$ with $n > 1$, have intensities equal to 0 in the case of an harmonic oscillator in q (no terms in q^3, q^4, etc. in $V(q, Q_s)$). Overtones of ν_s, particularly those corresponding to $0 \rightarrow 2$ transitions in the 6000–6500 cm^{-1} region, however often appear in experiments (1). Their intensities remain weak, weaker than the intense $0 \rightarrow 1$ transition by several orders of magnitude. In the adiabatic separation of q (fast vibration) and Q_s (slow vibration), the q, Q_s anharmonic coupling cannot be at the origin of the presence of these overtones, however weak they may appear, because it is harmonic in q for Q_s fixed. It means that in addition to the strong anharmonic q, Q_s coupling included in eq. (6.1), the potential $V(q, Q_s)$ must also contain small terms in q^3, q^4, etc. This type of anharmonicity is called 1D anharmonicity because, in opposition to the preponderant anharmonic q–Q_s coupling, it only concerns the q mode and does not imply couplings of q with other modes. Such 1D anharmonic terms imply that for a fixed value of Q_s, $V(q, Q_s)$ departs from a (harmonic) parabola when q appreciably differs from q_0. For weak and moderately strong H-bonds, they take on the form displayed in Figure 6.1, which is that of a "Morse potential" in q, which is a good representation of the actual potential when this 1D anharmonicity is not too strong:

$$V(q, Q_s) = \frac{\hbar\omega(Q_s)}{\delta(Q_s)}\left\{1 - \exp\left[-\sqrt{\frac{m\omega(Q_s)\delta(Q_s)}{2\hbar}}(q - q_0)\right]\right\}^2 + 1/2M\Omega_s^2(Q_s - Q_0)^2 \tag{6.2}$$

We have already encountered this type of potential in eq. (4.6) where it governed Q_s intermonomer modes. It tends towards the harmonic potential of eq. (6.1) when the parameter δ, characteristic of this 1D anharmonicity in q, tends towards 0. The vibrational potential experienced by the slow Q_s mode when the rapid mode q is in its nth excited state, is then equal to

$$E_n(Q_s) = \left(n + \frac{1}{2}\right)\hbar\omega(Q_s)\left[1 - \frac{\delta(Q_s)}{4}\left(n + \frac{1}{2}\right)\right] + 1/2M\Omega_s^2(Q_s - Q_0)^2 \tag{6.3}$$

This potential is that, adapted to a Morse potential in q, of the Hamiltonian defined in eqs. (5.5) and (5.A38) in the case of a harmonic potential in q. The Morse potential in q that enters eq. (6.2) is represented in Figure 6.1, together with the harmonic potential that has at its minimum at $q_0 = 1$ Å the same curvature, defined by the parameter ω. We should remember that these potentials are represented for a fixed value of $Q_s = Q_0$. The effect of this 1D anharmonicity (parameter δ) is to shift the average value of q towards higher values, increase its amplitude of vibration, and also shift all $0 \rightarrow n$ transitions towards lower wavenumbers

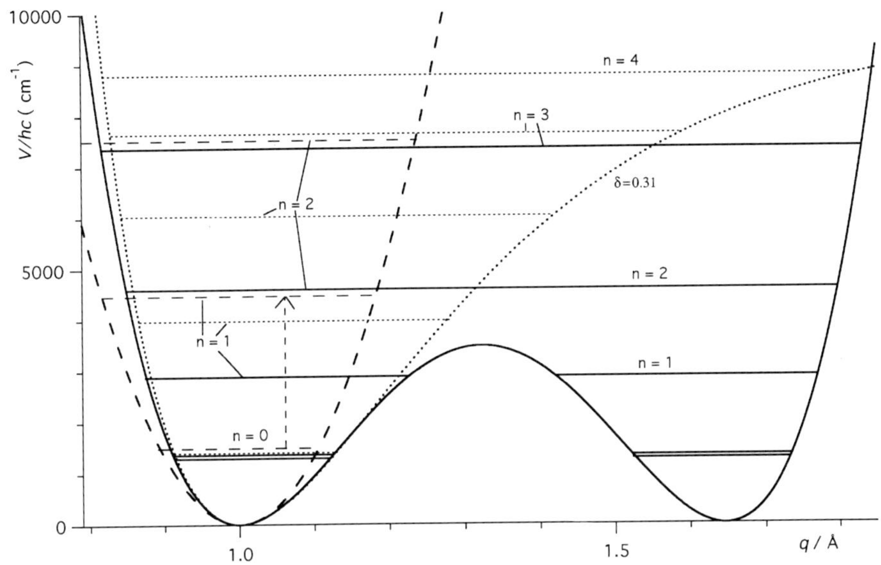

Figure 6.1 Harmonic (dash), Morse (dots, $\delta = 0.31$) and double-well potentials for the O–H stretch ν_s (coordinate q) motion in an O–H···O bond with a fixed O···O distance $Q_s = 2.65$ Å. Eigenvalues corresponding to observable values of vibrational energies, and defined by indices $n = 0$, 1, etc., are shown for each potential. All three potentials have the same curvature at $q = 1$ Å, defined by the parameter ω which is such that $\omega/2\pi c = 3000\,\text{cm}^{-1}$ (c velocity of light in cm sec^{-1}). With this value the $0 \rightarrow 1$ transition of the harmonic potential, represented by a vertical arrow, falls at $\tilde{\nu} = 3000\ \text{cm}^{-1}$.

(eq. (6.3)). This last shift remains nevertheless small as compared to that due to the q, Q_s anharmonic coupling.

Although the existence of such a 1D anharmonicity looks quite general, it may be considered in most cases a small correction, with no real importance, in the same way as we have considered Fermi resonances in the $0 \rightarrow 1$ transition in mid-IR spectra. This correction is well taken into account by the Morse potential defined above and shown in Figure 6.1. In the case of an H-bond such as the one represented in the upper part of Figure 6.2, this is, however, no longer true as when one starts from the left configuration of this drawing and extends the O–H distance, one reaches a symmetrical configuration where the H^+ proton is equidistant from both the O-atoms. If left to this symmetrical position this proton may either come back to its original position, or reach another position with same energy, represented by the right configuration of the upper drawing. It means that the Morse potential in q is no longer a good approximation, and should be replaced by a symmetrical double-well potential, such as that drawn in Figure 6.1. Such a potential gives the proton much greater amplitude. In terms of chemical mechanisms, it corresponds to a transfer of the proton H^+ between the two H-bond configurations drawn in the upper drawing of Figure 6.2. In the case of malonaldehyde, shown in the lower part of Figure 6.2, we find a similar, but not identical, situation—in the case of acid carboxylates or carbonates, shown in the upper part of this figure, the transfer leads to a final state that has the same energy as the initial one, and is consequently also described by a similar symmetrical double-well potential in q. The transfer

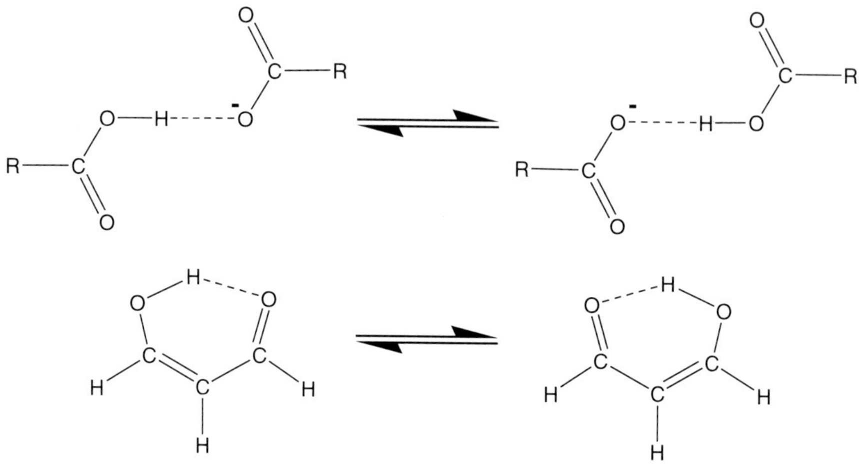

Figure 6.2 H-bond configurations of potassium hydrogen bistrifluorocarboxylate(2) ($R=CF_3$) or bicarbonate(3) (R=OH) in the upper drawing (K^+ not represented), and in malonaldehyde(4) in the lower drawing.

that occurs in such an intramolecular H-bond is, however, a transfer of an H-atom, not of a proton. We shall see later in this chapter the importance of this distinction. These simple considerations on anharmonicity show that in H-bonds great amplitude motions of the proton or of the H-atom may easily occur, in opposition to what occurs in covalent bonds.

PROTON TRANSFERS IN AN H-BOND NETWORK

Beyond these very particular symmetrical cases, we may wonder whether such transfers of protons are of practical importance in chemistry, biology or physics. The object of this section is to show that the answer to such a question is, yes they are important, and we often have to deal with them, without however always recognizing them. In order to support such an assertion, we shall examine some well-known mechanisms in which they are central and which clearly show that they may even be fundamental; and it will also clearly appear that we still have a lot to learn about them. In reality, they are not found in such simple systems we have described in the preceding section. They scarcely occur in single H-bonds; and most of the time they occur in systems that display a well-developed "H-bond network", such as liquid water or aqueous media, which are more complex systems with an appreciable number of closely lying and interacting H-bonds.

Ionization mechanism of an acid or a base

Dissociation of an acid or a base in water is certainly one of the most spectacular manifestations of proton transfers and is largely documented in classical chemistry. We thus very well know that on mixing hydrogen chloride with liquid water, we completely dissociate

HCl molecules into Cl^- and H_3O^+ ions strongly diluted among H_2O molecules, which clearly indicates that protons have been transferred from HCl molecules to water molecules. In this case, the H-bond network is exceptionally dense (liquid water is characterized (Ch. 9) by a number of H-bonds as great as that of covalent bonds), and the whole reaction that leads to dissociation of HCl requires at least two steps, with two different kinds of transfer of protons: ionization and diffusion of ions. In this subsection, we examine the first step, which is ionization, before examining diffusion in the next step, and keeping in mind that recent experiments based on nonlinear time-resolved IR spectroscopy have even shown the existence of three steps for this process (5).

In order to know how ionization proceeds in some details, and in order to have an idea of the various parameters that influence it, the study of simple systems is instructive. The first question that comes to mind is how many water molecules are necessary for ionization of an HCl molecule H-bonded to a small number of water molecules to proceed. In other words, does a molecule as that of HCl spontaneously transfer its proton to a single H_2O molecule with which it has established an H-bond, or is it necessary that a more developed H-bond network be present around it for such a transfer to occur? Experiments do not directly give an answer to this question. However, some of them give important indications. Thus, IR spectra of mixtures of gaseous HCl and H_2O molecules in various proportions trapped in solid matrices of a noble gas at a low temperature (6), show that with a single H_2O molecule only complexes of the type Cl–H···OH_2 are present, with well-marked ν_s(Cl–H···) bands and absence of bands due to hydronium ion (H_3O^+). Rotational microwave spectroscopy (7) of an expanded mixture of gaseous HCl and H_2O in an equal proportion leads to the same conclusion (see Ch. 3 for microwave spectroscopy of supersonic expanded gases). Applied to a related system, a mixture of HCl and NH_3, it clearly demonstrates the presence of neutral dimers Cl–H···NH_3 and the absence of corresponding ionic species Cl^-···H^+NH_3 (8). The NH_3 molecule being a stronger proton acceptor than H_2O, or as said in a language of classical chemistry the NH_3 molecule being a more basic molecule, it clearly shows that in a mixture of HCl and H_2O a greater number of H_2O molecules is necessary for such a strong acidic molecule as HCl to be ionized. How many? Experiments have not yet given any definitive answer to this question, even if IR studies of pulsed supersonic jet-cooled mixtures of HCl or HBr with H_2O indicate that four to five H_2O molecules are enough to ionize these acids (9). Furthermore, IR studies of mixtures of HCl and NH_3 trapped in solid matrices (10, 11) reveal that the proton of HCl involved in the Cl–H···NH_3 bond, without being transferred, approaches the N-atom of NH_3 more when the matrix is Ar, Kr or N_2, than when the matrix is Ne. In other words, HCl has a greater tendency to share its proton with the N-atom of NH_3 when in these matrices than when in an Ne matrix. This result may be understood as follows: because Ne is a hardly polarizable molecule, while Ar and Kr are more polarizable, and N_2 may even accept H-bonds, these molecules are not as inert as they are supposed to be. They are indeed at the origin of a small interaction with the H-bonded Cl–H···NH_3 dimer that weakly favours a partial ionization.

Such a result points to the high sensitivity of the proton position in these H-bonded dimers to the surrounding. On the other hand, theory suggests that the addition of one to two H_2O molecules to this Cl–H···NH_3 dimer is sufficient for a complete proton transfer to occur (12, 13). It also stresses the sensitivity of transfers of protons to the environment, and particularly to the presence of other H-bonds or, as we have put it at the beginning of this section, to the presence of the surrounding H-bond network. Another simple model system that gives precious indications on ionization and leads to the same conclusion is that of isolated amino acid molecules, which are neutral molecules in the gas phase and in the absence of H_2O molecules, but become "zwitterions" in the presence of H_2O molecules (14, 15), with a significant stabilization energy of about 30 kJ mol^{-1} in the case of glycine (16). Photoelectron spectroscopy coupled with mass spectroscopy gives a minimum number of five H_2O molecules to ionize glycine into its zwitterion (14). IR spectra of glycine molecules trapped in solid matrices at low temperatures show that the neutral form is the stable form when less than three H_2O molecules are present, whereas the zwitterion is in the stable form with four H_2O molecules (15) or more. Thus with three H_2O molecules no zwitterionic form is detected; and calculation shows glycine to adopt the configuration drawn in the left part of Figure 6.3. With four H_2O molecules, the most stable structure is the zwitterion drawn in the right part of Figure 6.3, with an NH_3^+ cationic part and COO^- anionic part, which means that a proton has been transferred from the carboxylic neutral group COOH to the amine neutral group NH_2. The four H_2O molecules, however, do not form a "water wire" connecting COOH to NH_2. This is done by three water molecules, and the fourth one extending the H-bond network of the whole system by establishing an H-bond with the other O-atom of the carboxyl COO^- group that is not directly H-bonded to the first three H_2O molecules. It strongly suggests that the formation of the H-bond with this fourth H_2O molecule is responsible for the stability of the zwitterion. Back to ionization of HCl, an *ab initio* molecular dynamics study (17) of the HCl···glycerol system shows that ionisation of HCl starts when the Cl-atom of HCl accepts at least two H-bonds from glycerol and its H-atom has established an H-bond with O-atoms. It clearly demonstrates that the H-bond network around the neutral molecule strongly influences ionization.

Incidentally, let us note that these questions are not purely academic in the sense that experiments and analyses described above have immediate implications; the answers to the questions raised on a possible proton transfer are precious for our knowledge of the reactions

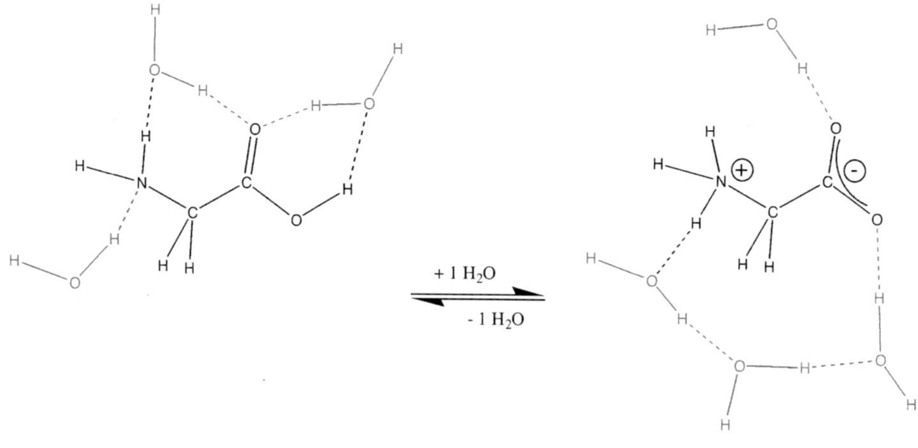

Figure 6.3 Stable configurations of glycine and water molecules in the gas phase. Left: glycine + $3H_2O$ (neutral form). Right: glycine + $4H_2O$ (zwitterionic form). From R. Ramaekers *et al.* (15).

that occur in the atmosphere. Traces of such gases as HCl, H_2SO_4, HNO_3, NH_3, etc. can be detected in the atmosphere (12, 13, 18) and it is certainly important to know whether these molecules are then as reactive as when they are in liquid water, that is whether they already behave in the atmosphere as strong acids or bases. We shall encounter other situations where atmospheric studies encounter H-bonds and water molecules (e.g., Ch. 8).

On the basis of these experimental and theoretical results, the proton transfer that occurs during ionization of HCl in liquid water may be represented as in Figure 6.4. This figure is schematic, as the extended H-bond network due to liquid water in which HCl is embedded, is represented by a single H_2O molecule around the Cl–H···OH_2 complex. The thermal fluctuations of this H-bond network, sketched in this figure by various configurations this single H_2O molecule adopts, induce variations of the vibrational potential $V(q, Q_0)$ for the stretching mode ν_s of HCl. When this single H_2O molecule adopts the configuration shown in the left part of the figure, $V(q, Q_0)$ has the form of a parabola centred around $q_0 \approx 1$ Å, with a hint of a 1D anharmonic distortion on its right-hand side indicating a slight tendency of some proton sharing. When this H_2O molecule adopts the configuration represented in the middle part of the figure, $V(q, Q_0)$ becomes flat. This configuration requires some energy to be attained, and the two minima of $V(q, Q_0)$ have higher energies than the minimum in the preceding configuration. This energy, however, lies within the range of thermal fluctuations when this reaction occurs in liquid water. Following Ando and Hynes (19), the first authors to propose such a mechanism, the potential in this configuration is sufficiently flat for the ground vibrational level of ν_s to be higher than its maximum. Consequently, it allows for a back-and-forth

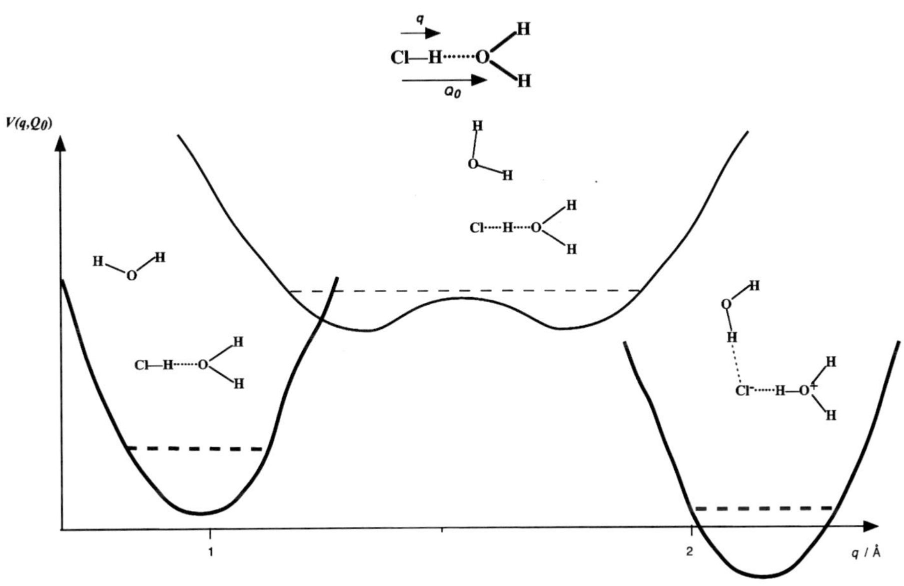

Figure 6.4 Influence of the H-bond network on the vibrational potential $V(q, Q_0)$ that rules the ν_s vibration of a Cl–H···O–H_2 complex (19). This H-bond network is schematically represented by the H_2O molecule that establishes various H-bond configurations with respect to this complex. The ground levels corresponding to these potentials are drawn as dashed lines.

quantum motion of the proton between the two atoms Cl and O, which is nevertheless not a tunnelling. If the single H_2O molecule adopts the configuration shown in the right-hand side of the figure with the establishment of an H-bond with the Cl-atom, the corresponding potential has a lower maximum than the initial one and represents a situation where we have ionization of HCl and appearance of a $Cl^-\cdots H_3O^+$ complex. This mechanism has been called "adiabatic proton transfer" by Borgis and Hynes (20). It means that the ν_s vibration of the proton is sufficiently rapid with respect to the fluctuations of the H-bond network around the $Cl^-\cdots OH_3^+$ or $Cl{-}H\cdots OH_2$ complex, for the potential that governs it to immediately adapt itself to any new state of this H-bond network. In other words, the ν_s vibration adapts itself immediately to the various potentials shown in Figure 6.4, which pass slowly and continuously from one form to another one under the influence of the surrounding H-bond network. As seen in Ch. 5, this adiabatic decoupling of two motions, one slow and one rapid, is the same as that on which relies the Born–Oppenheimer separation between electronic and nuclear motions in a molecule.

For such a strong acid as HCl, the overall reaction is exothermic, which corresponds to the vibrational potential for the ionized complex, shown on the right-hand side of Figure 6.4, having a ground state level lower than that for the neutral complex (left-hand side). For weaker acids such as HF, the ground state level of the ionized complex $F^-\cdots OH_3^+$, has energy higher than that of the neutral complex $F{-}H\cdots OH_2$. Let us note, however, that this ionization step is not the only one in the process of dissociation of an acid or a base in water, and that this gain or loss in enthalpy is only part of its gain in free energy. The other part is that due to diffusion of H_3O^+ (or OH^- for a base), which we examine in the next subsection and which may be at the origin of an entropic gain.

We can now realize that proton transfers at the origin of the ionization of an acid or a base are certainly not as simple as imagined when considering isolated H-bonds. In particular, the role of the surrounding H-bond network is predominant. Nevertheless, we have the impression that we have begun to understand them, and the theory of an "adiabatic type" transfer looks quite adapted. Before leaving this subsection, let us note that we may transpose all what has been described above on ionization of acids to ionization of bases. In that case, we write the equation that describes this mechanism, analogue to ionization of HCl represented in Figure 6.4, as

$$RH_2N\cdots H{-}O{-}H \rightarrow RH_2N^+{-}H\cdots O{-}H^- \quad (6.4)$$

Diffusion of H_3O^+ and $O{-}H^-$ ions in liquid water

The second molecular mechanism encountered in this process of acid/base solvation is diffusion in water of the H_3O^+ or OH^- ions formed during the ionization step described previously. It is represented in Figure 6.5 for both H_3O^+ and $O{-}H^-$ ions. The mechanism of this proton transfer is not the same as that encountered in ionization. Following Ando and Hynes (19) this proton transfer is also adiabatic. This point is however a matter of discussion, as other authors (21) find that this may be true in supercooled water but at room temperature and above, a nonadiabatic tunnelling is responsible for this proton transfer. In this mechanism, the influence of the surrounding is less marked, and instead of having a potential $V(q, Q_0)$

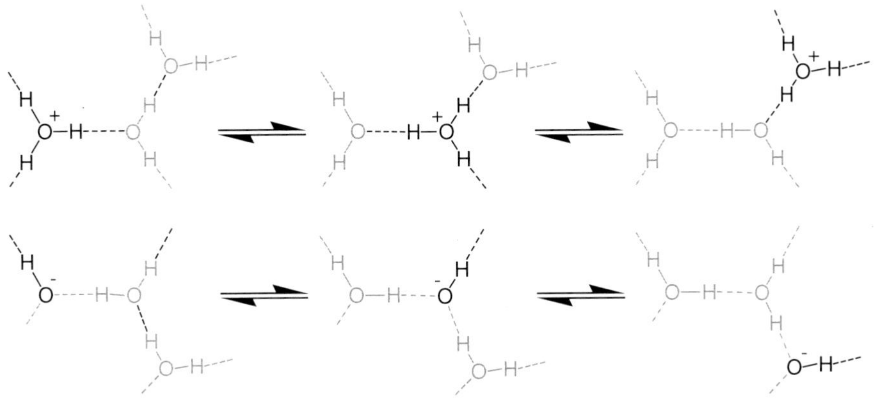

Figure 6.5 Diffusion of an H_3O^+ ion (upper drawing) or of an $O{-}H^-$ ion (lower drawing) in liquid water.

that slowly fluctuates with changes of the H-bond network around the proton that is to be transferred, as depicted in Figure 6.4, we have a potential which is, in the first approximation, scarcely sensitive to these fluctuations, but is a double-well potential similar to that depicted in Figure 6.1 (20). The mechanism at the origin of the transfer of protons is then tunnelling in this double well. It is called nonadiabatic because, when tunnelling occurs, nuclei take some time to adapt themselves to the new position of the proton. Let us note that in this mechanism, the electrons follow the proton adiabatically, that is, they are sufficiently rapid so as to adapt themselves to any change of the proton, tunnelling included. But this is not so of the nuclei. The vibrational double-well potential of the proton is therefore a potential obtained in the Born–Oppenheimer approximation of an adiabatic separation of electrons and nuclei. In other words, this tunnelling may result in a transition of the proton or other vibrations between various vibrational levels, but no transition to excited electronic levels appear. When temperature is varied this nonadiabatic tunnelling displays a behaviour that is in general non-Arrhenius. It also displays a strong H/D isotope effect, with the probability of transfer of D-atoms being generally much smaller than that for H-atoms, but still being not that which is usually expected, and is consequently to be analyzed with care (22, 23). These properties are criteria to distinguish this nonadiabatic proton transfer from the adiabatic proton transfer ruled by the potentials of Figure 6.4. We shall come back to this nonadiabatic proton transfer. Let us note that even if this tunnelling is more insensitive to the fluctuations of the H-bond network around the proton to be transferred, it is not completely insensitive to its environment. In particular, the H-bond distance Q will appear to be a parameter that strongly influences the shape of this double-well potential and, consequently, the rate of transfer. Also, the potential(s) that rules such transfers as those depicted in Figure 6.5 are symmetric and the initial and final states have same energy. It means that in the region of low temperatures where an adiabatic transfer looks predominant, the potentials shown on the left and right parts of Figure 6.4 are symmetric with respect to a vertical line (parallel to the $V(q, Q_0)$ axis) that cuts the q axis at $q \approx 1.35$ Å (the O···O distance is approximately 2.7 Å) and their minima have equal energies. At higher temperatures, the double well that governs nonadiabatic tunnelling at the origin of the transfer, such as represented in Figure 6.1, is also symmetric.

A direct consequence of this proton transfer, be it adiabatic or not, is that H_3O^+ and OH^- ions are much more mobile in water than any other ions. This results in a conductivity of acids and bases in water, which is at least one order of magnitude greater than for any salt (21). Ions other than H_3O^+ and OH^- are surrounded in water by several H_2O molecules from which they accept H-bonds (anions) or have an electrical interaction with the non-bonding electrons pairs of the O-atom comparable in energy to that of an H-bond (cations—see Ch. 9). They suffer ordinary mass diffusion and during this process they behave as a colloid, more or less keeping their H_2O dressing, which makes this mechanism somewhat slow. H_3O^+ and OH^- ions also obey this "ordinary" diffusion mechanism. In addition, they also exhibit the other type of diffusion due to the stepwise proton transfer shown in Figure 6.5. It is a much faster process that involves transfer through an H-bond of a single proton, with no transfer of water molecules, in opposition to diffusion of an ion dressed by several water molecules. The corresponding hopping time is some picoseconds (psec) (21), a short time for such a process. This high proton conductivity of water is greatly favoured by the existence of the exceptionally developed tridimensional H-bond network of liquid water. This process is also at the origin of an appreciable gain in entropy in the case of liquid water. Its exceptionally dense H-bond network offers H_3O^+ or OH^- the possibility to occupy many different sites with equal energies. The gain in free energy for the dissociation of HCl in water is of about 75 kJ mol^{-1} (24). As Ando and Hynes (19) give a gain in enthalpy due to the ionization of HCl of about -30 kJ mol^{-1}, this gain in entropy is of the order of -45 kJ mol^{-1}, about 50% greater in absolute value than the gain in enthalpy. It is this consequent entropic gain that allows weaker acids to be nevertheless dissociated in water, despite their endothermic ionization step that occurs with a loss of enthalpy.

We thus see that this type of proton transfers, which is at the origin of the efficient diffusion of H_3O^+ and OH^- ions in water, is not yet so well understood. It is different for the transfer that occurs during ionization of an acidic or basic molecule, and is consequently sensitive to other parameters that are presently not clearly defined.

PROTON TRANSFERS IN THE ELECTRONIC EXCITED STATE

Photoacids

To know more on these various proton transfers, the study of proton transfers that occur after excitation of an electronic level, display some interest of their own with respect to proton transfers in the ground electronic state. Such transfers are often labelled ESPT (excited states proton transfers) and the origin of the interest they provoke is to be found in the possibilities they offer in designing sensors. They occur as a relaxation mechanism of the electronic excitation of a molecule that follows absorption of a visible or UV photon. A good representative of these ESPTs are photoacids. They consist of molecules of phenol or naphtols, that is aromatic rings (naphtol consists of two aromatic rings sharing a common C–C bond) where one H-atom attached to a C-atom has been replaced by an alcohol group O–H. When immersed in an H-bond acceptor solvent, typically water, this O–H group establishes an H-bond with the O-atom of a solvent molecule. Upon excitation of a phenol molecule, for instance, with visible or UV light, an electronic excited state may be reached where, relative to the ground state, a displacement of electrons occur from the O–H group to the aromatic ring. Within some femtoseconds (fsec) it induces a change in the strength of the H-bond established by the O–H group of the phenol, which may result, after some ps (25), into a proton transfer. The whole reaction may be written as

$$\phi-\mathrm{O}-\mathrm{H}\cdots\mathrm{OH_2}\cdots+\mathrm{h}\nu\rightarrow\phi-\mathrm{O}-\mathrm{H}^{*}\cdots\mathrm{OH_2}\cdots\rightarrow\phi-\mathrm{O}^{-*}\cdots\mathrm{OH_3}^{+} \qquad (6.5)$$

where ϕ symbolizes an aromatic or biaromatic ring, and the asterisk characterizes an electronic excited state of the (bi)aromatic-water complex. This ϕ–O^- part in its electronic excited state behaves as the anion of a strong acid, whereas the solvated H_3O^+ may diffuse in water within some ps, the same way it does in ordinary acidic solutions we have already seen (Figure 6.5). Let us note two points. First, the same type of reaction occurs when the solvent is an amine R-NH_2 instead of water. We then have the same ϕ–O^- ion linked to an ammonium R–$NH_3{}^+$ ion. Second, photobases exist as well as photoacids. They are well represented by aromatic ketones where the C–OH group of a photoacid is replaced by a carbonyl C=O group. The corresponding photoreaction is as follows:

$$\phi=\mathrm{O}\cdots\mathrm{H}-\mathrm{O}-\mathrm{H}\cdots+\mathrm{h}\nu\rightarrow\phi=\mathrm{O}^{*}\cdots\mathrm{H}-\mathrm{O}-\mathrm{H}\rightarrow\phi-\mathrm{O}^{+}-\mathrm{H}^{*}\cdots\mathrm{O}-\mathrm{H}^{-} \qquad (6.6)$$

The interest of studying these systems is that the dynamics of proton transfer is more accurately probed using such methods as time-resolved spectroscopy, than for proton transfers in the ground state; excitation starts with a laser flash that excites the molecule at time $t = 0$, which can then be defined with a resolution better than psec. Time-resolved IR spectroscopy or pump-probe spectroscopy or any other "two-photon spectroscopy" (Ch. 3) may then be applied to convey original information on the dynamics of the proton transfer and the various parameters that govern it. This is not possible for ordinary proton transfers in the ground state. Thus, the intramolecular proton transfer responsible for enol–keto isomerization in the excited state of conjugated aromatic rings has been shown to be modulated by a well-identified mode of the molecule in its excited state (26, 27). This system is not exactly a photoacid, but falls in the neighbouring class of the zwitterions of aminoacids we encountered above. The proton transfer has been shown to occur on a timescale of about 30–50 fsec after electronic excitation. Also, proton transfers to water have been shown to occur some psec after excitation (28) in a series of "superphotoacids", anthocyanins, which are found in common pigments of numerous red flowers and fruits. It has been deduced that water reorientation is an important parameter that controls these transfers and that, furthermore, these proton transfers in anthocyanins are suspected to be an efficient protection for these flowers and fruits against deleterious and excess solar radiation, particularly in the blue and green regions of the visible light. These examples show the interest of these ESPTs, if we wish to know more accurately about the parameters that govern proton transfers at a molecular level, particularly all those that concern their dynamics.

ESPT's in biology: photosynthesis and vision mechanisms

So far, we have seen proton transfers are fundamental molecular mechanisms in ordinary chemistry, particularly that part of chemistry that deals with acid/base reactions. In a subsequent section, we shall see transfers of different types, which are basic mechanisms in the reactivity of biomedia and are consequently fundamental and general in biology. Before describing them, we presently keep to the topics of ESPT to examine one particular type that also occurs in biology and plays a fundamental role, even if this role is not general, being limited to a very specific kind of reactions. These are the mechanisms of photosynthesis and vision. As usual in biology, they involve a great number of different steps with relatively low barriers between them, so that reactions can proceed in an aqueous medium at room temperature. They are consequently more complex mechanisms than ESPTs encountered in chemistry. The methods to study them are also different and involve using various tools that give partial results, and later recombining all these results into a consistent mechanism, the usual procedure in biochemistry. For these complex mechanisms, we shall stress, in this section, the part that deals with proton transfers. The result of photosynthesis is storage of energy after absorption of a photon in the visible or UV region. This energy is later used to synthesize ATP (adenosine triphosphate), which in turn is transformed into ADP (adenosine diphosphate) when energy is required for metabolism. The energy liberated by this ATP $\rightarrow$ ADP reaction is, used, for instance, to synthesize mono or polysaccharides, and also for the metabolism of the cell in which it occurs. In vision mechanism, GTP (guanine triphosphate) is synthesized instead of ATP. It will be used to trigger an electric signal that will be analyzed by the brain and interpreted in terms of points (pixels in the language of electronics) in an image.

Both mechanisms, photosynthesis and vision, are very similar. We shall concentrate on the one that has been the most thoroughly studied, that is photosynthesis in bR (bacteriorhodopsin), occasionally pointing to the differences between these mechanisms. In both mechanisms, a photon is absorbed by a molecular pigment embedded in a protein. In order to be efficient in the visible region, the region where the solar emission is most intense, this pigment is a somewhat large molecule with many conjugated double covalent bonds. The most familiar of these pigments is chlorophyll, found in plants and algae. It is a porphyrin with magnesium as a ligand, represented in Figure 6.6. The porphyrin nucleus is the tetrapyrrole structure characterized by the four N-atoms linked to Mg. It may accommodate various metals as ligands. In haemoglobin, for instance, the ligand is Fe. The Mg porphyrin pigment of chlorophyll is transparent in the green region of the visible spectrum, absorbs in the blue and red regions, and consequently scatters the green light. Let us recall (Ch. 3) that absorption of light (a one-photon or first-order process) is predominant over scattering (a two-photon or second-order process). Even if plants appreciably contribute to photosynthesis, the greatest part of photosynthesis does not occur in plants. Most of the photosynthesis that results in absorption of atmospheric CO_2 and rejection of O_2 occurs on sea surface and is performed by cyanobacteria, commonly called "blue algae". They have been the organisms at the origin of the production of O_2, about 2.5–3.5 billion years ago, well before plants existed (32), and also before oxygen could escape in the atmosphere (it was then mainly trapped in oxidized rocks and oceanic iron).

The mechanism of photosynthesis in chlorophyll is somewhat complex and is not as well known as the one that occurs via another chromophore (or pigment), a carotenoid, also shown in Figure 6.6. It requires a smaller number of steps and is simpler. It is also encountered in simpler and much easier to handle species, the purple membranes of special archaebacteria *Halobacterium salinarum*, which are found in great quantities in salt-saturated waters,

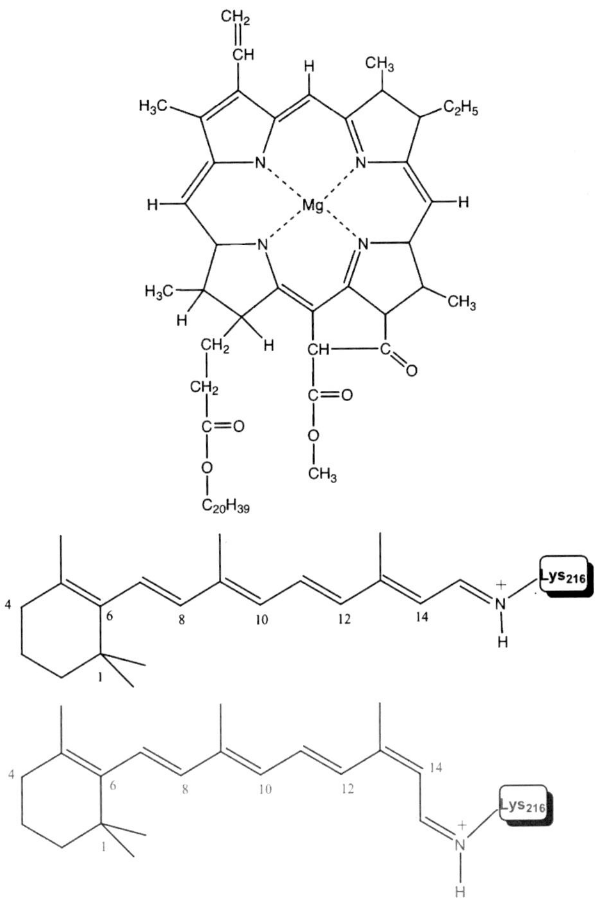

Figure 6.6 Two photosynthetic pigments: chlorophyll (upper diagram), and the bR retinylidene chromophore, which consists of a carotenoid, retinal, linked to a Schiff-base, itself linked to a lysozyme residue of bR (29–31). Retinal is in its *all-trans* conformation in the intermediate diagram, and in its 13-*cis* conformation in the lower diagram. Conventional atom numbering of retinal is shown.

such as in Lac Rose (Sénégal), in the far end of the San Francisco Bay, where they give the shallow water its purple colour. These bacteria do not absorb CO_2 and do not reject O_2, which makes them simpler systems. Photosynthesis nevertheless helps them survive in their medium in extreme conditions by providing them with enough energy for their metabolism. The primary and secondary structures of bR (Ch. 2) are discussed in Ref. (31). The "light harvesting system", called antenna in radio technology, of these purple membranes, is bR, which is a transmembrane protein. The proteins of rods in retins, for vision mechanism, are transmembrane proteins of a similar structure, rhodopsines, which can also be collected in great quantities from bovine retins and has also been widely studied. The primary and secondary structures of bovine rhodopsines are discussed in Ref. (33). In both types of protein, bR and rhodopsine(s), the chromophore is a carotenoid, called retinal, shown in Figure 6.6, and consists of a partially saturated six C cycle at one end, which is attached to a polyene and bound to a "Schiff base", also called imine group, at the other end. A Schiff base is an amine group with a double C=N bond. It is covalently linked to a lysine group of the protein on its other side. In the ground state of the protein, this Schiff base takes on the form of a protonated cation (33), with a COO^- counterion, most probably that of a glutamate group in its vicinity. In this form, the chromophore absorbs in the visible region at wavelengths greater than 440 nm, whereas, in its neutral form, it is transparent in the visible region, hardly absorbing wavelengths greater than 400 nm. Let us note that the absorption band of retinal in the visible region is also sensitive to the protein of which it is a part. It is always a somewhat broad band that exhibits a maximum at 568 nm when this protein is bR. This maximum varies from 425 nm in human cone cells of the retin that are sensitive in the blue, to 530 nm for those sensitive in the green and to 560 nm for those sensitive in the red (31). These three kinds of cone cells, which are responsible for bright-light colour vision, have the same retinal + Schiff base chromophore, but slightly different "opsins", the rhodopsin to which the protonated Schiff bases are linked. The rhodopsin of the rod cells, responsible for dim-light vision, is also somewhat different, the absorption in these cells occurring around 500 nm (33). It shows that optical spectroscopy in the visible region is a precious tool to characterize these very similar proteins, which are parts of the system that perform their own particular task.

As schematically shown in Figure 6.7, the retinal chromophore is deeply embedded inside the protein itself, to which it is linked by covalent bonds on the side chain of a lysine residue, itself being a part of an α-helix. In its ground state the retinal of bR takes on its *all-trans* conformation, shown in the middle diagram part of Figure 6.6. Upon absorption of a visible

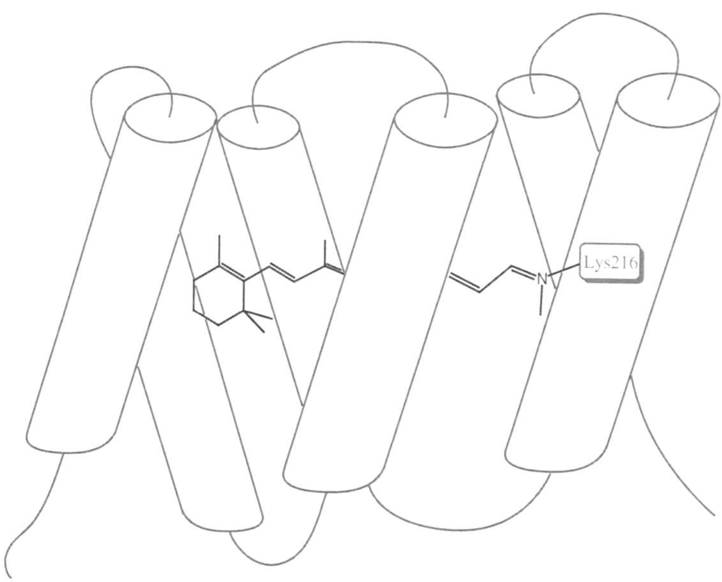

Figure 6.7 Schematic structure of bR, showing the retinal chromophore and its six transmembrane α-helix regions represented as cylinders.

photon, it suffers a photoisomerization; and the stable conformation of the retinal in its excited state that follows absorption of a visible photon, is shown in the lower part of Figure 6.6, a 13-*cis* conformation. This conformation then allows an H-bond to be established by the N^+–H group of the Schiff base on the O-atom of the carboxylic COOH group located on the side chain of an aspargine residue of bR. The appearance of this H-bond is depicted in the upper and middle page, right-hand side drawings of Figure 6.8. It triggers a series of proton transfers that starts in the newly established N^+–H···OH-bond, proceeds in the cytosol (the aqueous medium inside the cell) through a series of connected H-bonds forming a "proton wire". As described in Ch. 11, the structure and role of H_2O molecules within this proton wire have been firmly established by X-ray crystallography (34, 35). It ends with the appearance of a cation, represented in Figure 6.8 as H_3O^+, which is bound to the membrane, at the outer side of the cell, in the cytoplasm. After that, other steps allow the protein to retrieve its initial conformation. These steps that close the loop are not shown in Figure 6.8, except for the bottom-left drawing that represents the energetic aspect of the process. The energy gain is positive, even after the retinal has retrieved its original *all-trans* configuration, having consequently broken the H-bond it had established with its aspargine neighbour, and has been reprotonated, and is hence ready for a new cycle. This energy is stored before being used to synthesize ATP, or saccharides in the case of plants. A very similar mechanism occurs for vision, with slight differences; in the case of rhodopsin, the stable conformation of retinal is an 11-*cis* conformation, which is only slightly different from the 13-*cis* conformation shown in the lower right part of Figure 6.6 for bR. After absorption of a photon, the photo-induced conformation is *all-trans* (33), the opposite situation of retinal in bR.

We see that these proton transfers that occur in photosystems exhibit some similarities with the mechanism we have seen for photoacids. Thus the relaxation of the structure established just after absorption of the photon towards a more stable structure with a lower energy is due to proton transfers in both kinds of systems. Differences also appear. In photoacids, H-bonds are already established when the photon is absorbed, which is not the case in bR or rhodopsines. Also, the H_3O^+ ions, which appear in the courses of both mechanisms, behave differently: in photoacids, they freely diffuse in the solvent, which is most of the time water, and can consequently modify its ionic equilibrium (its pH in the case of water). In photosynthetic or vision membranes, they remain bound to the membrane and generally do not escape into water, which would have the result of modifying this equilibrium. The energy storage is then due to the appearance of a proton-based electric potential gradient across the membrane. Another difference lies in the relative complexity of the mechanism of proton transfer in these membranes. We point here to the systematically different situations and methodologies that are encountered in physics and chemistry on one hand, and biology on the other hand. In the case of a photoacid, two types of proton transfers occur, ionization of the phenol (naphtol) photoacid and diffusion of H_3O^+ in water. In bR, many more types of proton transfer appear that require the existence of a "proton wire", made of successive H-bonded O–H and N–H groups of water, alcohol, amine, etc. groups. We shall encounter this concept of a proton wire in Ch. 10 devoted to water and macromolecules, where we shall see that the H_2O molecule is indeed the ideal molecule to allow these proton transfers.

The relative complexity of the proton transfer mechanism in bR or other photosensitive proteins explain why the mechanism of photosynthesis in plants is much less understood than that performed by *H. salinarum*, a bacterium that is much less familiar than green leaves,

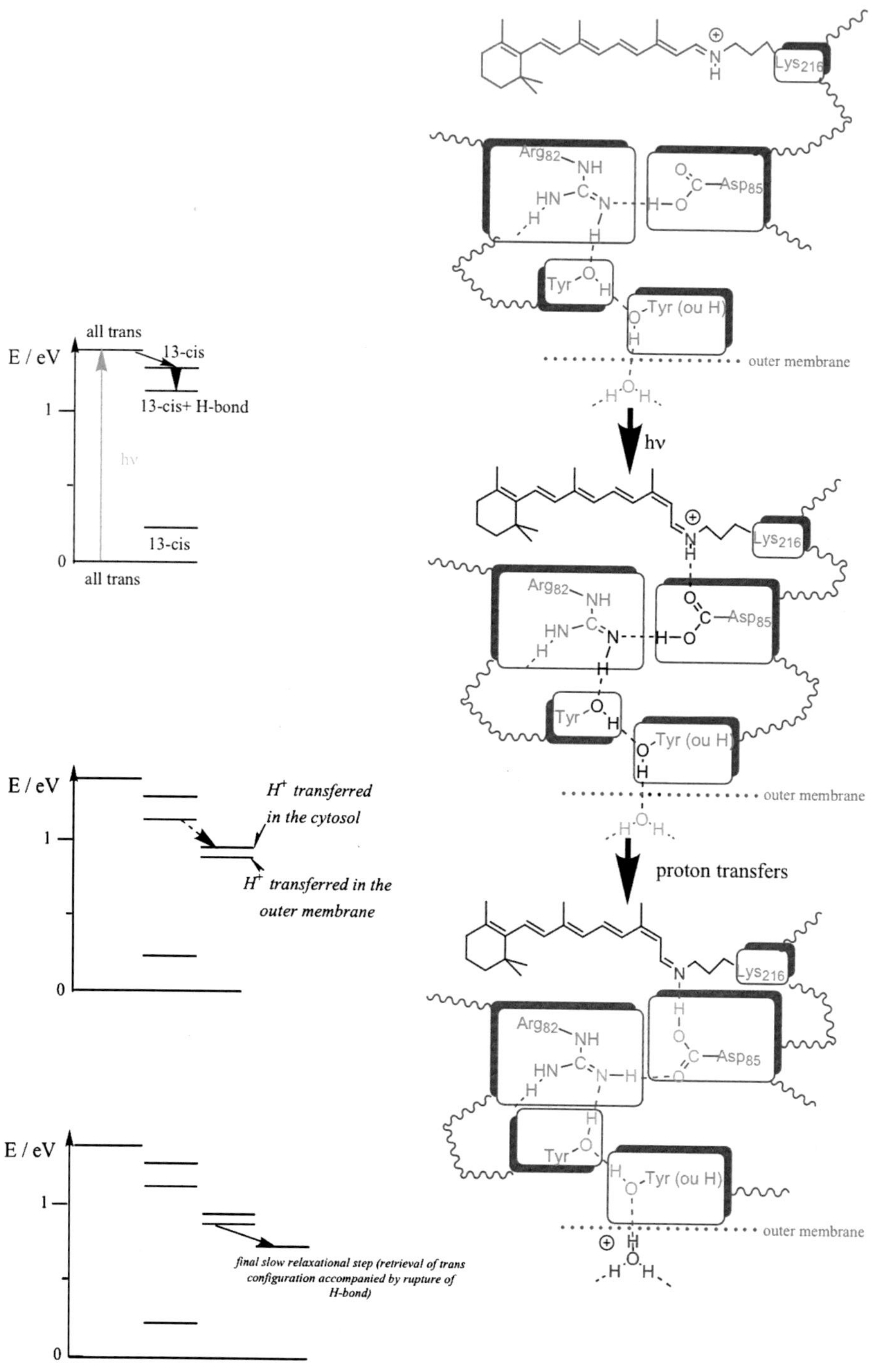

Figure 6.8 The proton transfer in bR that follows absorption of a visible photon ($h\nu$). Upper right drawing: scheme of bR before absorption. Middle drawing: bR just after absorption. Lower drawing: final state after proton transfers. The full retinal part of bR is drawn, whereas only residues of bR that take part in the proton transfer are shown within frames. Undulations represent part of the backbone of the protein. The energetic developments of the reactions are shown in a qualitative way in the left part.

but is simpler and much easier to handle. Compared to mechanisms that occur in chemistry, it is fairly well complicated, and we come to understand why the quest for most simple systems is a necessity. We may wonder how these various steps can be put into evidence. As already stated, the photosynthetic scheme in bR could be proposed after a series of various experiments, based on various methodologies and the use of various tools. The characterization of all steps, part of them being displayed in Figure 6.8, is basically performed by spectroscopy in the visible-UV region. We have seen that the absorption bands of retinal in the visible region are sensitive to the protein (the type of opsin) they are bound to. For a given protein they also depend on the conformation of the protein. It allows us to follow the various steps. These steps are transient, and in order to be characterized, they have to be isolated and stabilized. This is done by varying such parameters as the pH of the aqueous solution, or its temperature. At low pH, for instance, the last step, transfer of the proton in the outer solution will be more difficult than in neutral pH. After illumination by a flash that starts the photosynthesis, this last step does not occur, and the bR is consequently stopped in the conformation found before this last step. This conformation can then be characterized by IR spectrometry (29), or X-ray (33, 36) scattering. These two methods convey the most useful structural information. The analysis of IR spectra, which we have seen (Ch. 4) to be highly sensitive to H-bonds, mainly allows us to follow the proton along the "proton wire", while X-rays allow us to define which amino acids are involved in it. Illuminating bR at various temperatures, which fall between 77 K and room temperature, has also the effect of quenching the bR membrane into various intermediate states. At very low temperatures, only the first step, isomerization of retinal, occurs. Increasing the temperature allows us to go further into the process. Also, IR spectroscopy, X-ray scattering and occasionally other techniques allow us to characterize and analyze these numerous intermediate states.

The determination of this somewhat complex mechanism is presently one of the most accomplished results of biochemistry and biophysics. This is the reason why we described it in some detail. It shows that proton transfers may be of importance in biology. It will also raise many questions in case we want to extend these results to other biomechanisms, which most certainly will turn out to be even more complex. First, questions concern the structure of the proton path itself. Is the proton wire made of O–H and N–H groups of amino acids, such as tyrosine, which has been proposed to be part of it (37), or is it mainly a "water wire" made of O–H groups of H_2O molecules, some of which have been localized by X-ray experiments (38) despite the intrinsic difficulties of X-ray to observe such small and labile molecules as H_2O (Ch. 11), or the difficulty of IR to distinguish an H-bond established by the O–H group of an H_2O molecule from that established by an alcohol group, or both of them? Is this proton wire specific to bR and other rhodopsines, or is it a generally established structure of aqueous biomedia? How does the flexibility of this proton wire influence these proton transfers, and how could we predict the effects of temperature on these proton transfers, particularly the position of the quenched protonated structure when T is varied? Many questions concerning the dynamics of the proton also appear unanswered. One of them is whether the proton transfer is stepwise, or is a concerted transfer of several protons on various adjacent sites, or a mixture of both these mechanisms. This question has been theoretically studied in the case of such enzymatic reactions as H_2 ablation from alcohol in liver (39). The conclusion was that the transfer was stepwise, but a high level is required for the treatment of the electronic structure of the molecule to give such an answer (40). It means that this important question will certainly not receive any clear answer soon.

More generally, the mechanism of these proton transfers along proton wires is yet unknown. Most probably they are nonadiabatic proton transfers, in opposition to the adiabatic proton transfers we have seen in the case of ionization of a strong acid. This nonadiabaticity means that the basic mechanism of proton transfer is a tunnelling of the proton through the H-bond. The barrier height of the corresponding double-well potential remains sufficiently high even if modulated by solvent fluctuations or O···O distances of the H-bond and so the vibrations require some time to adapt themselves to the new position of the proton after tunnelling.

H-BONDED FERROELECTRICS

In the preceding subsections, we have encountered proton transfers that are at the origin of fundamental mechanisms of chemistry. We have also seen ESPTs, at the origin of photoacids in chemistry, which have a special importance in biology, as they constitute the central mechanism of photosynthesis or vision. In this subsection, we consider a transfer that has far fewer general consequences, but these consequences are somewhat original and have immediate applications, in electronics for instance. These transfers occur in crystals of potassium dihydrogen phosphates, KH_2PO_4, often abbreviated as KDP, or related crystals. KDP is ferroelectric at temperatures below 122 K. It means that it displays a spontaneous electrical polarization $\vec{\mu}$ that can be reversed by applying an electric field in the opposite direction (41). This spontaneous polarization is not a direct consequence of H-bonding, but the transition between the ferroelectric and paraelectric phases at 122 K is an order–disorder transition induced by proton transfers through the H-bonds of this crystal, which themselves induce a small displacement of both the anions and cations that are responsible for the polarization. It is indeed one of the first manifestations of proton transfers that have been put into evidence. The key role played by these proton transfers appears in the great difference of the temperature of the ferroelectric/paraelectric phase transition, which increases up to 222 K when H-atoms are replaced by D-atoms.

This KDP crystal consists of an assembly of $H_2PO_4^-$ anions that are linked together by H-bonds and of K^+ cations that are positioned between these anions. Along the $\vec{c}$ axis, anions and cations alternate. The four O-atoms of an anion occupy the four summits of a tetrahedron around the P-atom (Figure 6.9). Two H-atoms are covalently bound to two of the four O-atoms of a tetrahedron, and establish H-bonds with two O-atoms of two neighbouring tetrahedra. In the ferroelectric phase, both H-atoms of each PO_4 tetrahedron are covalently linked to the two O-atoms of the tetrahedron that have largest component along $\vec{c}$, as represented in the left part of Figure 6.9. In a first approximation, the O–H bonds are perpendicular to this $\vec{c}$ axis. The two other O-atoms of a tetrahedron accept H-bonds from two neighbouring tetrahedra. The coordinate along $\vec{c}$ of the P-atom is slightly displaced from that of the centre of the tetrahedron. That of the K^+ cation that lies between two tetrahedra is also displaced, in an opposite direction, from that of the middle point between these two tetrahedra. It results in an electrical dipole moment $\vec{\mu}$ parallel to $\vec{c}$.

Tunnelling of protons through their H-bonds is at the origin of disorder. This tunnelling is phonon-assisted, that is, they appear more easily when vibrational modes of the crystal, called phonons, are excited. When temperature increases from 0 K, such low-frequency phonons

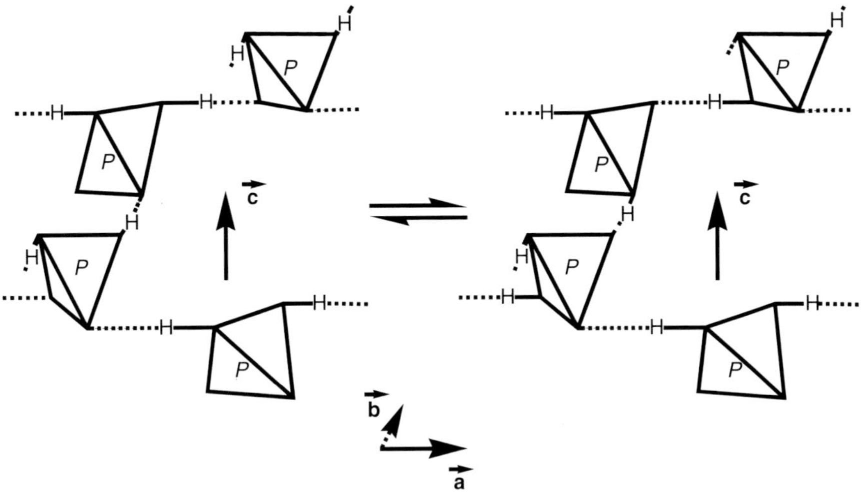

Figure 6.9 The H-bond configuration in the ferroelectric phase of KDP (left). New configuration after transfers of protons (right). Cations are not represented. O-atoms are neither represented, but they are positioned at all summits of tetrahedra. Axis $\vec{c}$ is the vertical axis in the plane of the figure. Axis $\vec{a}$ is horizontal and also in the plane of the figure, while axis $\vec{b}$, perpendicular to the plane of the figure, is drawn in perspective.

may become excited. Transfers of protons are therefore favoured by an increase of temperature. They transform the ferroelectric H-bond configuration with a permanent $\vec{\mu}$, shown in the left part of Figure 6.9, into another configuration, shown in the right part of the same figure. This new configuration, as any other possible configuration, obeys a law similar to the "ice rule" defined by L. Pauling for the distribution of H-atoms around each O-atom in ice (Ch. 8), also known as "Slater rule". Presently, this rule states that two H-atoms are covalently bound to two of the four O-atoms of each anion $H_2PO_4^-$, and that each of the two O-atoms that is not covalently bonded to an H-atom accepts one H-bond from the H-atom of a neighbouring anion. In this configuration, the coordinates along $\vec{c}$ of the centres of charges of the anionic tetrahedra coincide with those of their geometrical centres, and also with those of the K^+ cations that lie close to them and suffered a slight translation, a consequence of the changes of the positions of the H-atoms. In this new configuration, no permanent electrical dipole moment appears. The crystal may thus suffer a ferroelectric/paraelectric phase transition. Let us note that in the ferroelectric phase, reversing the permanent electrical moment by applying a strong enough electric field also implies transfers of protons, more precisely of all protons.

As already noted, various crystals similar to KDP, where the K-atom or the P-atom of the anion is replaced by homologous atoms, display a similar ferroelectric behaviour. Some of them are antiferroelectrics (41). Recently, some other types of ferroelectrics have been discovered, where the origin of their ferroelectric behaviour is due to the position of the proton itself, which is not the case for KDP-type ferroelectrics where H-bonds are roughly perpendicular to the permanent electric dipole moment $\vec{\mu}$ in the ferroelectric phase and do not directly contribute to the permanent electric moment. Thus such a crystal as

1-4 diazabicyclo[2.2.2]octane perrhenate (42) displays chains with N–H···N groups arranged in a zigzag chain and separated by rigid assemblies of saturated cycles that exhibit a permanent electrical dipole moment, because they are all oriented in the same direction. After transfers of all protons through their H-bonds, which may be induced by a strong enough electric field, and which we may represent as a transition from the N–H···N configuration to the N···H–N configuration for all N–H groups, this permanent moment is reversed, giving the crystal ferroelectric properties.

HYDROGEN ATOM TRANSFERS BY TAUTOMERISM

Besides the proton transfers that we have seen in the preceding sections, which are at the origin of such fundamental reactions as acid/base reactions in chemistry, or photosynthesis and vision in biology, there exists another type of transfer through H-bonds, which is the transfer of H-atoms by "tautomerism". The distinction between these two types of transfers is not yet often highlighted, and the object of this section is to do so, because it has important consequences. The two "tautomer" configurations of a carboxylic acid dimer, or of a molecule such as formamidine, are shown in Figure 6.10. We have already met such an intramolecular tautomerization, due to transfer of an H-atom, in the case of malonaldehyde (Figure 6.2). H-atom transfers by tautomerism neither create charges, in opposition to proton transfers that cause ionization of an acid (Figure 6.4) or a base, nor transfer charges as proton transfers may do, for instance by diffusing H_3O^+ and OH^- ions (Figure 6.5). They may be viewed as transfers of protons that are, however, accompanied by a simultaneous transfer of electrons. In the case of malonaldehyde (Figure 6.2) or carboxylic acids (Figure 6.10), this transfer of electrons occurs between the two CO-bonds, the C=O bond becoming a C–O(H) bond and vice versa. In the case of formamidine, this transfer of electrons exchanges C=N with

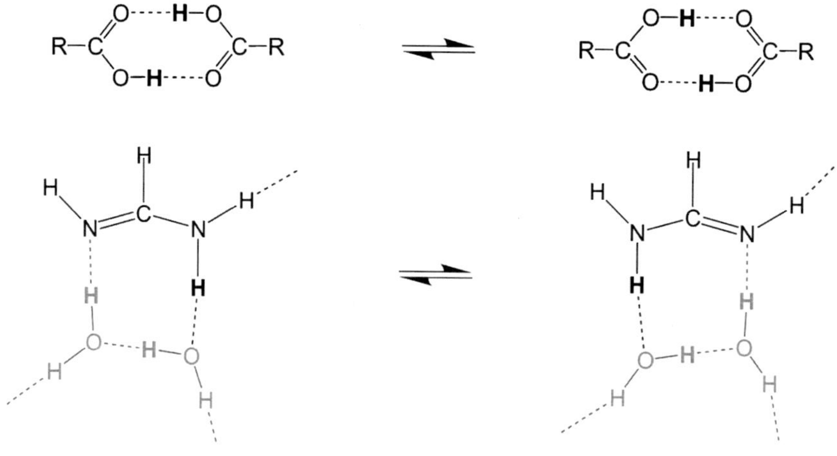

Figure 6.10 The two tautomer forms of a carboxylic acid dimer (upper drawing) and of formamidine H-bonded to two H_2O molecules (bottom). One passes from one form to the other one by H-atom transfers.

C–N(H) and vice versa. This simultaneous transfer of electrons requires that such an H-atom transfer (or transfers) occurs within a cyclic H-bonded structure. It points to the first difference between H-atom transfers and proton transfers, for which no such requirement exists.

These H-atom transfers by tautomerism are common mechanisms. They occur in liquid water at neutral pH (=7), as one easy-to-perform experiment shows: on mixing water with heavy water, we most rapidly obtain a mixture of H_2O, HDO and D_2O molecules, which can all be precisely characterized by IR spectroscopy. At this pH, the molar concentration of H_3O^+ and of OH^- ions is about 2×10^{-9}, which is nearly zero. Consequently, these ions cannot play any role in these exchanges. Such an H/D exchange furthermore occurs in other H-bonded liquids such as alcohols, where the ionic concentration is orders of magnitude smaller than in water. Furthermore CH groups, which, as discussed in Ch. 1 do not establish H-bonds, hardly become CD groups in a deuterated liquid. Consequently, such H/D exchanges in H-bonded liquids proceed via H-atom transfers through H-bonds by tautomerism, not via proton transfers. At pH 7, H-atom transfers are efficient mechanisms, whereas proton transfers are not efficient because of the absence of a sufficient concentration of H_3O^+ or OH^- ions. Proton transfers may be at the origin of the ionization of an acid or a base, which is accompanied by the appearance of such H_3O^+ or OH^- ions. In that case, they strongly modify the value of the pH, which consequently most often departs from 7 after such transfers. Another experiment also shows the efficiency of these H-atom transfers—at an ice surface, made of ordinary H_2O molecules, a D_2O molecule that links by establishing or accepting an H- or a D-bond does not exchange (43); it remains a D_2O molecule, whereas it exchanges when it is introduced into bulk ice thus becoming an HDO molecule. This is because no cyclic H-bonded structure exists on the ice surface, whereas plenty of them are present in the bulk, as appears in Figure 8.3, making these transfers at the origin of H/D exchanges efficient. In Ch. 10, these H-atom transfers by tautomerism are shown to be as fundamental mechanisms as proton transfers in chemistry, but in a different domain, biology, which occurs as in the above case of H/D exchange, in an aqueous medium at pH 7.

We have seen H-atom transfers require the presence of a cyclic H-bonded structure. This is a severe steric requirement, which may be fulfilled by a very few number of molecules, such as carboxylic acid, for instance, represented in Figure 6.10, which are found in vapours of carboxylic acids, and which have the exact required geometry to fulfil this condition in their dimer form. Such molecules are exceptional. It means that these steric conditions are so stringent that such a mechanism is impossible for a great majority of molecules. There exists, however, one exception, which has its own importance (44): the water molecule, H_2O, which we shall see in Chs. 9 and 10 is unique to form small clusters that develop around them such a dense and flexible H-bond network, that this condition is almost automatically fulfilled in any aqueous media. Biomedia are aqueous media, and this *sine qua non* condition of the existence of cyclic H-bonded structure for H-atom transfers to occur is no point in these media.

We presently know very little about these H-atom transfers. They have been experimentally put into evidence by sophisticated two-photon methods in the visible-UV region (45–47), completed by NMR and neutron scattering methods (Ch. 3), applied to cyclic dimers found in benzoic acid crystals (48). These dimers are those of the upper drawing of Figure 6.10, with R being a benzene ring. Impurities of thioindigo molecules, which occupy the same volume as a dimer, can be introduced in these crystals with no perturbation of the surrounding dimers. They act as sensitive probes for these H-atom transfers, which may be observed by UV spectroscopy. These experiments have clearly shown a significant dependence of these transfers on temperature (49), an indication that the low-frequency modes of these dimers, most certainly intermonomer modes of the H-bonds (Ch. 4), strongly influence the tunnelling of the H-atoms. The frequency of these transfers has also been directly measured, and found equal to about $2 \times 10^8\,\mathrm{sec}^{-1}$ at 20 K and about $10^{11}\,\mathrm{sec}^{-1}$ at 300 K (48). This is apparently the only measurement that has been performed of this crucial quantity. The existence of these transfers and their fundamental role require knowing them in more detail. An important question about these transfers in such simplified model is whether the multiple transfers are simultaneous or simply correlated? INS (inelastic neutron scattering) experiments (3, 50) seem to indicate they are not correlated, in these most studied carboxylic acid dimers, while *ab initio* studies found these transfers to be strongly correlated, even if somewhat delayed by some fsec (51). In the case of other dimers that model the A-T (adenosine-thymine, see Ch. 2) H-bond in DNA the H-atom transfer has also been found by *ab initio* methods to be concerted but delayed, in the absence of water, but stepwise with a zwitterionic intermediate in water (52). These examples illustrate how far we still are from being able to make any prediction concerning these mechanisms, even in these simple models. Theoretical approaches have been carried out to get a deeper understanding of them. They treat the great majority of these transfers as nonadiabatic, that is, following Borgis and Hynes (20), their basic mechanism is a tunnelling governed by a double-well potential with a barrier height much greater than kT at room temperature. A small number of them, intramolecular transfers in the excited state that might be called "phototautomerism", are adiabatic (53). We have seen that the tunnelling is strongly modulated by intermonomer modes, which we have represented by the coordinate Q, and also certainly by other modes of the surrounding medium. As we see it, because of the numerous parameters that influence it (we have seen some of them) it is not so easy to handle. The problem is that, in order to convey original information, these theoretical methods should give accurate results. Adiabatic transfers are quantum in essence. Full quantum treatment is possible for a maximum of five atoms. This is insufficient for most chemical systems, not to speak of biosystems, and approximations should consequently be introduced to describe these transfers. New methods, based on approximations such as the "instanton approach", seem promising (54, 55). As defined by Benderskii *et al.* in 1993 (56), the instanton path, depicted in Figure 6.11, is the least action trajectory, which consists of reaching, with the help of thermal fluctuations, some excited vibrational state from which tunnelling occurs. It is an optimum path, intermediate between a direct tunnelling and a classical path. The direct tunnelling occurs between the two configurations defined by $q = q_0$ and $q = q_1$, with no change in Q, and occurs mainly when the energy E_f of the second minimum is smaller or equal to that of the first minimum, taken as 0 in Figure 6.11. The barrier height for this tunnelling, E_t, is, for most systems, so high, or equivalently, the distance $q_1 - q_0$ is so great, that this direct tunnelling has a very low probability to occur. The classical path, drawn as a dash line, consists of reaching the pass with energy slightly smaller than E_c. It is also unlikely, because E_c is most often far greater than thermal fluctuations. Let us note that the energy E_c required to make this classical path feasible is smaller than the barrier height E_t of the direct tunnelling. This is a consequence of the strong q, Q anharmonic coupling at the origin of the most spectacular effects H-bonds display in IR spectroscopy, which modifies the shapes of the equipotential lines. With pure harmonic vibrations, these lines would all be ellipses centred around q_0 and q_1 with axes

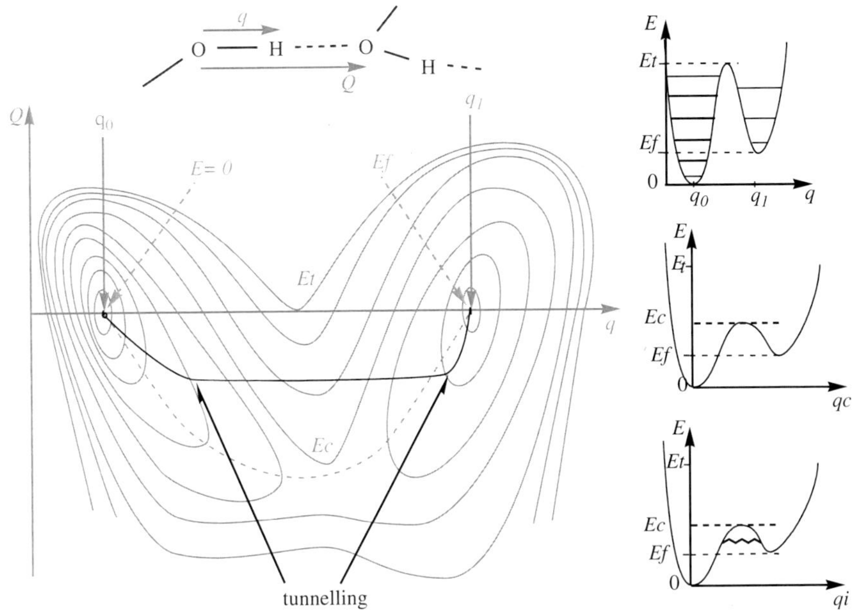

Figure 6.11 Vibrational equipotential curves for the H-bond defined by its two degrees of freedom q and Q in the left diagram. The classical path is drawn as a dashed line, and the instanton path as a bold line. For this path tunnelling occurs between the two arrows. Energies of some equipotential curves are indicated. Energetics of the different paths is shown on the right part. Upper right: path along q, Q remaining equal to Q_0. Middle right: classical path (reaction coordinate q_c). Bottom right: instanton path (reaction coordinate q_i), with tunnelling part indicated with a zigzag line.

parallel to q and Q axes. E_c would then be equal to E_t. The instanton path consists of a tunnelling from an already excited state. The tunnelling distance is markedly shorter than $q_1 - q_0$, with a smaller barrier height, and becomes consequently feasible. Meanwhile, the energy of the excited state from which the tunnelling starts may now be within reach of thermal fluctuations, which makes the whole transfer possible. Note that such a mechanism predicts a strong H/D isotope effect, as a D-atom with mass $m_D = 2m_H$ will require much longer time to tunnel through a potential barrier than a H-atom. This is sufficient to explain why life occurs in ordinary water, but is hardly possible in heavy water, as described in Chs. 7 and 10.

The short description of these H-atom transfers given above, clearly shows the need to know more about them. Detailed studies of H/D exchanges in H-bonded liquids such as water, alcohol, etc. by IR spectrometry described in Ch. 11 might be an interesting source of information on them. Such experiments have not yet been reported. Also, tautomerism in the excited state, which exists in much the same way as ESPT, at the origin of photoacids we have seen above, might be interesting mechanisms to study. They occur in such cyclic dimers as found with 7 azaindole, a model for the H-bonded DNA base pairs and their photosensitivity. This "phototautomerism" could be experimentally shown to be a cooperative mechanism (57), in opposition to the stepwise one we have seen for proton transfers, by showing that deuteration of one H-bond or of both H-bonds considerably modified the transfer rate.

CONCLUSION

The third fundamental property of hydrogen bonds, the possibility they offer to transfer protons or H-atoms between the two molecules they link, is a fundamental molecular mechanism in chemistry and the basis of the reactivity of aqueous media at pH = 7, particularly biomedia. Incidentally, it also appears in physics, being for instance responsible for the ferro/paraelectric transitions in some ferroelectric materials as KDP. In ordinary chemistry, transfers of protons are at the origin of acid/base reactions in liquid water, a vast domain. They also appear more occasionally, as for instance, in the chemistry of the atmosphere. In biology, transfers of H-atoms by tautomerism appear as the molecular mechanism at the origin of bioreactivity, particularly enzyme catalysis, as discussed in more detail in Ch. 10. It may be viewed as a transfer of several protons accompanied by simultaneous electron transfers that avoid appearance of space charges and make it a mechanism that can occur in a neutral (pH 7) medium and does not modify the value of this pH. It requires a cyclic H-bonded structure to occur, such as drawn in Figure 6.10, a most severe sterical condition for nearly all molecules, except for that small, labile and versatile water molecule, H_2O, which displays an exceptional ability to develop around it an extended H-bond network inside which this condition is almost automatically fulfilled.

Despite their importance, transfers of protons or of H-atoms still remain poorly known mechanisms at molecular level. In such a simple system as that made of a single HCl molecule mixed with some H_2O molecules, the question as to how many H_2O molecules are necessary to ionize the HCl molecule, has not yet received a definitive answer. The study of model systems shows that these transfers are sensitive to many different parameters. Two of them appear general: the intermolecular distance Q of the H-bond (Figure 6.11, for instance) and the structure of the surrounding H-bond network, which may influence this transfer in various ways. Theoretical approaches of these transfers have distinguished two extreme classes—adiabatic transfers, the first class, where the tunnelling barrier is lower than the energies of thermally populated vibrational states. The transfer is a simple relaxational vibrational mechanism, which is highly sensitive to the surrounding H-bond network. The influence of this surrounding H-bond network has a slow enough dynamics for ν_s and other vibrations to adapt themselves immediately to the changes it induces. Hence, the term "adiabatic". Ionization of HCl by water molecules, drawn in Figure 6.4, is a good representative of this category. The second class, nonadiabatic transfers, is one where the basic transfer mechanism is a tunnelling, which is ruled by a potential having a barrier height significantly greater than the energies of thermal fluctuations. Generally, this potential strongly depends on Q, and may also depend on the H-bond network surrounding the hydrogen bond. In this nonadiabatic mechanism, the ν_s vibration and all other vibrations require some time to adapt to the new positions of the atoms after tunnelling has occurred. Much work still remains to be done on these transfers in chemistry or physics, before they are understood in the much more complex biosystems. Thus simple experiments that permit the measurement of their probability, such as time-resolved (IR) spectroscopy, are strongly needed, before their role, which is often crucial in biology, as described in Ch. 10, can be fully evaluated. Also spectroscopic studies, mainly by IR, of mechanisms such as H/D exchanges in aqueous media, or hydrolysis of esters or of oligopeptides may convey original information on H-atom transfers.

In this chapter, we have evocated proton transfers in chemistry (acid/base ionization) and in biology (photosynthesis). It gives us the opportunity to compare the conditions encountered and methodologies used in both these fields. In chemistry, or physics, we have to do with relatively well-defined systems, which depend on a limited number of parameters, such as the concentration of acid (base) in water. A chemical reaction is very similar to walking from one valley to another, with elevation taking the role of energy. In ordinary chemistry, the most direct way is chosen, a pass between these two valleys. Relatively, precise measurements can then be performed, using a small number of techniques, which can be fairly well exploited, often in a quantitative manner. The conditions in biosystems do not allow reaching such a pass, which requires too much energy for an aqueous medium at room temperature. Reactions then comprise resting on the valleys, where the elevation (energy) does not change very much, and passing through the vicinity of the junction of both valleys. This way, used by biosystems is not direct and much longer and depends consequently on a greater number of parameters, which cannot all be controlled and even known. Thus, even if the primary and secondary structures of bR, the "light harvesting system" in photosynthesis, are known, the tertiary structure of bR is still elusive, the position of water molecules can only be guessed, etc. In these conditions, establishing a mechanism of photosynthesis or vision, requires performing a far greater number of experiments, using a great variety of techniques. Each measurement conveys limited information, as in all these various experiments the parameters do not remain unchanged and cannot all be controlled. It limits the exploitation of each experiment. On the other hand, the great number of experiments that can be carried out compensates this lack of precise measurements. It requires having the capacity of assembling a great number of results, which should be analyzed in such a way that a consistent scenario emerges from these experimental results.

REFERENCES

1. C. Sandorfy, in Schuster, G. Zundel and C. Sandorfy (Eds.), The Hydrogen Bond: Recent Developments in Theory and Experiments, Vol. II, Ch. 13, North Holland, Amsterdam, 1976.
2. F. Fillaux and J. Tomkinson, Chem. Phys., 158 (1991) 113.
3. F. Fillaux, J. Mol. Struct., 615 (2002) 45.
4. X. Krokidis, V. Goncalves, A. Savin and B. Silvi, J. Phys. Chem. A, 102 (1998) 5065.
5. M. Rini, D. Pines, B.-Z. Magnes, E. Pines and E. T. J. Nibbering, J. Chem. Phys., 121 (2004) 9593.
6. B. S. Ault and G. C. Pimentel, J. Phys. Chem., 77 (1973) 57.
7. A. C. Legon and L. C. Willoughby, Chem. Phys. Lett., 95 (1983) 449.
8. N. W. Howard and A. C. Legon, J. Chem. Phys., 88 (1988) 4694.
9. M. Farnik, M. Weiman and M. A. Suhm, J. Chem. Phys., 118 (2003) 10120.
10. L. Andrews, X. Wang and Z. Mielke, J. Phys. Chem. A, 105 (2001) 6054.
11. A. J. Barnes, T. R. Beech and Z. Mielke, J. Chem. Soc., Faraday Trans. 2, 80 (1984) 455.
12. R. A. Cazar, A. J. Jamka and F.-M. Tao, J. Phys. Chem. A, 102 (1998) 5117.
13. M. Biczysko and Z. Latajka, Chem. Phys. Lett., 313 (1999) 366.
14. S. Xu, M. Nilles and K. H. Bowen, J. Chem. Phys., 119 (2003) 10696.
15. R. Ramaekers, J. Pajak, B. Lambie and G. Maes, J. Chem. Phys., 120 (2004) 4182.
16. A. Fernandez-Ramos, Z. Smedarchina, W. Siebrand and M. Z. Zgierski, J. Chem. Phys., 113 (2000) 9714.

17. W. Zhuang and C. Dellago, J. Phys. Chem. B, 108 (2004) 19647.
18. L. J. Larson, A. Largent and F.-M. Tao, J. Phys. Chem. A, 103 (1999) 6786.
19. K. Ando and J. T. Hynes, J. Phys. Chem. B, 101 (1997) 10464.
20. D. Borgis and J. T. Hynes, J. Chem. Phys., 94 (1991) 3619.
21. B. Cohen and D. Huppert, J. Phys. Chem. A, 107 (2003) 3598.
22. P. M. Kiefer and J. T. Hynes, J. Phys. Chem. A, 108 (2004) 11793.
23. P. M. Kiefer and J. T. Hynes, J. Phys. Chem. A, 108 (2004) 11809.
24. B. R. Ringeisen, A. H. Münter and G. Nathanson, J. Phys. Chem. B, 106 (2002) 4988.
25. N. Agmon, J. Phys. Chem. A, 86 (2001) 13.
26. C. Chudoba, E. Riedle, M. Pfeiffer and T. Elsaesser, Chem. Phys. Lett., 263 (1996) 622.
27. M. Rini, A. Kummrow, J. Dreyer, E. T. J. Nibbering and T. Elsaesser, Faraday Discuss., 122 (2002) 27.
28. P. F. Moreira, Jr., L. Giestas, C. Yihwa, C. Vautier-Giongo, F. H. Quina, A. L. Maçanita and J. C. Lima, J. Phys. Chem. A, 107 (2003) 4203.
29. A. Maeda, J. Sasaki, Y. Shichida and T. Yoshizawa, Biochemistry, 31 (1992) 462.
30. J. Wang and M. El-Sayed, Biophys. J., 80 (2001) 961.
31. S. Hayashi, E. Tajkhorsid, E. Pebay-Peroula, A. Royant, E. M. Landau, J. Navarro and K. Schulten, J. Phys. Chem. B, 105 (2001) 10124.
32. L. Schalchli, La Recherche, 360 (2003) 91.
33. R. Vogel and F. Siebert, Biopolymers (Biospectroscopy), 72 (2003) 133.
34. H. Lücke, B. Schobert, H. T. Richter, J. P. Cartailler and J. K. Lanyi, J. Mol. Biol., 291 (1999) 899.
35. H. Belrhali, P. Nollert, A. Royant, C. Menzel, J. P. Rosenbusch, E. M. Landau and E. Pebay-Peroula, Structure, 7 (1999) 909.
36. J. K. Lanyi and B. Schobert, J. Mol. Biol., 328 (2003) 439.
37. B. Brzezinski and G. Zundel, Faraday Discuss., 103 (1996) 363.
38. J. Baudry, E. Tajkhorshid, F. Molnar, J. Phillips and K. Schulten, J. Phys. Chem. B, 105 (2001) 905.
39. Q. Cui, M. Elstner and M. Karplus, J. Phys. Chem. B, 106 (2002) 2721.
40. Q. Cui and M. Karplus, J. Phys. Chem. B, 107 (2003) 1071.
41. V.H. Schmidt, in Schuster, G. Zundel and C. Sandorfy (Eds.), The Hydrogen Bond: Recent Developments in Theory and Experiments, Vol. III, Ch. 23, North Holland, Amsterdam (1976).
42. M. Szafranski, A. Katrusiak and G. J. McIntyre, Phys. Rev. Lett., 89 (2002) 215507.
43. M. A. Everest and C. J. Pursell, J. Chem. Phys., 115 (2001) 9843.
44. Y. Maréchal, J. Mol. Struct., 700 (2004) 207.
45. J. M. Clemens, R. M. Hochstrasser and H. P. Trommsdorff, J. Chem. Phys., 80 (1984) 1744.
46. C. Rambaud, A. Oppenländer, M. Pierre, H. P. Trommsdorff and J.-C. Vial, Chem. Phys., 136 (1989) 335.
47. A. Oppenländer, C. Rambaud, H. P. Trommsdorff and J. C. Vial, Phys. Rev. Lett., 63 (1989) 1432.
48. M. Neumann, D. F. Brougham, C. J. McGloin, M. R. Johnson, A. J. Horsewill and H. P. Trommsdorff, J. Chem. Phys., 109 (1998) 7300.
49. L. Torres, M. Moreno and J. M. Lluch, Chem. Phys. Lett., 340 (2001) 591.
50. F. Fillaux, M. H. Limage and F. Romain, Chem. Phys., 276 (2002) 181.
51. H. Ushiyama and K. Takatsuka, J. Chem. Phys., 115 (2001) 5903.
52. Y. Podolyan, L. Gorb and J. Leszczynski, J. Phys. Chem. A, 106 (2002) 12103.
53. S. Lochbrunner, A. J. Wurzer and E. Riedle, J. Phys. Chem. A, 107 (2003) 10580.
54. Z. Smedarchina, W. Siebrand and A. Fernandez-Ramos, J. Chem. Phys., 112 (2000) 566.
55. A. Fernandez-Ramos, Z. Smedarchina and J. Rodriguez-Otero, J. Chem. Phys., 114 (2001) 1567.
56. V. A. Benderskii, D. E. Makarov and P. G. Grinevich, Chem. Phys., 170 (1993) 275.
57. K. Sakota and H. Sekiya, J. Phys. Chem. A, 109 (2005) 2722.

– 7 –

H/D Isotopic Substitution in H-Bonds

THE H AND D ATOMS: SIMILARITIES AND DIFFERENCES

The deuterium nucleus D, or deuteron, labelled ^{2}H when using the general isotope nomenclature, is a stable isotope of the proton that is encountered in a proportion that falls in the vicinity of 10^{-4} that of protons or H-nuclei. H-atoms that establish H-bonds can be easily exchanged into D-atoms because, as seen in Ch. 6, protons or H-atoms can be relatively easily transferred through H-bonds. Immersing a sample that contains O–H or N–H groups into liquid heavy water consequently often provokes an O–H/O–D or N–H/N–D substitution on most of these groups of both sample and water. The H-atoms of groups that can establish H-bonds with water molecules or with molecules of other solvents are thus called exchangeable H-atoms, in opposition to H-atoms that cannot establish H-bonds. C–H groups, for instance, are much more difficult to transform into C–D groups, as they do not establish H-bonds. The H-atoms of these groups are nonexchangeable H-atoms. This opportunity that only H-bonds offer makes such a substitution an interesting tool in molecular studies. It requires, however, precise knowledge of the effects of such a substitution on the properties of H-bonds or, in other words, what the differences are between H-bonds and D-bonds. This is the object of this chapter where the interest of performing such an exchange and the kind of information it conveys when observed using appropriate experimental methods is also examined.

The origin of the differences between H-bonds and D-bonds lies in the structure of the nuclei of H- and D-atoms, which is a single proton for the H-atom, and a proton plus a neutron for the D-atom. This has several consequences. First, the mass of the deuteron is twice that of the proton. Second, the spin of the deuteron, made of two particles, a neutron and a proton that have each a spin ½, has a spin 1 different from the spin ½ of a proton. The electric charge of a deuteron is conversely identical to that of the proton. It implies that the electronic structure of the D-atom, which is sensitive to the charge of the D-nucleus but not to its spin, is identical to that of the H-atom. This result itself implies that the electronic structure of a D-bond is identical, *for the same geometry of the nuclei*, to the electronic structure of its equivalent H-bond. Does it imply that a D-bond is identical to an H-bond? No, because if some properties of H-bonds mainly depend on their electronic structures, and are consequently barely sensitive to the doubling of the nuclear mass that is encountered when passing from an H-bond to a D-bond, other ones, mainly dynamic properties, are much more sensitive to such a doubling and may consequently be at the origin of important changes upon an H/D substitution. The object of the next sections is consequently to examine first the properties of H-bonds that are weakly sensitive to this mass effect, second those that display important changes after such a substitution, before examining those that display dramatic changes. Some experimental methods moreover probe the spins of the nuclei. For them a D-bond is consequently different from an H-bond. How this point can be exploited is examined in the last section before the conclusion.

GEOMETRIES AND THERMODYNAMICS OF H-BONDS AND D-BONDS

In order to understand how the mass m_H of the H-atom influences the geometry, the thermodynamics and the vibrational properties of H-bonds, and how it modifies them when it is changed into $m_D = 2m_H$, we develop the basic equations that rule the mechanics of H-bonds in the appendix of this chapter. We exploit them below in this section and in the next one. This organization should allow a reader who is more interested in established results than in demonstrations to read the subsequent sections without missing important points. The reader who does not accept only results but expects more rigorous explanations may find them in this appendix. These equations explicitly take into account the characteristic and strong anharmonic modulation by intermonomer distances of the stretching vibration of the H-atom that establishes the H-bond. The basis of this modulation, together with some of its implications, has already been established in the appendix of Ch. 5, which we often refer to in the next sections.

Geometries of H-bonds and D-bonds

The simulations of bandshapes of ν_s modes of H-bonds, such as displayed in Figure 5.2, clearly demonstrate that the dynamics of H-atoms of H-bonds is well taken into account if one simply supposes that the frequencies of their stretching vibrations are modulated by slow intermonomer modes. This modulation is well accounted for by the vibrational potential (eq. (7.A5)) in the appendix of this chapter, which is the potential that enters the vibrational Hamiltonian that has been defined in eq. (5.2) when the only considered intermonomer mode was the stretching mode Q_s. The consequences of this modulation appears in eq. (7.A11) in this appendix, which clearly shows that the equilibrium X···Y distance is not Q_0, the distance at which the electronic energy of the H-bond is minimum for both an H-bond and its corresponding D-bond. It is $\bar{Q} = Q_0 - \delta Q_0$, with δQ_0 given by eq. (7.A12). Let us note that as $\partial\omega/\partial Q_s$ is positive, as indicated by Novak's curve of Figure 4.5, δQ_0 is positive, which implies that the equilibrium distance is somewhat shorter than Q_0. This shift, δQ_0, of the equilibrium distance of the H-bond is a direct consequence of this specific anharmonic modulation. It arises from the average (square) amplitude of the stretching vibration of the H-atom that establishes this H-bond in its ground state. As this amplitude is mass dependent and is consequently not the same after the H-atom of the H-bond has been replaced by a D-atom, this shift is different in an H-bond and in its corresponding D-bond. This difference, following eq. (7.A12), is

$$\delta Q_0^{\mathrm{H}} - \delta Q_0^{\mathrm{D}} = \frac{\hbar\left\{\dfrac{\partial\omega^{\mathrm{H}}}{\partial Q_s} - \dfrac{\partial\omega^{\mathrm{D}}}{\partial Q_s}\right\}}{2M_s\Omega_s^2} = \delta Q_0^{\mathrm{H}}\left(1 - \frac{1}{\sqrt{2}}\right) \qquad (7.1)$$

where indices H and D have been written to distinguish H- and D-bonds, and eq. (7.A7) has been taken into account to write the last equation which indicates that H-bonds are somewhat shorter than corresponding D-bonds.

This isotopic difference is small and has been scarcely measured. A difference of –0.02 Å for O···O distances in the H-bonds and D-bonds of formic acid dimers in the gas phase has been measured by electron diffraction (1) as covered in Ch. 3. These dimers are excellent models of H-bonds. Their structure is drawn in Figures 1.6 and 4.4. From the analysis of the shape of their ν_s bands, developed in Ch. 5 and from which the equations developed in the appendix of this chapter are deduced, a value for the dimensionless parameter b that characterizes the anharmonic coupling between ν_s and intermonomer modes and is proportional to $\partial\omega/\partial Q_s$, following the equation

$$b = \frac{1}{\Omega_s}\sqrt{\frac{\hbar}{2M_s\Omega_s}}\frac{\partial\omega}{\partial Q_s} \tag{7.2}$$

has been found equal to 0.7 (2). With $\Omega_s = 220\ \mathrm{cm}^{-1}$, $M_s = 11m_H$ (3) and the value of the Planck's constant as given in the appendix of Ch. 1, a value of 0.06 Å is found for δQ_0^H. This value is not at all small and emphasizes the important geometrical consequence of the anharmonic modulation of the frequency of ν_s by intermonomer modes (4). It gives a value of 0.017 Å for $\delta Q_0^H - \delta Q_0^D$ (eq. (7.1)), a value close to 0.02 Å, measured by electron diffraction. It shows that the theoretical description of the anharmonic coupling of ν_s and intermonomer modes of H-bonds, presented above, not only quite precisely reproduces spectra, but also reproduces such small geometrical effects, which are measured by completely independent methods. The experimentally verified correlation this theoretical description establishes between IR bandshapes and isotopic length variations $\delta Q_0^H - \delta Q_0^D$ of H-bonds constitutes a strong support of its validity. Values of $\delta Q_0^H - \delta Q_0^D$ can be in principle also measured by X-ray diffraction of H-bonded samples in crystals. This existence of slightly different values of X···Y distances has thus been originally measured using this technique by Ubbelohde and Gallagher (5) and has been sometime called the "Ubbelohde effect". More recent measurements have been performed on cyclic dimers of dicarboxylic acid in crystals, showing, for instance, a difference of –0.02 Å, the same as in formic acid dimers in the gaseous phase seen above, for the O···O distance between deuterated and hydrogenated oxalic acid dihydrates (6). Due to difficulties in ensuring a complete deuteration of the crystals, added to the difficulty of measuring such small distance differences, no new data, however, have been given with this technique (7). In the case of ice, a very small difference of $\delta Q_0^H - \delta Q_0^D$ has been measured. It implies a value for $\partial\omega/\partial Q_s$ weaker than in carboxylic acid dimers and justifies classifying H-bonds established by water molecules as "weak H-bonds", as has been made in Figure 1.6.

As can be seen from eq. (7.A11), intermonomer modes in Q_θ and Q_φ are not at the origin of any geometrical H/D isotopic difference, as the equilibrium positions of these two modes remain 0. This might, however, not be exactly true for the very small H_2O molecule in liquid water, for which bending intermonomer vibrations, called librations, have exceptionally high frequencies in the 700 cm^{-1} region and are seen in Ch. 9 to display unusually great amplitudes. The harmonic approximation that is implicit for the intermonomer modes in eqs. (7.A8) and (7.A11) might consequently be insufficient.

Enthalpies of H-bonds and D-bonds

The differences in enthalpies of formation between H-bonds and D-bonds are a consequence of this same anharmonic modulation that is at the origin of the geometric differences defined by eq. (7.1). Establishing them requires some mathematical development that we perform in this subsection, because the following equations can be written with no difficulties. The reader who is not interested in the detail of these equations may directly skip to the resulting eq. (7.7) and consequent results established in eqs. (7.9) and (7.10).

Vibrational energies $E_0^{N_s,N_\theta,N_\varphi}$ of the isolated H-bond, which have been defined in eq. (5.7) when only the stretching intermonomer mode Q_s was considered and which are the eigenvalues of $H_0(Q)$ defined in eq. (7.A11) that now includes all three intermonomer modes, are equal to

$$E_0^{N_s,N_\theta,N_\varphi} = V(q_0,Q_0,0,0) + \frac{\hbar\omega(Q_0,0,0)}{2} + \left(N_s + \frac{1}{2}\right)\hbar\Omega_s + \left(N_\theta + \frac{1}{2}\right)\hbar\Omega_\theta' + \left(N_\varphi + \frac{1}{2}\right)\hbar\Omega_\varphi' - \frac{1}{2}M_s\Omega_s^2\delta Q_0^2 \tag{7.3}$$

with

$$\Omega_\theta'^2 = \Omega_\theta^2\left\{1 + \frac{\hbar}{2M_\theta\Omega_\theta^2}\frac{\partial^2\omega}{\partial Q_\theta^2}\right\} \tag{7.4}$$

and a similar equation for $\Omega_\varphi'^2$. N_s, N_θ and N_φ are positive or null integer numbers. With the help of eq. (7.A12) that defines the shift δQ_0 of the equilibrium distance in Q_s due to this anharmonic modulation, we may write

$$\frac{\hbar\omega(Q_0,0,0)}{2} - \frac{1}{2}M_s\Omega_s^2\delta Q_0^2 = \frac{\hbar\omega(Q_0-\delta Q_0,0,0)}{2} \tag{7.5}$$

so that the enthalpy ΔH of formation of H-bonds, as defined in eq. (1.5) and derived from eq. (7.3) above writes

$$\Delta H = V(q_0,Q_0,0,0) - V(q_0,\infty,0,0) + \frac{\hbar[\omega(Q_0-\delta Q_0,0,0)-\omega_f]}{2} + \bar{E}_Q - \frac{3}{2}kT \tag{7.6}$$

where the first two terms in V are the differences of electronic energies between the complex X–H···Y and the "free" separated molecules X–H and Y represented by $Q_0=\infty$. With respect to eq. (1.5), these terms include the effects of Q_θ and Q_φ that appear in the form of the two zeros in V. The following fraction of eq. (7.6) is the energy difference of ground vibrational ν_s modes in H-bonded and "free" molecules with $\omega_f = \omega(\infty, 0, 0)$. The last two terms are the average energies of intermonomer modes, $\bar{E}_Q$ for the X–H···Y that becomes $\frac{3}{2}kT$ for the free molecules, as these intermonomer modes are then pure translational modes with frequencies and force constants equal to zero. Neglecting the small differences of Ω'_α of eq. (7.4) in X–H···Y and X–D···Y, taking into account that Ω_s is the same in these two complexes, we find the enthalpy difference of formation of an H-bond with respect to the corresponding D-bond:

$$\Delta H^{\mathrm{H}} - \Delta H^{\mathrm{D}} \simeq \frac{\hbar[\omega^{\mathrm{H}}(Q_0 - \delta Q_0^{\mathrm{H}},0,0) - \omega^{\mathrm{D}}(Q_0 - \delta Q_0^{\mathrm{D}},0,0) - \omega_f^{\mathrm{H}} + \omega_f^{\mathrm{D}}]}{2} \tag{7.7}$$

The physical origin of this difference is a pure quantum effect that comes from the gain in "zero vibrational energy" of the ν_s vibration between the H-bonded state and the "free" state that enters the enthalpy of formation of the H-bond and is different in H- and D-bonds. This quantity is negative because, as illustrated in Novak's curve of Figure 4.5, the difference between terms 1 and 3 of eq. (7.7) is negative: the wavenumbers of ν_s bands, which are related to the ω's by the following eq. (7.8) that takes into account eq. (1.A3),

$$\omega = 2\pi\nu = 2\pi 10^2 \tilde{\nu} c \tag{7.8}$$

with c the velocity of light, are found at 3600 cm^{-1} for X–H and at lower wavenumbers for X–H···Y. The corresponding terms 2–4 are of opposite sign and, in a first approximation, equal to those for H-molecules (terms 1–3), divided by $\sqrt{2}$. H-bonds are consequently a little stronger than D-bonds, as $\Delta H^{\mathrm{H}} - \Delta H^{\mathrm{D}}$ is negative. An estimate of this difference of enthalpies $\Delta H^{\mathrm{H}} - \Delta H^{\mathrm{D}}$ is easily made. For a weak O–H···O bond, the value of the centre of the ν_s band is typically found at 3400 cm^{-1} and $\omega^{\mathrm{H}}(Q_0 - \delta Q_0^{\mathrm{H}},0,0)$ is related to this quantity by eq. (7.8). $\omega^{\mathrm{D}}(Q_0 - \delta Q_0^{\mathrm{D}},0,0)$ falls in a close vicinity of $\omega^{\mathrm{H}}(Q_0 - \delta Q_0^{\mathrm{H}},0,0)/\sqrt{2}$. The same is true for free ν_s O–H bands that are found at wavenumbers equal to 3600 cm^{-1} in O–H and at that value divided by $\sqrt{2}$ in O–D. For such a weak H-bond, we consequently find, using eqs. (1.A2) and (1.A4) to convert cm^{-1} into kJ mol^{-1}:

$$\Delta H^{\mathrm{H}} - \Delta H^{\mathrm{D}} = \frac{(3400-3600)}{2}\left(1-\frac{1}{\sqrt{2}}\right)\frac{96.3}{8054} \simeq -0.35 \text{ kJ mol}^{-1} \tag{7.9}$$

This value is to be compared with that of ΔH^{H} that is somewhat smaller than –20 kJ mol^{-1} for water dimers given in Ch. 1. It consequently amounts to only some percents of the total enthalpy of formation, much less than we would calculate by erroneously applying the same Novak's curve to H-bonds and D-bonds, as sometime appears in the literature. Let us note that in the case of ice, with its centre of the ν_s band at 3250 cm^{-1} (Figure 9.2), the value for $\Delta H^{\mathrm{H}} - \Delta H^{\mathrm{D}}$ is then found equal to –0.6 kJ mol^{-1}. This value is appreciably greater than that experimentally measured (8), which corresponds to a shift of –50 cm^{-1} only, instead of –350 cm^{-1} used to calculate this value (350 = 3600 − 3250). It once more shows that in the case of these small H_2O molecules, one should incorporate more precisely the effects of intermonomer modes Q_θ and Q_φ. The ν_s centre of a typical medium-strength H-bond falls around 3000 cm^{-1}. Its enthalpy ΔH^{H} is typically of –30 kJ mol^{-1}. The difference between enthalpies of H-bonds and D-bonds is consequently:

$$\Delta H^{\mathrm{H}} - \Delta H^{\mathrm{D}} = \frac{(3000-3600)}{2}\left(1-\frac{1}{\sqrt{2}}\right)\frac{96.3}{8054} \simeq -1 \text{ kJ mol}^{-1} \tag{7.10}$$

also some percents only of their enthalpies, a result already seen in Ch. 1. Such small differences are hard to measure. It supports the assertion formulated in Ch. 1, that enthalpies are not the best quantities to characterize H-bonds.

It may be concluded that the geometry and the enthalpy of formation of a D-bond are slightly different from those of the corresponding H-bond. H-bonds are shorter than D-bonds by somewhat less than 0.05 Å for a medium-strength H-bond, a distance that is in principle measurable, but has been actually scarcely measured, and by <0.05 Å for a weak H-bond. They have enthalpies greater than enthalpies of corresponding D-bonds by a few percents. These are small differences that are hard to put into evidence due to the relative imprecision of corresponding measurements.

DYNAMIC PROPERTIES OF H-BONDS AND D-BONDS

Vibrational spectra of H-bonds and D-bonds

Dynamic properties of H-bonds are much more sensitive to an H/D substitution than geometrical and thermodynamic properties, because the mass effect is then directly resented when moving an H-atom that doubles its mass on becoming a D-atom. The most familiar dynamic effects appear in the form of vibrations of H-atoms themselves. IR spectra of model systems of H-bonds, cyclic dimers $(CH_3COOH)_2$ and $(CH_3COOD)_2$ displayed in Figure 7.1 illustrate this mass effect. The structure of these dimers is displayed in Figures 1.6 and 4.4. The main band that is sensitive to this H/D substitution is the ν_s band, extensively discussed in Chs. 4 and 5, which has all its first moments, integrated intensity, average wavenumber and width modified by it. The integrated intensity of the ν_s band of these cyclic dimers is abnormally reduced when passing from the H-bonded to the D-bonded dimer (9, 10), more than the expected ratio $\frac{1}{\sqrt{2}}$. This effect that seems to be only present when the H-bonded complexes are cyclic has up to now not been explained. The average wavenumber and width of the ν_s bands of H-bonds have been calculated in Ch. 5, on the basis of same equations as those developed in the appendix of this chapter. From eq. (7.A7), a variation of these quantities upon such an H/D substitution is predicted, particularly an important shift towards lower wavenumbers of the centres of these bands. It is clearly visible on the spectra of Figure 7.1, with the centre of ν_s of $(CD_3COOH)_2$ appearing (10) at 2960 cm^{-1} and that of $(CD_3COOD)_2$ at 2275 cm^{-1}. We may note, however, that such a shift also appears in the spectra of free monomers with no H-bonds, with the centre of ν_s of CD_3COOH monomers appearing at 3578 cm^{-1} whereas that of CD_3COOD appears at 2642 cm^{-1}. The shift of the ν_s band is not consequently due to changes between H- and D-bonds, but due to the change upon doubling the mass of the H-atom of the frequency of ν_s when passing from X–H to X–D, independently of whether this group establishes an H-bond or not. This great shift provides an easy way to distinguish H-bonds and D-bonds. The width of ν_s of $(CD_3COOH)_2$ is conversely appreciably greater than that of $(CD_3COOD)_2$, more precisely greater by a factor $\sqrt{2}$ following eqs. (5.A47) and (7.A7) of the appendix of this chapter, a result that does not apply to monomers.

Figure 7.1 furthermore shows that other modes, such as $\nu_{C=O}$ modes, also display some shift of their centres. These shifts are detectable in free monomers as well as in H-bonded

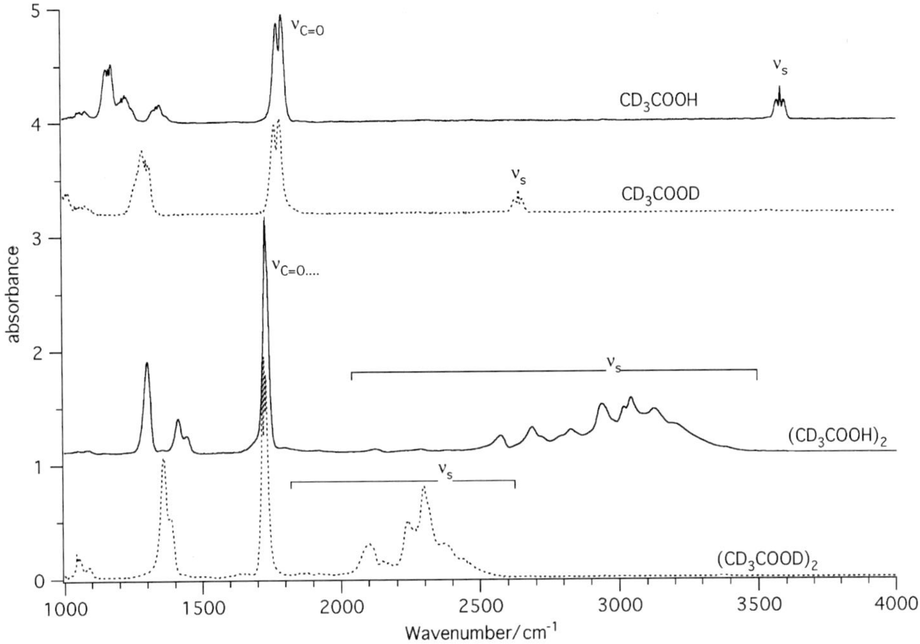

Figure 7.1 IR spectra of cyclic H-bonded dimers of carboxylic acids $(CD_3COOD)_2$ and $(CD_3COOH)_2$ (bottom) and of corresponding not H-bonded "free" monomers (upper spectra). Spectra are offset for clarity. The structure of these dimers is shown in Figure 4.4.

dimers: 1783 cm^{-1} for the $\nu_{C=O}$ mode in CD_3COOH monomer versus 1775 cm^{-1} in the free CD_3COOD monomer, to be compared to 1735 cm^{-1} in the H-bonded $(CD_3COOH)_2$ dimer versus 1728 cm^{-1} in the D-bonded $(CD_3COOD)_2$ dimer. This small shift may be assigned to a small modification of the harmonic decomposition of the $\nu_{C=O}$ mode on local vibrations when passing from an O–H group to an O–D group. It is not due to the H-bond itself. The effect of the H-bond appears in the shift of the centres of these $\nu_{C=O}$ modes in free monomers with respect to H-bonded dimers, which are $\simeq$1780 cm^{-1} for monomers and $\simeq$1730 cm^{-1} for dimers, an effect that is discussed in Ch. 4 and illustrated in Figure 4.10. It should be, however, noted that this shift is partly due, in these cyclic dimers, to a supplementary harmonic interaction between the two $\nu_{C=O}$ vibrations of the dimer, which is more precisely defined below in eq. (7.11). The value of this harmonic interaction can be deduced from the measurement of the shift of the same $\nu_{C=O\cdots}$ bands in Raman spectra. In these centro-symmetric dimers, IR sees only antisymmetric modes labelled *u* that are schematized in Figure 4.4, while Raman sees symmetric modes labelled *g* that do not destroy the symmetry of the dimer. These symmetric and antisymmetric modes appear at wavenumbers separated by a quantity proportional to twice this interaction.

Such a harmonic modification of the decomposition of the normal modes on their local vibrational components, which is not a specific property of H-bonds, is even more apparent on the $\nu_{C\text{–}OH}$ and $\delta_{C\text{–}O\text{–}H}$ or $\delta_{C\text{–}O\text{–}D}$ modes: in the H-bonded dimers (bottom spectra in Figure 7.1) the $\nu_{C\text{-}OH}$ and $\delta_{C\text{-}O\text{-}H}$ modes appear as bands at 1300 and 1415 cm^{-1} for $(CD_3COOH)_2$. These two modes have both important components on both C–OH stretch and C–O–H bending local vibrations that are not far from resonance (the frequencies of these isolated vibrations fall in a close vicinity). This is not so with the C–OD stretch and C–O–D bending vibrations in $(CD_3COOD)_2$ that display more separated $\nu_{C\text{-}OD}$ and $\delta_{C\text{-}O\text{-}D}$ modes that appear at 1360 and 1030 cm^{-1} respectively. It may be concluded that the comparison of IR spectra of isotopically substituted H-bonded systems is full of information on the H-bond itself but that one should carry a precise analysis before ascribing the results to H-bonds themselves.

This H/D substitution may be most interesting in the case of amide I modes of amide, peptides or proteins, which are widely studied bands that are mainly composed of C=O stretching vibrations. Replacing N–H groups in these species by N–D groups has the effect that bending C–N–D vibrations fall further apart from C=O stretch vibrations than C–N–H vibrations do. $\nu_{C=O}$ and $\delta_{C\text{-}N\text{-}D}$ vibrations are consequently no longer resonant vibrations, with the consequence that $\nu_{C=O}$ modes of D-substituted amide groups have more important components on C=O stretch vibrations than $\nu_{C=O}$ modes of normal amides. This property has been used in Figure 4.9 to put into evidence the effect of H-bonds on C=O stretching vibrations in proteins.

Partial H/D substitution and isotopic dilution

As can be seen in Figure 7.1, an H/D substitution on all H-atoms that establish H-bonds has important effects on the positions of bands such as ν_s that directly imply vibrations of H-atoms. We have seen that the important shift of this band such a substitution induces is, however, not entirely due to it. It also has a part that is due to the harmonic interaction between the two proximate and resonant H-atoms that establish H-bonds in these dimers. Suppressing this harmonic interaction can be achieved by a partial deuteration of these H-atoms. Introducing both acetic-H CD_3COOH and acetic-D CD_3COOD molecules in a gas cell gives a mixture of monomers and H-bonded cyclic dimers. Two kinds of monomers are present in this mixture, CD_3COOH and CD_3COOD molecules, and three kinds of H-bonded cyclic dimers are also present: $(CD_3COOH)_2$, $(CD_3COOD)_2$ and mixed dimers $CD_3COOH\text{–}CD_3COOD$. The spectra of these latter dimers can be isolated from the experimental spectra by subtracting the spectra of the monomers and of the homodimers $(CD_3COOH)_2$ and $(CD_3COOD)_2$ represented in Figure 7.1. Their ν_s bands are shown in Figure 7.2, together with the ν_s bands of homodimers of Figure 7.1 for a comparison. Other bands of the mixed dimers do not exhibit significant differences with those of homodimers, the $\nu_{C=O}$ band being hardly distinguishable from that of the homodimers, whereas the $\nu_{C\text{-}OH}$ and $\delta_{C\text{-}O\text{-}H}$ bands of the mixed dimer exhibit only slight shifts from the corresponding bands of $(CD_3COOH)_2$, and the $\nu_{C\text{-}O\text{-}D}$ and $\delta_{C\text{-}O\text{-}D}$ bands of the mixed dimer much resemble those of $(CD_3COOD)_2$. The ν_s bands shown in Figure 7.2 conversely exhibit a clearly marked change of their centres, with the centre of $\nu_s(O\text{–}H\cdots)$ of homodimers falling some 110 cm^{-1} above that of mixed dimers and the centre of $\nu_s(O\text{–}D\cdots)$ of homodimers falling some 80 cm^{-1} above that of mixed dimers. As already mentioned, this shift is due to the harmonic interaction term, defined below in eq. (7.11), between the two ν_s vibrations in these dimers. This interaction term has but a negligible effect in mixed dimers

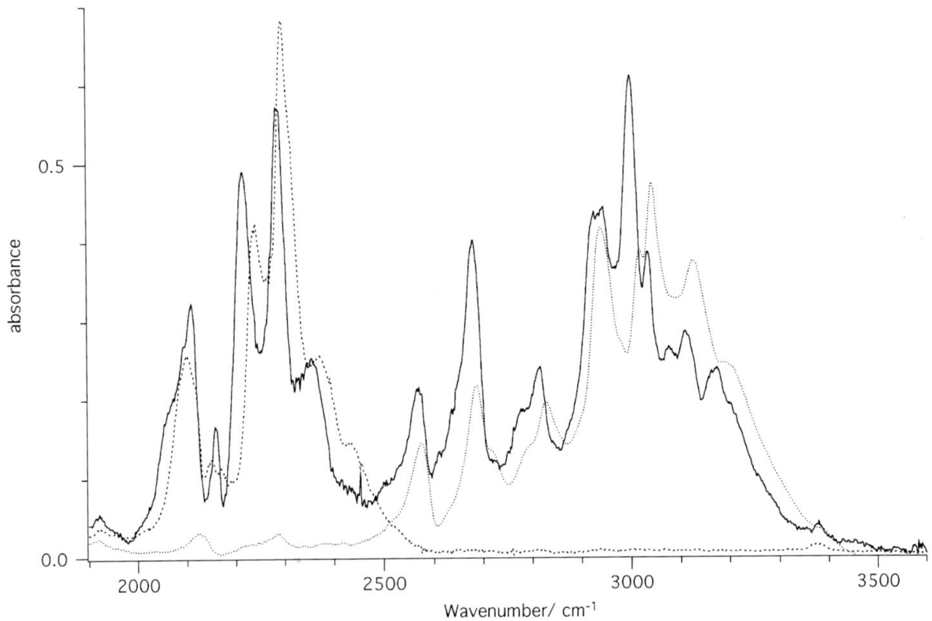

Figure 7.2 ν_s bands of cyclic H-bonded dimers of carboxylic acids: mixed dimers CD_3COOH–CD_3COOD (solid line) and homodimers $(CD_3COOH)_2$ and $(CD_3COOD)_2$, same as in Figure 7.1 (dotted lines). The numbers of mixed dimers is approximately twice that of each homodimer, so that ν_s(O–H···) and ν_s(O–D···) bands have comparable intensities. The structure of all these H-bonded cyclic dimers is that represented in Figure 4.4.

where the two ν_s(O–H···) and ν_s(O–D···) vibrations are far from resonance, as they respectively vibrate around 3000 and 2250 cm^{-1}. It has an appreciable effect in homodimers where the two ν_s bands are fully resonant because they are identical. This harmonic interaction may be easily quantified. The Hamiltonian $\breve{\mathrm{H}}$ that governs the two ν_s vibrations, 1 and 2 of such a homodimer, is equal to the sum of the Hamiltonians H defined in eq. (7.A8) for each H-bond of the dimer, plus a harmonic interaction term q_1q_2.

$$\breve{\mathrm{H}} = H(q_1, Q_s^1) + H(q_2, Q_s^2) + m\omega^2(Q_0)\breve{V}_0 q_1 q_2 \qquad (7.11)$$

This last interaction term is harmonic because it is quadratic in the vibrational coordinates and harmonic terms are by definition those that are of degree less or equal to 2. Q_0 is defined in eq. (7.A5) of the appendix. The effect of this harmonic interaction term is to shift the centre of the band by a quantity equal in wavenumber to $-(\omega(Q_0)\breve{V}_0/2\pi c)$, with c the velocity of light. In the case of a mixed dimer the effect of this last term is, as already mentioned, negligible, because the two ν_s vibrations are not resonant. The magnitude of this term can consequently be taken equal to 110 cm^{-1} in the case of $(CD_3COOH)_2$ and 80 cm^{-1} in the case of $(CD_3COOD)_2$. These are values that are consistent with eq. (7.A7). The comparison of the ν_s bands of these mixed dimers with those of homodimers consequently conveys useful information on these stretching vibrations of H-bonds, more particularly on their harmonic interactions.

Dimers such as those mentioned above are excellent models of H-bonds. They are, however, not often found in studies of H-bond-containing species that are observed with other goals than studying H-bonds for themselves. H-bonds in solids and liquids are conversely commonly encountered. In these species, the equivalent way to eliminate interactions between resonant vibrations in view of gathering information on the dynamics of H-bonds is to use isotopic dilution. It is often used in the case of liquid water or of aqueous media. In these media it consists of mixing a small quantity of heavy water in normal water, or, less commonly the inverse, mixing a small quantity of normal water in heavy water. The two molecules, H_2O and D_2O, rapidly exchange their H- and D-atoms and if the concentration of D-atoms that originates from D_2O molecules of heavy water is small in comparison to that of H-atoms, more D-atoms are found in HDO molecules, which remain in a small number compared to H_2O molecules. In these conditions, the vibrations of O–D parts of HDO behave as if these O–D groups were isolated, because closely lying O–H groups vibrate at different frequencies and only weakly influence the O–D vibrations (they are far from resonance).

This method of isotopic dilution provides a way to avoid saturation of bands, a real problem in the case of spectra of liquid water or aqueous solutions, as described in Ch. 11. It consists of observing O–D vibrations of diluted groups in the transparency region of normal liquid water, between 1800 and 3000 cm^{-1}. As shown in Figure 7.3, which displays the spectrum of HDO molecules diluted in heavy water, the ν_s(O–H$\cdots$) band of HDO appears in this region around 3400 cm^{-1} and can therefore be studied in detail if the concentration of HDO molecules is low. It implies that the concentration of H_2O molecule is often negligible. This method has for long been the main method to record spectra of liquid water before the advent of more general recent methods described in Ch. 11. It is still much used in nonlinear time-resolved IR spectroscopy studies of liquid water described in Ch. 4. The spectra displayed in Figure 7.3 are the spectra of a mixture of heavy water and normal water in quantities initially proportional respectively to $1-C$ and C, giving a final relative proportion of C^2 H_2O, $2C(1-C)$ HDO and $(1-C)^2$ D_2O molecules ($0 \leq C \leq 1$). In Figure 7.3 these relative proportions are consequently 0.04 H_2O, 0.32 HDO and 0.64 D_2O molecules. When C tends towards zero, HDO molecules establish H-bonds mainly on D_2O molecules. As a consequence, the ν_s(O–H$\cdots$) band of HDO molecules then tends towards that of an isolated O–H$\cdots$ vibration in liquid water for which the effect of the solvent is absent. This clearly appears in Figure 7.3 where, even with a concentration $C = 0.2$ that is not so small, the ν_s(O–H$\cdots$) band of HDO molecules (full line) around 3400 cm^{-1} appears at higher wavenumbers than that of H_2O molecules in normal liquid water (dashed line). This shift of the ν_s(O–H$\cdots$) band of H_2O molecules towards lower wavenumbers is a consequence of harmonic interactions between each O–H stretching vibrations of an H_2O molecule with surrounding O–H vibrations of the same H_2O molecule or of the four surrounding ones. These interactions have the same form as the last term of eq. (7.11). It has a vanishing effect for ν_s(O–H$\cdots$) vibrations of HDO molecules mainly surrounded by D_2O molecules, once more because of off-resonance. This harmonic interaction also broadens the ν_s(O–H$\cdots$) band of H_2O molecules as compared to that of HDO molecules. These features do not appear in the ν_s(O–D$\cdots$) band of HDO molecules with peak around 2500 cm^{-1}, because the ν_s(O–D$\cdots$) vibration of these molecules still harmonically interacts with other resonant ν_s(O–D$\cdots$) vibrations of the numerous D_2O molecules with which this HDO molecules establishes H-bonds. The ν_s(O–D$\cdots$) bands of both HDO and D_2O molecules are

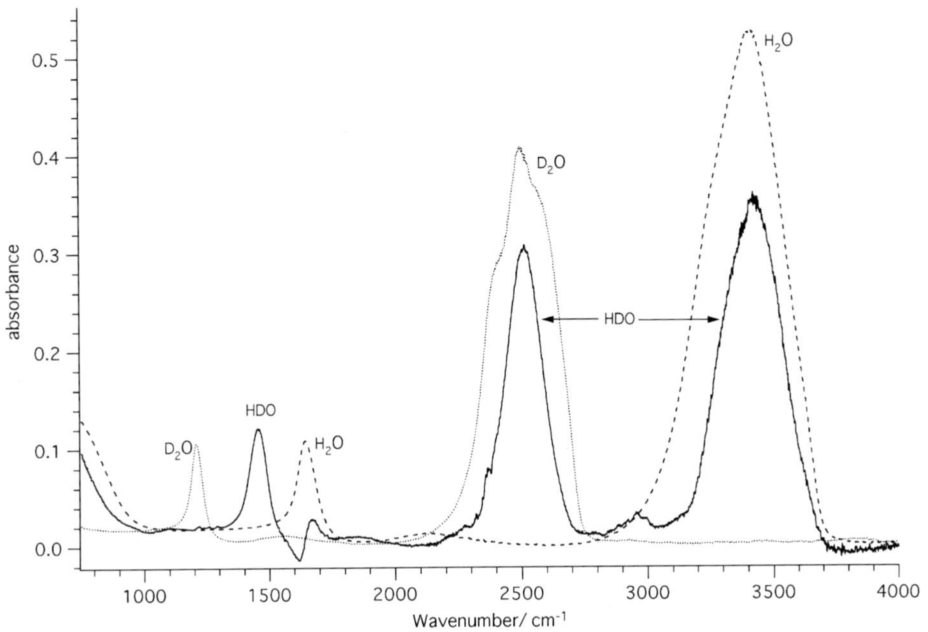

Figure 7.3 IR spectra of HDO molecules diluted in heavy water (full line) and of normal and heavy waters (dashed line and dots). The intensity of each spectrum is normalized so as to correspond to the same number of H_2O, HDO or D_2O molecules. The original mixing corresponds to a concentration $C = 0.2$ (0.8 heavy water for 0.2 normal water).

consequently similar. Only their Fermi resonance shoulder are different: it appears around $2400\,cm^{-1}$ for D_2O, at twice the wavenumber of the bending band $\delta_{D\text{-}O\text{-}D}$ that appears at $1200\,cm^{-1}$, and around $2900\,cm^{-1}$ for HDO, at twice the wavenumber of the bending band $\delta_{H\text{-}O\text{-}D}$ that appears around $1450\,cm^{-1}$. This $\nu_s(O\text{–}H\cdots)$ band of HDO molecules diluted among D_2O molecules consequently keeps all the properties of that of H_2O molecules in normal water, with the exception of its harmonic interactions with neighbour molecules. The properties of H-bonds in liquid water can consequently be precisely studied on this band. Its interest lies in the low concentration of HDO molecules that offers a simple way to avoid saturation problems with an ordinary absorption setup. Let us note that the $\delta_{H\text{-}O\text{-}D}$ band around $1450\,cm^{-1}$ can also be precisely studied. It appears to be accompanied with a difference band around $1650\,cm^{-1}$, the wavenumber of $\delta_{H\text{-}O\text{-}H}$ in H_2O molecules. It shows that this $\delta_{H\text{-}O\text{-}H}$ band appears at a slightly different wavenumber when H_2O molecules are surrounded by other H_2O molecules, as in normal liquid water, than when H_2O molecules are surrounded by D_2O and a few HDO molecules, as in the solid spectrum of Figure 7.3. In the latter case, H_2O molecules are surrounded by D_2O and HDO molecules that display nonresonant $\delta_{D\text{-}O\text{-}D}$ and $\delta_{H\text{-}O\text{-}D}$ vibrations, whereas in normal liquid water $\delta_{H\text{-}O\text{-}H}$ vibrations are in resonance with $\delta_{H\text{-}O\text{-}H}$ vibrations of neighbour moleculess. The $\delta_{H\text{-}O\text{-}H}$ band falls consequently at a somewhat different wavenumber. The result is the difference band at $1650\,cm^{-1}$ in the spectrum of HDO molecules that arises from the subtraction of the spectrum of liquid water from the spectrum of the preceding mixture, a subtraction performed in view of obtaining the spectrum of HDO molecules in Figure 7.3. Let us finally note that, as described in Ch. 11, this technique of isotopic dilution is used in medicine to measure the total amount of body water of a person.

H/D substitution in biology: a dramatic effect on reactivity

The doubling of the mass caused by an H/D substitution in H-bonds has still more marked effects on transfers of protons or transfers of H-atoms, defined in Ch. 6, than on vibrations examined above. These transfers are a specificity of H-bonds and their existence has been declared the third fundamental property of H-bonds. Transfers of protons are at the origin of acid/base chemistry and of protonic conduction. Important changes of conductivities are found in protonic conductors when protons are changed into deuterons. Exchanging protons into deuterons may consequently serve to decide whether conductivity is electronic or protonic (11), because electronic conduction is insensitive to such an exchange. However, as argued in Ch. 10, transfers of protons are certainly not fundamental mechanisms in biology. H-atom transfers are conversely fundamental mechanisms. Little is known of these transfers for which the effect of deuteration has been scarcely measured. In benzoic acid dimers, the lifetime of a tautomer such as shown in the upper part of Figure 6.10 is some 30 times longer in the deuterated sample than in the hydrogenated one, as measured by precise optical techniques (12). In other words, the transition probability of H(D)-atoms between the two tautomers is decreased by a factor of 30 when passing from H-benzoic acid to D-benzoic acid. This is not so much surprising a result as tunnelling, which is for an important part responsible for such transfers, is well known to be hypersensitive to the mass of the transferred particle. It strongly suggests that an H/D substitution may have dramatic consequences in biology. And indeed life becomes very difficult in heavy water. It may be interesting to look in more detail to this point that has been reviewed (13) in detail, with, however, not the point of view of a molecular chemist adopted here but rather that of a biologist who attempts to define the toxicity of such an H/D exchange. It is thus true that the replacement of normal water by heavy water is lethal for nearly all living organisms, particularly evolved organisms. Prior to 1960, significant concentrations of deuterium were considered incompatible with life. Presently such a definitive conclusion cannot be reached and should be somewhat modified. Some very primitive organisms, such as green algae and bacteria, have thus been shown to be able not only to survive but also to grow within heavy water (14). Not very surprisingly, they however grow with a very much slowed-down metabolism. These "autotrophic" simple organisms can accommodate such a slow metabolism, even if they do it with some difficulties. Autotrophy is that feature of very primitive organisms that only need water, inorganic salts, CO_2, N_2 and light to live. All other organisms require having at their disposal amino acids, polysaccharides and other essential nutrients that have been preformed by other feeding organisms. This requirement strongly complicates the interpretation of the effect on them of H/D substitutions, which can be satisfactorily carried out only with such autotrophic organisms. When grown in heavy water, these algae perform photosynthesis at a rate that is depressed by at least a factor of 3 (15), which strongly supports the previous suggestion that the primary effect of such an H/D substitution is to slow down all bioreactivity. Some primitive algae and also primitive bacteria are able to adapt themselves to such a slowing down. It becomes so much difficult for more evolved organisms that it is indeed lethal.

This partial compatibility of life with heavy water suggests that a Darwinian evolution might have also succeeded if, instead of normal water, life would have started developing in heavy water some 3–4 billion years ago. What would be presently such a life? No answer can be given to this question. However, we may suspect that life would most likely still be in a much earlier stage of the evolution, because of the much slowed-down reactivity in heavy water. It might also have taken a completely different route because evolution, as history, strongly depends on contingency, which means it changes its course with the advent of fortuitous events that have an extremely low probability to be the same that the evolution we know has experienced.

Heavy water can be easily accepted at low concentration and, as explained in Ch. 11, can be injected with no problem in the blood of a person to measure his total body water. This is so as long as HDO and D_2O molecules remain in a sufficiently low concentration that they do not take the place of H_2O molecules in metabolic reactions. At higher concentrations, they become toxic, as by reducing rates of transfers of H-atoms between proteins and enzymes, for instance, they completely disorganize the synchronization of the sets of metabolic reactions these organisms require to perform continuously and coherently to survive. It becomes consequently lethal for most organisms, with the exception of primitive ones that apparently succeed finding and adapting new kinds of metabolisms.

H-BONDS AND D-BONDS AS SEEN BY METHODS SENSITIVE TO NUCLEAR SPINS

In this section the effect of the change of the nuclear spin after an H/D substitution in H-bonds is examined. Such a change of the nuclear spin does not induce any modification of the properties of H-bonds, as their electronic structures are not modified, but some experimental methods are sensitive to this change. NMR spectroscopy is one of them, which is sensitive to the spin of the nucleus. The proton has a spin ½, whereas the deuteron has a spin 1 that goes on with a relatively intense electric quadrupole moment that is zero for the proton. The deuteron is consequently, from the point of view of NMR, a completely different nucleus. A deuterated sample then becomes interesting to study with NMR spectroscopy, as it distinguishes H-bonds from D-bonds and allows studying only those bonds that are deuterated. Furthermore, deuterated bonds experience much less the effects of neighbour nonexchangeable H-atoms of for instance C–H groups. As these groups are often numerous, this may result in a considerable simplification of the spectra. This possibility is, however, limited by the increase of the numbers of NMR bands due to the spin 1 of the deuteron as compared to the spin ½ of the proton, and by the complication due to the presence of a quadrupole moment that makes spectra more difficult to interpret than those originating from protons. Such a substitution, in principle interesting, is consequently not a general method to study H-bonds, all the more so that NMR spectroscopy, in opposition to IR spectroscopy, is not especially sensitive to H-bonds.

Neutron scattering is another method that probes nuclear spins. Neutrons are scattered by spins of both protons and neutrons of nuclei. Scattering cross-sections of deuterons are consequently different from those of protons. As already seen in Ch. 3, coherent scattering by deuterated samples is at the origin of precise measurements of molecular distances by diffraction of neutrons, making neutron diffraction a method that is complementary to X-ray diffraction, a fundamental method in molecular physics. The strong point of neutron diffraction is that the deuteron coherent cross-section is comparable to that of other nuclei. It allows precise localization of D-atoms. X-rays, which are scattered by electrons, often meet difficulties localizing these H- or D-atoms that have around them a much smaller number of electrons than all other atoms. Neutron scattering is consequently an interesting method to determine the structures of D-bonds. The proton itself has a coherent cross-section comparable to that of the deuteron. It is, however, severely hampered by its much bigger incoherent cross-section that often masks its coherent scattering. Most samples have consequently to be deuterated for their structure to be determined by neutron scattering. This is not always an easy task. In some cases, such as liquid water or aqueous solutions, where one kind only of H(D)-atoms that all establish H(D)-bonds are present, neutron diffraction can then be informative. As seen in Ch. 11, one can then furthermore take advantage of the different signs of the coherent cross-sections of deuterons and protons to vary the contrast of these nuclei, a unique possibility offered by neutron scattering experiments. Deuteration of H-bonds is consequently interesting to carry out when performing coherent neutron scattering experiments that allow determining the positions of D-atoms and therefore decide whether they establish D-bonds or not. Such experiments require, however, heavy equipments and their implementation is consequently limited, even if most often profitable.

CONCLUSION

Replacing H-bonds by D-bonds has no effect on the electronic structure of H-bonds when nuclei are held at fixed positions. It nevertheless has a small, indirect effect on thermodynamic properties, because quantities such as enthalpies of formation of H- or D-bonds not only depend on their electronic structure, but also incorporate a significant vibrational component that is sensitive to this substitution. The result is a difference of enthalpies of formation of H- and D-bonds that amounts to only some percent of the total enthalpy of formation and remains consequently weak with respect to kT at room temperature for weak and intermediate strength H-bonds. It means that at room temperature this difference is not really apparent. Such a substitution consequently conveys almost no information for calorimetric methods. The geometry of the H-bond is also little affected by such a substitution that results in a small lengthening of the X···Y equilibrium distance in X–D···Y as compared to X–H···Y. It is an indirect, or second-order consequence of the strong and characteristic anharmonic modulation of the ν_s stretching vibration of the H-atom by the X···Y distance. It implies a weak effect with little information.

This H/D substitution has conversely appreciable effects on the dynamics of H-bonds that are sensitive to the doubling of the mass of the H-atom that establishes an H-bond. It appears particularly well in vibrations and can consequently be used in vibrational spectroscopy, particularly IR spectroscopy, to convey original information on H-bonds. Isotopic dilution techniques can then be most interesting either to decouple the ν_s vibration of a particular H-bond from that of surrounding H-bonds that display resonant vibrations, thus simplifying the interpretation of IR spectra. It can also be used to diminish the absorption mainly due to this vibration to avoid catastrophic saturation of bands. This is particularly useful in the case of liquid water or aqueous media. The ability of H-bonds to transfer protons or H-atoms is much more sensitive to such an H/D substitution that has then more important consequences, particularly in biomedia where it displays dramatic effects: it is lethal for almost all living organisms, except a few primitive ones that may not only survive when living in heavy water but may adapt themselves to such hostile conditions. Deuterium is consequently not, as previously thought, really a poison for life. It is toxic for nearly all organisms above some concentration that depends on these organisms but its toxicity is the result that H_2O molecules are necessary for life as we know it to proceed, and cannot be replaced by D_2O or HDO molecules that have the effect of severely slowing down the reactivity, completely disorganizing the synchronization of the set of metabolic reactions these organisms require to perform continuously and coherently to survive.

Besides these effects on the properties of H-bonds, H/D substitutions may be used to observe exchangeable X–H groups using experimental methods such as NMR spectroscopy or neutron diffraction that are sensitive to the change of the nuclear spin when passing from H- to D-atoms. Coherent neutron scattering experiments can thus provide a precise determination of the positions of D-atoms. As these methods are not especially sensitive to H-bonds they are not, however, general methods for the study of H-bonds.

APPENDIX

The basic equations that rule the mechanics of H-bonds are developed in this appendix. They have been already established in the appendix of Ch. 5 but take here a slightly different form that makes the role of the mass m of the H-atom more apparent, in view of predicting effects of an H/D substitution. The formation of an H-bond is the result of an electrostatic interaction between the electrons and the nuclei of two molecules X–H and Y. Molecules are quantum objects that are ruled by an Hamiltonian H that depends on the coordinates r of electrons, q of the H(D)-atom that establishes an H(D)-bond and Q that defines the relative positions of the two molecular components X–H and Y. r stands for all coordinates r_e of all electrons e. The relative coordinates q and Q of the nuclei are defined in Figure 2.1. Q stands for all three intermonomer coordinates Q_s, Q_θ and Q_φ defined in this figure. The quantum description is necessary for this H-bond, because a classical description fails to describe any chemical bond. This Hamiltonian H writes:

$$H(r,q,Q) = \sum_e \frac{\pi_e^2}{2\mu} + \frac{p^2}{2m} + \frac{P^2}{2M} + U(r,q,Q) \qquad (7.A1)$$

where the first three terms are the kinetic energy operators for all electrons e of mass μ, for H- or D-atom of the H-bond with mass m that will later be set equal to m_H or m_D ($= 2m_H$), and for intermonomer modes with (reduced) mass M. The momenta π_e, p and P of these three sets of coordinates are equal to $-i\hbar(\partial/\partial x)$ where x is one of the coordinates r_e of electron e, q of H-atom, or Q of an intermonomer distance. The eigenfunction of interest of this Hamiltonian, or in other words the wavefunction governed by this Hamiltonian that describes the whole H-bond, writes

$$\psi(r,q,Q) = \varphi_{\text{el}}(r,q,Q)\phi(q,Q) \tag{7.A2}$$

where $\varphi_{\text{el}}(r,q,Q)$ is the ground state electronic wavefunction and $\phi(q,Q)$ the vibrational wavefunction. Writing $\Psi(r,q,Q)$ in this form is the consequence of the "Born–Oppenheimer" or adiabatic approximation that is a very good approximation for nearly all molecular systems in their ground electronic state. It supposes that, due to their masses μ that are three orders of magnitude smaller than m, and at least four orders of magnitude smaller than M, the electrons are much more rapid than vibrations. They consequently adapt themselves immediately to any change of the nuclear configuration. Nuclei are conversely so slow that they cannot induce transitions between electronic states. Mathematically, this leads to each matrix element of momenta of nuclei p or P between ground electronic state $\varphi_{\text{el}}(r,q,Q)$ and any excited electronic state $\varphi_{\text{el}}^{*}(r,q,Q)$ to be equal to 0, which leads to the form written in eq. (7.A2) for $\Psi(r,q,Q)$. The ground state electronic wavefunction $\varphi_{\text{el}}(r,q,Q)$ is the wavefunction of lowest energy of the electronic Hamiltonian $H_{\text{el}}(r,q,Q)$ that enters $H(r,q,Q)$ of eq. (7.A1) and has the form

$$H_{\text{el}}(r,q,Q) = \sum_{e} \frac{\pi_e^2}{2\mu} + U(r,q,Q) \tag{7.A3}$$

with $U(r,q,Q)$ the total (electrostatic) interaction potential between electrons and nuclei of this H-bond. The energy $V(q,Q)$ of this ground electronic state is the eigenvalue of lowest energy of $H_{\text{el}}(r,q,Q)$, which means that the electronic wavefunction of this H-bond verifies the equation

$$H_{\text{el}}(r,q,Q)\varphi_{\text{el}}(r,q,Q) = V(q,Q)\varphi_{\text{el}}(r,q,Q) \tag{7.A4}$$

We have already seen in eq. (1.5), how this energy term $V(q_0,Q_0)$ at the distances q_0 and Q_0 of the nuclei of the H-bond for which $V(q_0,Q_0)$ is minimum, contributes to the value of the enthalpy of formation of this H-bond, together with the term $V(q_0,\propto)$, for the two molecules X–H and Y when they do not establish an H-bond, which is equivalent to having Q infinite. In our present description this term $V(q,Q)$ acts as the potential energy for the vibrations q of the H-atom and Q for intermonomer vibrations. We may develop it in powers of nuclear coordinates as

$$V(q,Q) = V(q_0,Q_0) + \frac{1}{2}m\omega^2(Q)(q-q_0)^2 + \frac{1}{2}M\Omega^2(Q-Q_0)^2 \tag{7.A5}$$

which is that of a harmonic vibration in q that is strongly anharmonically coupled to intermonomer vibrations Q via the Q dependence of ω. This coupling physically appears as a modulation of the frequency $\nu = \omega/2\pi$ of this vibration in q (the X–H distance) by intermonomer distances or modes. The Fermi resonance terms that appear in many IR spectra of H-bonds are neglected because they are pure spectroscopic effects that have no geometrical or thermodynamic consequences. The last term in eq. (7.A5) is the potential that governs the three intermonomer modes. It is written in the form of a harmonic potential in Q. Such a form has been established in eq. (5.2), where only the intermonomer stretching vibration Q_s was considered to modulate q and where also the term $V(q_0,Q_0)$ was not written because it was then irrelevant. In eq. (7.A5), Q represents the three intermonomer vibrations. In this last equation, other vibrations of the H-bonded complex X–H···Y are disregarded because they are independent of the establishment of the H-bond and are at the origin of an irrelevant constant term in the energy.

The mass m of the H- or D-atom in X–H(D)···Y explicitly appears in eq. (7.A5). However, $V(q,Q)$, the energy of the ground electronic state, is independent of this mass, as the potential $U(r,q,Q)$ in eq. (7.A3) only depends on the charges of nuclei, not on their masses. This is also true of the electronic wavefunction $\varphi_{el}(r,q,Q)$. It implies

$$m_H \omega_H^2 (Q) = m_D \omega_D^2 (Q) = K(Q) \tag{7.A6}$$

with K representing the "force constant" for the q motion. It only depends on Q and is independent of the isotopic constitution. As $m_D = 2m_H$, we deduce from eq. (7.A6):

$$\omega_D (Q) = \frac{\omega_H (Q)}{\sqrt{2}} \tag{7.A7}$$

This result is also valid when no H-bond exists, that is when the two components X–H and Y of the H-bond are very far apart, which is equivalent to Q_s being infinite and ω_D and ω_H being independent of Q. It is also valid for C–H groups that do not establish H-bonds. It implies that stretching bands of X–H groups appear at frequencies $\omega_H/2\pi$ that are completely different from frequencies $\omega_D/2\pi$ at which stretching bands of corresponding X–D groups appear. It allows us to unambiguously identify bands due to D-atoms of X–D groups. When this group does not establish an H-bond, these bands are narrow. In the case of broad bands that originate from groups that establish H-bonds, supplementary effects appear due to the Q dependence of ω_H and ω_D. They are described below.

The Hamiltonian that governs these intermonomer vibrations in the ground electronic state of the H-bond is $H(q,Q)$. It has been defined in eq. (5.2) and takes here the form:

$$\begin{aligned} H(q,Q) &= \frac{p^2}{2m} + \frac{P^2}{2M} + V(q,Q) = \frac{p^2}{2m} + \frac{P^2}{2M} + V(q_0,Q_0) \\ &+ \frac{1}{2} m\omega^2 (Q)(q-q_0)^2 + \frac{1}{2} M\Omega^2 (Q-Q_0)^2 \end{aligned} \tag{7.A8}$$

In Ch. 1, we have seen that the contribution of these vibrational terms represents some 25% of the total enthalpy of an H-bond. It is not negligible and stresses the importance of these vibrations. An adiabatic separation between the rapid mode q with wavenumbers around 3000 cm^{-1} and slow intermonomer modes Q with wavenumbers around 200 cm^{-1} is still valid, although less easily justified than the Born–Oppenheimer separation between electrons and nuclei. The wavefunctions of $H(q,Q)$ then takes the form defined by eq. (5.3). The energy of the rapid mode q is then equal to $\hbar\omega(Q)/2$ following eq. (5.4), with the ground state in q defined by $n = 0$ representing the only state of this mode that is populated at ambient temperature around 300 K (the first excited state of this mode with $n = 1$ has an energy corresponding to a band with wavenumber $\bar{\nu} \simeq 3000$ cm^{-1} giving a

Boltzmann factor for its relative population $e^{-(3000 \times 11{,}600/300 \times 8054)} \simeq 610^{-7} \simeq 0$, using eq. (1.A4) and following text of the appendix of Ch. 1). This energy, $\hbar\omega(Q)/2$, of the q mode acts as the potential energy that governs the intermonomer modes, so that the Hamiltonian for these modes, equal to that defined in eq. (5.5) where $V(q_0,Q_0)$ is omitted and Q limited to Q_s is then, following eq. (7.A5),

$$H_0(Q) = \frac{P^2}{2M} + V(q_0,Q_0) + \frac{\hbar\omega(Q)}{2} + \frac{1}{2}M\Omega^2(Q-Q_0)^2 \tag{7.A9}$$

Assuming that all three intermonomer modes modulate the frequency of the H-atom stretching mode q, we now have to specify their coordinates Q_s, Q_θ and Q_φ that we have up to now represented as Q. The values at which the electronic energy of the H-bond is minimum are Q_0 for Q_s and 0 for Q_θ and Q_φ. Then ω takes the form already displayed in eq. (5.A48) that writes

$$\omega(Q) = \omega(Q_0,0,0) + (Q_s - Q_0)\frac{\partial\omega}{\partial Q_s} + \frac{Q_\theta^2}{2}\frac{\partial^2\omega}{\partial Q_\theta^2} + \frac{Q_\varphi^2}{2}\frac{\partial^2\omega}{\partial Q_\varphi^2} \tag{7.A10}$$

With linear and quadratic terms in all three coordinates being regrouped, H_0 (eq. (7.A9)) writes:

$$\begin{aligned} H_0(Q) = {} & \frac{P_s^2}{2M_s} + \frac{P_\theta^2}{2M_\theta} + \frac{P_\varphi^2}{2M_\varphi} + V(q_0,Q_0,0,0) + \frac{\hbar\omega(Q_0,0,0)}{2} \\ & + \frac{1}{2}M_s\Omega_s^2\{(Q_s - Q_0 + \delta Q_0)^2 - \delta Q_0^2\} \\ & + \frac{1}{2}\left\{M_\theta\Omega_\theta^2 + \frac{\hbar}{2}\frac{\partial^2\omega}{\partial Q_\theta^2}\right\}Q_\theta^2 + \frac{1}{2}\left\{M_\varphi\Omega_\varphi^2 + \frac{\hbar}{2}\frac{\partial^2\omega}{\partial Q_\varphi^2}\right\}Q_\varphi^2 \end{aligned} \tag{7.A11}$$

with

$$\delta Q_0 = \frac{\hbar\dfrac{\partial\omega}{\partial Q_s}}{2M_s\Omega_s^2} \tag{7.A12}$$

REFERENCES

1. A. Almenningen, O. Bastiansen and T. Motzfeldt, Acta Chem. Scand., 24 (1970) 747.
2. P. Excoffon and Y. Maréchal, Spectrochim. Acta A, 28 (1972) 269.
3. Y. Maréchal, Chem. Phys. Lett., 13 (1972) 237.
4. N. D. Sokolov and V. A. Savele'ev, Chem. Phys., 181 (1994) 305.
5. A. R. Ubbelohde and K. J. Gallagher, Acta Cryst., 8 (1955) 71.
6. R. G. Delaplane and J. A. Ibers, Acta Cryst. B, 25 (1969) 2423.

7. G. A. Jeffrey, An Introduction to Hydrogen Bonding, Oxford University Press, Oxford, 1997.
8. J. P. Devlin, J. Chem. Phys., 112 (2000) 5527.
9. J. Bournay and Y. Maréchal, J. Chem. Phys., 59 (1973) 5077.
10. Y. Maréchal, J. Chem. Phys., 87 (1987) 6344.
11. J. E. Rupley and G. Careri, in C. B. Anfinsen, J. T. Edsall and F. M. Richards (Eds.), Advances in Protein Chemistry, Vol. 41, Academic Press, San Diego, 1991.
12. H. P. Trommsdorff, Proc. Acad. Sci. Estonian SSR, 37 (1988) 166.
13. J. J. Katz and H. L. Crespi, in C. J. Collins (Ed.), Isotope Effects in Chemical Reactions, Vol. 167, Ch. 5, Van Nostrand Reinhold, New York, 1970.
14. D. J. Kushner, A. Baker and T. G. Dunstall, Can. J. Physiol. Pharmacol., 77 (1999) 79.
15. S. Bose, H. L. Crespi, M. I. Blake and J. J. Katz, Plant Cell Physiol., 8 (1967) 545.

Part II

THE WATER MOLECULE

– 8 –

The H_2O Molecule in Water Vapour and Ice

H_2O: AN EXCEPTIONAL MOLECULE

The water molecule, H_2O, is one of the simplest and most stable molecules, often appearing at the end of many chemical reactions. It is also one of the most familiar molecules, the unique component of liquid water, which we drink and use in many ways everyday. All this suggests that it should be a well known and a more or less inert molecule. This chapter and the following ones clearly show that this is far from true (1). It is a polar molecule, with a marked electric dipole moment of 1.83 D (1 D = 1 Debye $\approx 3.34 \times 10^{-30}$ C m ≈ 0.21 e Å) directed along its axis of symmetry. Its H–O–H angle is 104.5° and its O–H distance is 0.96 Å (2). It is depicted in Figure 8.1 where another of its properties is emphasized: it has two acceptor sites for H-bonds and two donor sites. This is due to its electronic structure, which consists of two σ orbitals that form the two covalent O–H bonds, and of two nonbonding orbitals that remain on its O-atom and are each occupied by two lone-pair electrons. The geometry of these four electronic orbitals around the O-atom is that of a tetrahedron. The two O–H groups are H-bond donors and the two lone-pair electrons are H-bond acceptors. This discrete and scarcely invoked property may look trivial, but it is indeed exceptional and has far-reaching consequences, one among them being that without this unique possibility of the H_2O molecule, life would most probably not exist and we would not be here to comment on it. We shall see in the following how it can be so. This structure provides this small molecule with four sites to establish H-bonds. It makes it a ubiquitous molecule that easily finds a configuration to establish H-bonds with any neighbour molecule that displays H-bond receptor or acceptor sites. It gives the H_2O molecule a unique ability to develop around itself a dense *H-bond network*, which is *an assembly of numerous, closely lying and interacting H-bonds*. This H-bond network ensures the stability of the various important biological configurations covered in Ch. 2, as, at room temperature, it is a source of flexibility that makes the steric conditions required for these configurations much less stringent than suggested in the figures there. We now think, based on arguments presented in Ch. 10, that the presence of such a dense and flexible H-bond network, which only the H_2O molecule can develop, is at the basis of bioreactivity and is necessary for bioreactions to proceed. The importance of this dense H-bond network that appears around H_2O molecules is described in this and the following two chapters.

The H_2O molecule is thus a paradoxical molecule: it is both a very simple molecule on the one hand, and one that has exceptionally rich and complex ways to bind to other molecules on the other hand. It consequently displays very subtle properties, which have vital consequences, but which we still do not completely understand nor are able to evaluate. It makes this most familiar molecule a still poorly known molecule. In this and the following chapters we consequently describe it in its various states and environment: first of all when it is isolated, as in water vapour, behaving as any other isolated molecule. Then in company of other H_2O molecules, in an ordered structure at relatively low temperature, ice (this chapter), where it encounters few possibilities to evolve, but where it already starts displaying peculiarities. Next in liquid water, where its exceptional possibilities fully appear, most of which are not yet really understood. Finally, when embedded in macromolecules, where exceptional properties also appear, are even less understood, but nevertheless give hints for answering the question: how is it that water is a necessary condition for life, and water only? After these chapters a last one on the water molecule follows, where experimental methods are described to study this molecule that is not so easy to observe, another paradox.

We thus see that from now on we depart from the objects of preceding chapters, which were mainly related to the properties of H-bonds. In these preceding chapters aqueous systems were occasionally discussed, when some of their properties were considered as representative of

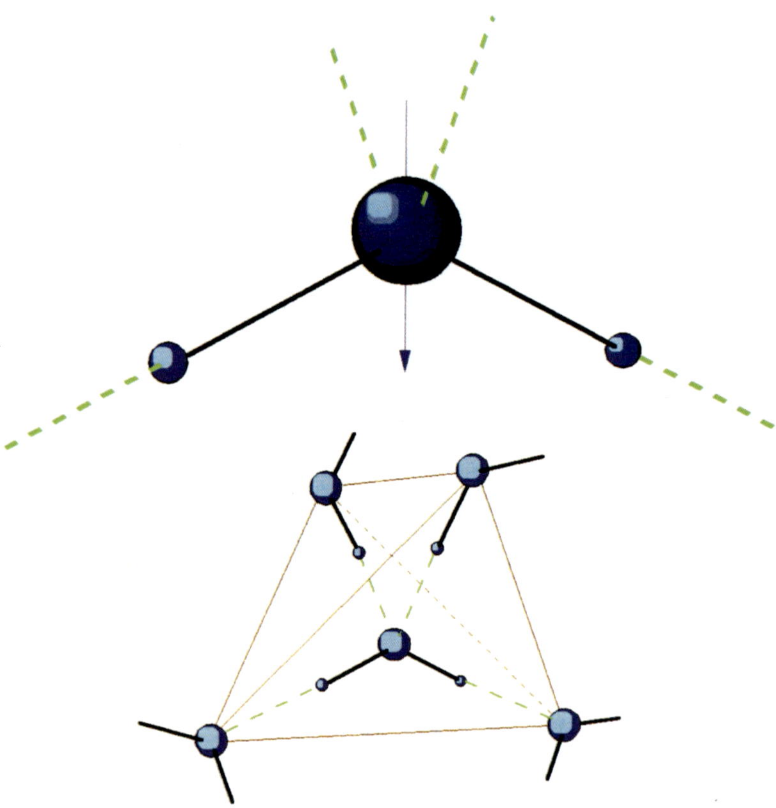

Figure 8.1 The water molecule (upper drawing) and its tetrahedral configuration in ice (lower drawing).

more general properties of H-bonds. We now concentrate on aqueous systems where a dense H-bond network always exists, with the exception, however, of the next section of this chapter, where isolated H_2O molecules are described.

WATER VAPOUR

The major greenhouse gas and its strong IR bands

The water molecule, which displays only σ bonds, exhibits no absorption bands in the visible region. Water vapour is consequently transparent in this spectroscopic region. It means that we do not see water vapour with our eyes, even if we often see white plumes in our surrounding atmosphere or clouds in the sky that we may be tempted to attribute to water vapour, but are in fact microdroplets of liquid water that do not absorb visible light, but scatter it. It shows that we have to carefully distinguish the various aspects of what we call "water" and in order to fix things, we always specify in this book whether we are considering water vapour, liquid water, a water molecule, etc. In the IR region, water vapour displays characteristic bands, shown in the IR spectrum of ambient atmosphere displayed in Figure 8.2, which makes H_2O molecules in the vapour phase most easily identified by IR

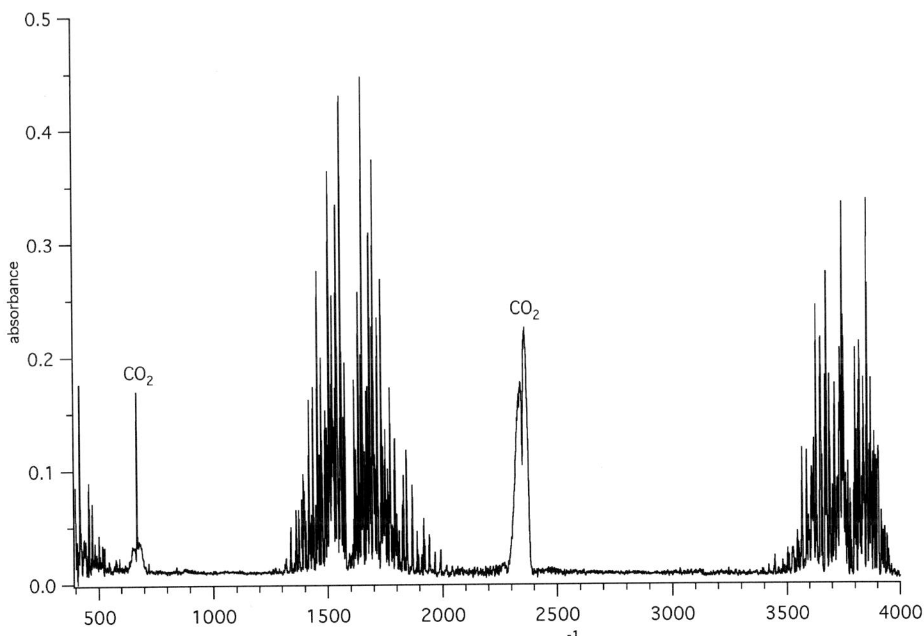

Figure 8.2 IR spectrum of a few centimetres of ambient atmosphere. The two bands due to atmospheric carbon dioxide are labelled CO_2.

spectroscopy. In this figure we see the bands due to the two main molecules at the origin of the "greenhouse effect", H_2O, responsible for more than two third of it, and CO_2, the molecule whose concentration is much less than that of H_2O, but has regularly increased in the atmosphere during the last 100–200 years and is thought to be responsible of the corresponding increase of the average Earth surface temperature. As these two molecules exhibit strong and broad absorption bands in the region 600–1800 cm^{-1}, they absorb an appreciable part of the radiation emitted towards the outer space by the Earth, a black body at temperature around 300 K. They are thus the two main gases of the atmosphere that keep this radiative energy at the Earth's surface. A characteristic feature of the H_2O molecules is their extremely rich rotational structure, which appears in the form of many very narrow bands with widths a few 1/100 cm^{-1}. This rotational structure is due to the presence of transitions between rotational levels that accompany vibrational transitions of gaseous molecules. As the resolution of the spectrum drawn in Figure 8.2 is 2 cm^{-1}, each narrow band encompasses several rotational sub-bands. Due to the very small moments of inertia of H_2O, these rotational sub-bands extend over a broad spectral region. No such detailed structure appears in the two vibrational bands of CO_2, because rotational bands are more compact, due to the moments of inertia of CO_2 being much bigger than those of H_2O. It makes individual rotational bands separated by much smaller distances. We thus only see, with this spectral resolution, the envelop of the rotational bands of CO_2, which extends over some 100 cm^{-1} for both its bands at 670 and 2350 cm^{-1}.

For H_2O molecules, we expect three internal vibration bands: two ν_s bands corresponding to the two O–H stretching vibrations and one bending $\delta_{H\text{-}O\text{-}H}$ band that corresponds to vibrations of the H–O–H angle around its equilibrium position of 104.5°. The $\delta_{H\text{-}O\text{-}H}$ band is the band centred around 1595 cm^{-1}, with rotational sub-bands that make it extend over 700 cm^{-1}. The two ν_s bands are centred at 3657 and 3756 cm^{-1} (2–4) and their overlapping rotational sub-bands extend over some 500 cm^{-1}. They correspond respectively to the symmetric and antisymmetric O–H stretching modes of the isolated H_2O molecule, called ν_1 and ν_3 respectively. The $\delta_{H\text{-}O\text{-}H}$ band has been called ν_2 by the first spectroscopists who recorded the spectrum of the water molecule. As nowadays ν stands for stretching bands, we keep its more recent labelling, $\delta_{H\text{-}O\text{-}H}$, with δ standing for bending bands. We may note that only individual H_2O molecules that do not establish H-bonds exhibit a rotational structure that extends over a broad spectroscopic region. These H_2O molecules keep a very small moment of inertia, which becomes markedly bigger when they form dimers, trimers, etc. H_2O molecules in n-mer then display rotational structures that extend over a much narrower spectroscopic region. Clearly bands with such rotational signatures are not visible in Figure 8.2. We conclude that in water vapour at ambient pressure a very small proportion of H_2O molecules establishes H-bonds on other H_2O molecules. Indeed the concentration of dimers in saturating water vapour is of about 5×10^{-3} only at 40 °C (5). Their roles in the thermal radiative balance of the Earth and also in many atmospheric processes, seem nevertheless more important than their low concentration suggests, a point that is much discussed in the meteorological scientific community (6). In these dimers ν_1 and ν_3 can still be identified (7), because the H-bond acceptor H_2O molecule of the dimer still keeps its two O–H groups free. In bigger n-meres, this is no longer the case, and two kinds of ν_s are seen: narrow ν_s(O–H) around 3700 cm^{-1} and much broader ν_s(O–H$\cdots$) at markedly lower wavenumbers, meaning that IR sees H-bonded O–H$\cdots$O groups and free

O–H groups. These two groups of configurations exist at the surface of liquid water, and are clearly distinguished in sum-frequency (8) vibrational spectra described in Ch. 4. The harmonic interaction that is at the origin of the 100 cm^{-1} splitting of ν_1 and ν_3 modes in isolated H_2O molecules is no longer visible, even if it is still present: the two O–H vibrations of a same H_2O molecules, which are resonant when this molecule is isolated, are no longer resonant when this H_2O molecule establishes H-bonds because of the great dispersion of the frequencies of ν_s(O–H···), seen in Chs. 4 and 5, and the separation of frequencies of ν_s(O–H) and of ν_s(O–H···) modes.

Formation of raindrops

The isolated H_2O molecule thus behaves as all other simple small molecules. It does not mean it is inert. In the atmosphere it is a central molecule that is strongly implicated in the physics and chemistry of the troposphere—that part of the atmosphere situated between 10 and 15–20 km from the Earth surface. In the troposphere and lower atmosphere, molecules such as H_2SO_4 or NH_3, the only significant base in the atmosphere (9), may establish H-bonds with it, forming aerosols that constitute nuclei for condensation of water vapour into water microdroplets. These water microdroplets are small. It means that their surface to volume ratio S/V is sufficiently great so that they remain sustained in the atmosphere, just as dust particles do. Owing to their strong absorption in the IR region, they contribute to the global climate balance (10). They are spherical, a shape that minimizes the number of "dangling" surface O–H groups that cannot establish H-bonds and consequently make spherical aggregates the clusters with lowest energy. The radii of these spheres are of the order of the wavelength of visible light. They consequently scatter it efficiently and may thus be detected by the eye. They may, however, gain enthalpy when two or several microdroplets collapse giving birth to bigger droplets, which exhibit a smaller S/V ratio, and therefore a smaller proportion of surface O–H dangling groups. Colliding microdroplets may thus coalesce into more stable bigger spherical droplets. This enthalpic effect is counterbalanced by an entropic one that favours a greater number of droplets and the equilibrium between coalescence and dispersion depends on the equilibrium between these two trends, an equilibrium that is a function of temperature and pressure. However, bigger droplets with their weaker S/V ratio can no longer be sustained in the atmosphere above some value of their radii. The water molecules then fall on the soil, in the form of raindrops. In addition, falling raindrops that contain these foreign molecules on which H_2O vapour molecules have initially nucleated, scavenge them from the atmosphere by solving them and driving them towards soils.

ICE (S)

The peculiarity of the H_2O molecule lies in the presence of four sites to establish H-bonds that are partitioned into two donor sites and two acceptor sites. An H_2O molecule is thus able to establish four H-bonds around it. This is what it does when surrounded by other H_2O molecules, and this gives a unique species where the *number of H-bonds is then equal to that of covalent bonds*, a really exceptional property. It happens in ice, but this will be seen to be also the case of liquid water. In ice H_2O molecules build highly ordered arrangements that are such that the two H atoms of each H_2O molecule establish two H-bonds on neighbour H_2O molecules, while the two lone-pairs of this same molecule accepts two other H-bonds from two other H_2O molecules. The four H_2O nearest neighbours of each H_2O molecule are consequently positioned in a tetrahedral arrangement around it (Figure 8.1, lower drawing). This is possible, at the price of a small distortion of the H-bonds: the angle H–O–H of individual H_2O molecules is equal to 104.5° in vapour. In ordinary ice, obtained by freezing liquid water or condensing water vapour at ambient pressure and temperatures below 0 °C, the value of the H–O–H angle falls in the vicinity of 106°, which allows accommodating the tetrahedral angle of 109.5° of the O-atoms with such a slight distortion of H-bonds.

Ice Ih and ice Ic

Crystal structures

Ordinary ice is called ice Ih for hexagonal ice. It is the only form of ice that is found in natural conditions, with maybe an exception mentioned below. It is the form of snow crystals, ice of glaciers, ice cubes that come from refrigerators, etc. Its structure is drawn in Figure 8.3. The O-atoms are found on summits of nonplanar hexagons having a chair form. Projection of these hexagons on a plane perpendicular to axis $\vec{c}$ gives regular hexagons (lower drawing). Between two adjacent O-atoms on a same nonplanar hexagon lies a single H-atom, covalently bonded to one of these two O-atoms and establishing an H-bond on the other O-atom. From the heat of sublimation of ice, Whalley (11) calculated the energy of formation E_{HB} of these H-bonds as being equal to some 23 kJ mol^{-1}. Each O-atom is in addition alternatively linked to the O-atom of a hexagon either of the upper layer or of the lower layer. This link is of the form O–H···O or O···H–O. The O-atoms in ice Ih are thus regularly assembled in a tetrahedral environment. Such a hexagonal arrangement is not so much usual. It is, however, encountered also in the case of silica, where the Si–O–Si motif replaces the O–H···O motif of ice Ih. Let us note that the electric dipole moment of individual H_2O molecules in ice Ih is of the order of 3.1 D (12), that is 1.7 D greater than that of isolated H_2O molecules in vapour, which we have seen to be of 1.83 D at the beginning of this chapter. This is a nice illustration of the cooperativity of H-bonds we have defined in Ch. 4.

In opposition to O-atoms, H-atoms are only partially ordered. They are found between two adjacent O-atoms, closer from one of them than from the other one. Their distribution between these two possible positions is not completely random, but obeys the "ice rule", which states that each O-atom is covalently bound to two H-atoms and at the same time accepts on its lone-pair electrons two H-bonds from two neighbour O–H groups. As many configurations of H-atoms can fulfil this ice-rule, the H-atoms are somewhat disordered. It means that the ground state of ice, the only one populated at 0 K, is degenerate, consisting of numerous states with same energy, which correspond to the numerous possible positions of the H-atoms. This is a situation which is scarcely encountered and is at the origin of a residual entropy at 0 K, which reflects the disorder of H-atoms at this temperature. As

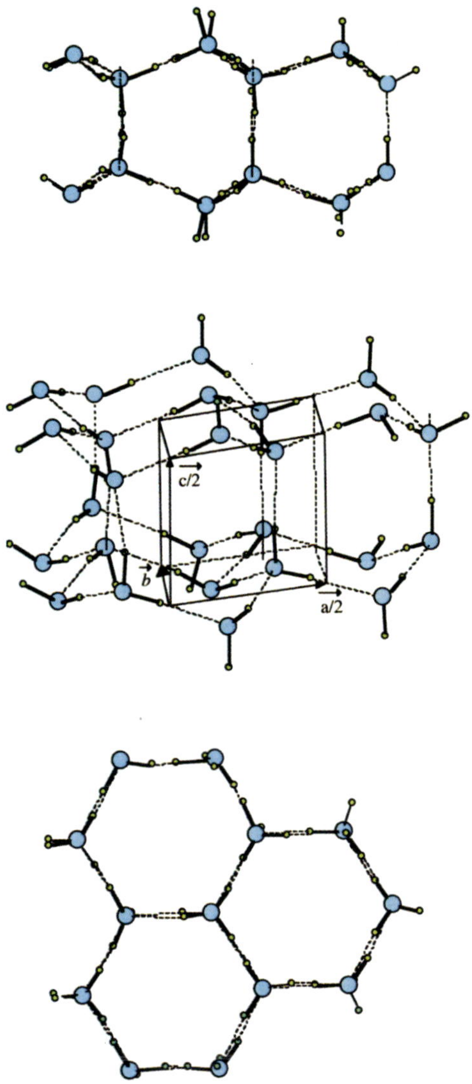

Figure 8.3 Hexagonal ice Ih: projection on the $\vec{a},\vec{c}$ plane (upper drawing) and on the $\vec{a},\vec{b}$ plane (lower drawing), where the hexagons are apparent. The middle drawing is an intermediate projection derived from the upper drawing by a rotation around $\vec{a}$ followed by a small rotation around $\vec{c}$.

L. Pauling has shown as early as 1935, the value of this residual entropy can be easily evaluated (13). For each O-atom, four energy boxes exist, which correspond to two covalent bonds and two H-bonds, and there are $C_4^2 = (4!/2!2!) = 6$ possibilities to distribute these two covalent and H-bonds among these four energy boxes (C_4^2 is the number of possible combinations of four objects taken two by two). With no ice-rule, which means with a complete random distribution of the H-atoms between the two positions they can take between two O-atoms, this number of possibilities would amount to $2^4 = 16$. For N H_2O molecules the number of configurations that obeys the ice-rule is consequently W, with

$$W = 4^N \left(\frac{6}{16} \right)^N = \left(\frac{3}{2} \right)^N \tag{8.1}$$

Using the Boltzmann's definition of entropy, we find the residual entropy S_r corresponding to this number of degenerate states for one mole of H_2O molecules to be, with R, the gas constant equal to 8.31 J deg^{-1},

$$S_r = R \ln \left(\frac{3}{2} \right) = 3.4 \text{ J deg}^{-1} \tag{8.2}$$

This is equal to the measured experimental value. Let us note that the Pauling's evaluation is a first approximation, as it treats all O-atoms as being independent and neglects correlations of configuration between neighbour O-atoms. This correlation enters for a small part only in the value of the residual entropy, but it has up to now not been calculated. The existence of such a residual entropy has no important consequences, except that if it is not properly taken into account, thermodynamical values may differ when measured from different experiments, thus lacking consistency.

On condensing water vapour at 193 K (−80 °C) a slightly different form of ice is obtained, which is cubic ice Ic. It is a metastable form with respect to ice Ih (11), which may coexist with it up to 273 K. Johari (14) gave rational explanations of the conditions of formation of ices Ic and Ih, based on the difference of surface energies of both forms. He concluded that small spherical clusters of ice Ic are the ice-stable form when radii of these spheres are smaller than some 15 nm and become metastable when radii are greater than some 15 nm. In that case, which is the most commonly encountered, the stable form is ice Ih. This 15 nm becomes 10 nm when such clusters are found in the form of films. Ice Ic thus appears preferentially to ice Ih when high pressure ice is decompressed towards atmospheric pressure at low temperatures (15). It takes on the cubic form of diamond, which is also that of crystals of silicium, germanium, etc. The basic arrangement of O-atoms is the same as in ice Ih: that of a tetrahedral ordering. The relative disposition of the various tetrahedrons is not the same. Thus the two tetrahedrons drawn in the two lower drawings of Figure 8.4 are the same. In these parts O-atoms 1, 2, 3 and 4 display the same relative configuration in both ice Ih and ice Ic. The tetrahedral configurations of atoms 5, 6, 7 and 8 are positioned differently with respect to atoms 1, 2, 3 and 4 in ice Ih (left) and ice Ic (right). This clearly appears in the projections perpendicular to the 1–5 axis that appears in the two upper diagrams. These projections clearly show that the configuration of the upper atoms 6, 7 and 8 of ice Ic is obtained from that of ice Ih by a 180° rotation around the 1–5 axis. This new configuration allows for a cubic elementary cell. In the left upper drawing of Figure 8.4, we recognize parts of the regular hexagons that appear in the $\vec{a}, \vec{b}$ plane of ice Ih (Figure 8.3). In the Ic crystal, H-atoms are disordered the same way as in ice Ih, with same consequent residual entropy at 0 K.

IR spectrum

In this chapter we do not display the IR spectrum of ice, as it conveys no new information that can be presently exploited, being consistent with the structure revealed by X-ray diffraction,

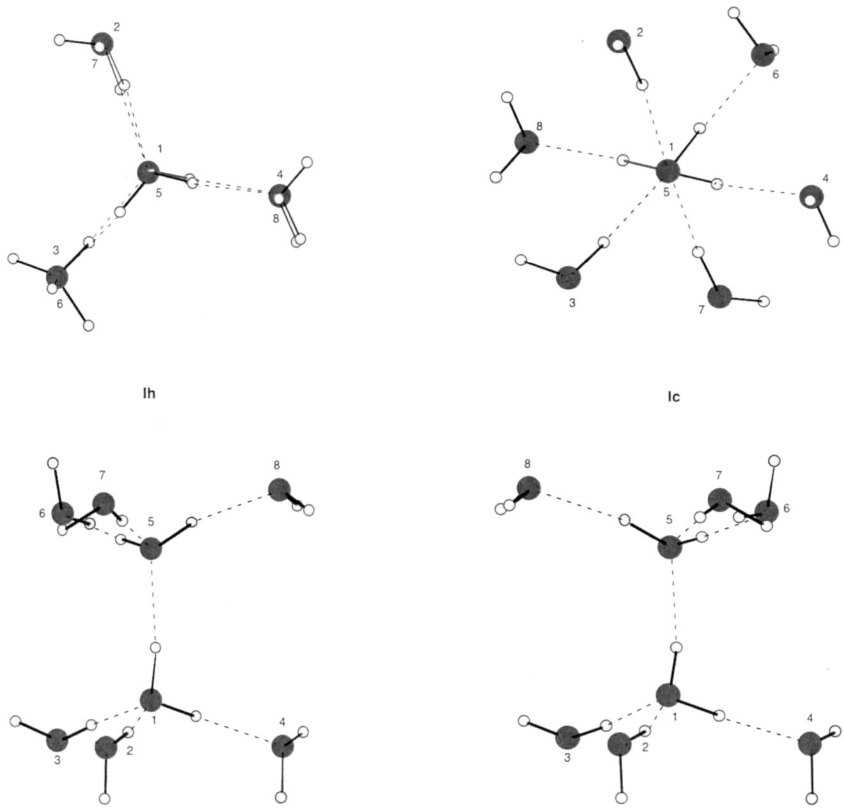

Figure 8.4 Tetrahedral orderings in ice Ih (left diagrams) and ice Ic (right diagrams). In both bottom drawings the axis defined by O-atoms 1 and 5 is the vertical axis of the paper sheet. The two upper diagrams are obtained from the lower ones by a 90° rotation around the horizontal axis of the paper sheet. The tetrahedra defined by O-atoms 1, 2, 3, 4 and 5 are the same in all drawings.

discussed previously. It fully agrees with the presence of a very dense H-bond network revealed by X-ray, displaying no band around 3700 cm^{-1} that would indicate the presence of free O–H groups. We nevertheless display this IR spectrum in Ch. 9 devoted to liquid water, as its comparison with the IR spectrum of liquid water will then allow to draw precise conclusion on the H-bond network of liquid water, which is far less precisely known than that of ice and for which X-ray or neutron scattering methods are much less informative. Concerning this IR spectrum of ice displayed in Figure 9.2, let us simply note two points. The first one is the value of the integrated intensity of the ν_s(O–H···O) band which is about 25 times that of the ν_s(O–H) band of water vapour (16), making H-bonds of ice appear as intermediate-strength H-bonds. The second point is the exceptional wavenumber at which the libration bands ρ_{H_2O} appear: around 830 cm^{-1}, its peak wavenumber, which is a value much greater than the wavenumber of the stretching intermonomer band, which appears at 214 cm^{-1}. Librations, also called hindered rotations, are the three relative rotations (intermonomer rotations) of individual H_2O molecules around their three main axes. Stretching intermonomer vibrations correspond to relative translations of individual H_2O molecules. We may realize how exceptional this appearance of the libration band at such high a wavenumber ($830\,cm^{-1}$) is, by comparing it to the bending intermonomer bands of other H-bonded systems: in the case of acetic acid dimers, which is a typical H-bonded system, the stretching intermonomer bands appear around $160\,cm^{-1}$, which compares quite well with the $230\,cm^{-1}$ of water made of lighter molecules. The intermonomer bending vibrations (17) appear well below $100\,cm^{-1}$, which is the rule for the great majority of other H-bonded systems. This value differs by one order of magnitude from the $830\,cm^{-1}$ of ice. This exceptionally rapid libration of H_2O molecules has especially important consequences in liquid water.

These librations are not free rotations, as in water vapour, but hindered rotations that are governed by force constants due to H-bonds between H_2O molecules. The fact that ρ_{H_2O} appears at such high a wavenumber is due to the particularly small moment of inertia of the H_2O molecule. In ice these librations have a limited amplitude, and may consequently be considered as vibrations. This will not be the case in liquid water, where librations will be seen to display great amplitudes that are at the origin of the dramatic differences between the properties of ice and liquid water.

Structures of ice surfaces

The crystalline structures of ices Ih and Ic described above are those found in the bulk. What happens of the organization of the H_2O molecules at the surfaces of these crystals? In ordinary crystals these surface molecules are not numerous enough to be detected and separated from the molecules in the bulk. However, Devlin and coworkers (18) developed elegant methods to grow nanocrystals of ice with diameters ranging from 2 to 100 nm. As seen above, these nanocrystals are Ic crystals. In these nanocrystals the numbers of H_2O molecules on the surface become appreciable as compared to that of H_2O molecules in the bulk, the ratio of the number of surface molecules to the number of bulk molecules varying as the inverse of the diameters of a spherical crystal. It gives a mean to isolate IR spectra due to bulk H_2O molecules and spectra due to surface H_2O molecules by comparing the spectra of clusters of different sizes. It consequently allows differentiating their structures. It has thus been found that the first layer of H_2O molecules on the surface is disordered and the transition between this layer and the organized crystalline bulk occurs through a "subsurface" layer made of more ordered H_2O molecules establishing bent H-bonds which extend towards about 3 nm from the surface. This structure implies that crystalline structure requires a minimum number of at least 300 H_2O molecules to appear. Below this number the ice clusters are made of disordered H_2O molecules on the surface and of more ordered H_2O molecules having the structure of the transition layer. Not so much surprising the libration bands ρ_{H_2O} of surface H_2O molecules appear at lower wavenumbers than those of bulk H_2O molecules (19): $750\,cm^{-1}$. The reason is that they are less H-bonded than bulk H_2O molecules and are consequently ruled by a smaller force constant. In these clusters, the "dangling" O–H groups at the surface of ice, more often called free O–H groups, are clearly visible (Figure 9.1) at $3694\,cm^{-1}$ (20), a value intermediate between the wavenumbers 3657 and $3756\,cm^{-1}$ of the symmetric and antisymmetric ν_1 and ν_3 modes of isolated H_2O molecules in the vapour phase. As explained in Ch. 9, this intermediate value is due to the rupture of symmetry between the two O–H groups of a same H_2O molecule that has one of it H-bonded and the other one free (dangling).

Infrared spectroscopy on thin samples of ice about 10–20 nm thick have also shown that the H-bond network on the surface of ice in contact with water vapour is not exactly the same as in the bulk, but more liquid-like, indistinguishable from that of liquid water (21). This transition layer extends over more than 10 H_2O layers at 0 °C but rapidly shrinks (22), becoming hardly a single layer at -10 °C. The structure of this layer on the surface of ice is not precisely known. Thus X-ray diffraction indicates that its density is more in the vicinity of 1.2 g cm^{-3} than around the expected value: 1 (23). It at least shows that the arrangements of H_2O molecules on the ice surface are definitely different from those in the bulk.

Other crystalline phases of ice

In these two previous crystalline forms, ices Ih and Ic, the H_2O molecules, with their H–O–H angle of some 106°, accommodate the tetrahedral arrangement of neighbouring molecules, with an angle of 109.5°, at the price of a small distortion of the O–H···O H-bonds. When pressure is applied to ice, the corresponding strain due to these not perfectly matching angles increases and at some values of the pressure strain becomes unbearable. The O-atoms are then reorganized to take different positions. It is at the origin of numerous various stable crystalline arrangements, described by the very rich phase diagram shown in Figure 8.5. This diagram is that initially proposed by Whalley and coworkers (25), completed of phases that are more recent. In 1984, Whalley and coworkers recognized nine various phases. In 1999, this number amounted to 10, which became 12 in 2002 (26) and 13 in 2004. One of this phase, ice X, with its H-atoms positioned at an equal distance of the two O-atoms to which it is bounded by one covalent bond and one H-bond, has not yet been experimentally put into evidence, but has been predicted from a theoretical model (27). Its domain of stability is that of pressures greater than 50 GPa = 500 kbar, which is a still difficult domain. In all these phases the O-atoms are tetracoordinated and the H-atoms partitioned in between O-atoms, obeying the ice-rule. For some of these ice forms (ices II, VIII and IX and hypothetical symmetrical X) the protons are fully ordered. Low-pressure ices Ih, II, III, IV, V and IX are open structures, while high-pressure ices VII, VIII and X are made of interpenetrating lattices which are not connected by H-bonds (28), but make the numbers of nearest neighbours of each H_2O molecules greater than four. The densities of these latter forms of ice are consequently higher than that of ice Ih. A somewhat ironical point is that the structure of most these various phases are more precisely known than that of ordinary ice Ih, where the positions of the O-atoms lack precise determination despite the numerous refinements that have been introduced in their analyses. A consequence is the relatively great uncertainty that characterizes the value of the H–O–H angle in ice Ih: around 106°. It indicates that our knowledge of the structure of ordinary ice is likely to be missing an elusive fundamental point.

Let us note that at the bottom of glaciers, where ice is in contact with rocks that support a great part of the weight of the glacier and may oppose its downwards flowing, the pressure may be locally sufficient for formation of ice IV or V (Figure 8.5). These may consequently be considered other natural forms of ice, even if they are encountered in exceptional situations.

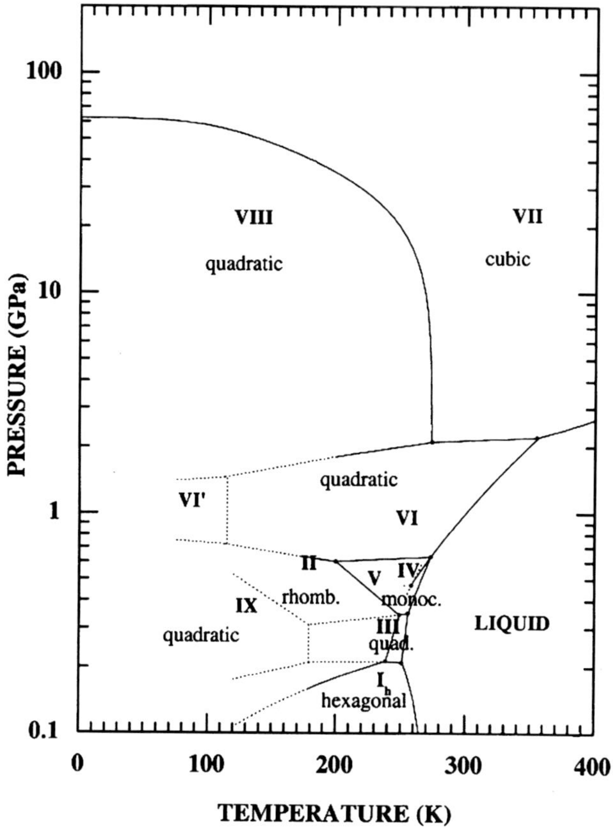

Figure 8.5 Semilog phase diagram of crystalline ice. Reproduced from (11) and (28) (courtesy of Dr Ph. Pruzan). The LDA ⇔ HDA transition is not shown. It appears around 0.2 GPa in the absence of hysteresis (24).

Ice Ih/liquid water interface

Figure 8.5 clearly shows that the line that separates ice Ih from liquid water displays a negative slope. It corresponds to a well-known property of ordinary ice that is highlighted by an also well-known experiment. It consists of straddling over an ice cube a wire on which weights are suspended. The wire slowly penetrates the ice cube. If we look at Figure 8.5, this line with negative slope indicates that applying a sufficient pressure on ice Ih at a temperature above −20 °C transforms it into liquid water. This is what the weighted wire does. This is an exceptional property, due to the presence in ice of an exceptionally dense H-bond network. This can be understood on a molecular basis: as shown by neutron scattering experiments (29) pressure on ice first destroys the tetrahedral symmetry adopted by O atoms. It means that pressure first bends H-bonds, weakening them and consequently increasing the fluidity of ice up to the point of transforming it into a liquid. Compressing H_2O molecules, that is reducing their O···O distances, appears as a secondary effect of pressure, whereas it is the primary effect on all other species that do not possess such a dense H-bond network. When pressure is applied on liquid water, where an as dense H-bond network is also found (see Ch. 9), it has a similar effect: at temperatures in the vicinity of 0 °C, the viscosity of liquid water decreases (or its fluidity increases) when pressure increases, a reaction in an opposite direction to that of all other liquids. It is also due to the same effect: bending of H-bonds upon pressure.

Skaters fully use this effect of pressure on the exceptional H-bond network of ice. At molecular level, the possibility of skating on ice is, however, more subtle. It has for long been thought that the pressure of the blade of a skating shoe is sufficient to transform ice Ih into liquid water. In reality it is far from sufficient, if one looks at values displayed in Figure 8.5. However, we have seen above that the surface of ice in contact with water vapour is not exactly the same as in the bulk, but more liquid-like. Thus if skating does not really occur on liquid water, this may not be far from it, as it is strongly favoured by the presence of this liquid-like fluid surface layer. We have also seen that this layer extends over more than 10 H_2O layers at 0 °C but rapidly shrinks (22), becoming hardly a single layer at −10 °C. It implies that skating is best practised on ice near its melting point at 0 °C than on colder ice, a point that is common knowledge among skaters. The existence of such a quasiliquid layer on the surface of a crystalline solid is not a specificity of ice, but also occurs on the surface of many other solids in equilibrium with their corresponding vapour. However, applying pressure on ice greatly favours sliding, as bending H-bonds makes this layer more fluid. This is not the case for other solids, where pressure, by shortening distances between molecules, makes it more viscous, that is less fluid, and consequently hampers sliding. The possibility of skating consequently owes much to the exceptional H-bond network of ice.

Amorphous phases of ice

Beside these stable crystalline phases described above, there exist metastable phases of ice. We have already seen the metastable crystalline cubic ice Ic. In addition, there exist various amorphous metastable forms. Some of them are obtained by an ultra rapid cooling of water vapour or microdroplets of liquid water down to temperatures in the range 77–200 K. Biologists have for long attempted to obtain noncrystalline ice, as crystalline ice has the inconvenience of having a volume 10% greater than that of liquid water from which it is produced, leading to destruction of biomaterials when ice appears, thus strongly hindering the study of these hydrated biomaterials at low temperatures. Vitreous ice may have a density close to that of liquid water, thus avoiding this drawback. The problem is that the H_2O molecule, with its particularly small moments of inertia and its exceptional ability to establish numerous H-bonds around it, most rapidly takes the optimum position of ice Ih or ice Ic with respect to neighbour molecules when it arrives in the vicinity of other H_2O molecules at ambient pressure. In other words, crystallization is hard to avoid when water is cooled under its freezing temperature. A lot of work has been done to obtain solid water in amorphous states. It is described in an especially clear review by Angell (30). Various methods to obtain such amorphous phases have been devised, and the various amorphous solids obtained have been given various names, which often refer to the way water was processed. Thus deposition of water vapour on a plate below 163 K gives amorphous solid water (ASW). Applying a pressure greater than 1 GPa (10 kbar) on ice Ih or ice Ic at 77 K (liquid nitrogen temperature) results in the formation of high-density amorphous ice (HDA) (25), which becomes a low-density amorphous ice (LDA) when heated around 120 K at atmospheric pressure. Quenching microdroplets of liquid water in a matrix or depositing it at a supersonic speed on a cryoplate gives hyperquenched glassy water (HGW) (31). The relation between these various forms are not clear (32, 33). Thus the densities of ASW, HGW and LDA are the same, around 0.94 g cm^{-3} at 136 K, and X-ray or elastic neutron scattering experiments are unable to distinguish between them. Methods that are more sensitive to their H-bond networks, such as IR absorption, Raman scattering or inelastic neutron scattering, allow to detect some small differences between them (34, 35), but no interpretation of these differences has yet been given. Furthermore, only small vibrational and thermodynamical differences exist between LDA, for instance, and crystalline ice. This is quite unusual for glasses that nearly always exhibit marked differences with crystals. Also, the density of LDA is thought to fall in the vicinity of 1.17 g cm^{-3}, but a value of 1.3 g cm^{-3} has recently been reported (30), which suggests that this density may depend on the route used to obtain LDA. Progress is thus necessary to clarify the structures and properties of these various amorphous forms of ice.

Despite the difficulty encountered to obtain them, these amorphous metastable ices are thought to be the most abundant form of water in the Universe, as they appear in this form on interstellar dust particles (30). In view of the temperatures found on most these dust particles, lower than 30 K, HDA is the most likely form to exist in these conditions. In laboratory conditions it appears when depositing vapour at temperatures below 30 K. These amorphous ices are also a matter of interest in exobiology, which beyond classical biology studies possible forms of life and the origin of life in the universe. They are thought to have played a fundamental role in bringing such light atoms as C, N and O at the surface of telluric planets, the four planets nearest to the Sun (Mercury, Venus, the Earth and Mars), which are the only ones to retain on their surface these atoms that are necessary for life to occur. These light atoms are thought to have appeared by nucleosynthesis some 10 billion years ago, before the formation of the solar system some 4.5 billion years ago. The Sun itself has not a big enough mass to synthesize these elements (36). However, remnants of this period of nucleosynthesis are found on the confines of the solar system, which are thought to be especially rich in these elements. One hypothesis nowadays is that they have been transported from these confines of the solar system to these telluric planets by comets that regularly accomplish back and forth trips between these borders of the solar system and the vicinity of the Sun. They would have been included inside amorphous ices, a possibility that crystalline ice does not offer, as foreign atoms or molecules are expelled from it during crystallisation. This is one of the reasons why satellites are launched towards comets, in order to precisely analyse the forms of their surface ice (s) and their compositions.

Reactivity of ice

Ice in the atmosphere

We have seen above that foreign molecules are expelled during crystallization of ice, the exception being the scarce molecules that occupy the same volume as an H_2O molecule and can thus be substituted to it. Ice (Ih) is consequently expected to have but a low reactivity. This is not exactly what experiments tell us. In fact, if ice indeed hardly reacts in its bulk, this is not so for its surface, which we have seen above to have H_2O molecules organized differently from the bulk and are disordered, thus displaying numerous sites for accepting or establishing H-bonds. These are respectively lone-pair electrons of O-atoms or free O–H groups, often called dangling O–H groups that point towards the external part of the solid. They may therefore establish H-bonds with foreign molecules that come in the vicinity of this ice surface. Thus ice surface adsorbs not only such acidic molecules as HCl, HNO_3, but also preacidic molecules such as SO_2. It makes ice-clouds efficient scavengers of polluting gases, all the more so that cirri, the ice clouds of the upper atmosphere, permanently cover a good proportion of the Earth surface, some 30%, and are stable clouds. As a consequence, various types of organic species, such as acetic, hydrochloric, nitric acids, etc., and also ethanol are later found on ice and snow deposited on the Earth surface (37).

Ozone depletion and polar stratospheric clouds

The importance, however, of ice clouds comes from the discovery around 1985 that the type of polar stratospheric clouds (PSC in the following) that consist of ice Ih, called type-II PSC, play a key role in the destruction of the ozone (O_3) stratospheric layer that protects life from harmful energetic UV radiations with wavelengths between 280 and 320 nm. The stratosphere is that part of the atmosphere that lies above the troposphere and extends some 25 km over it. This ozone destruction occurs in the early Antarctic spring, around September, and balloon sonde data have shown (38) that at this period of time most of the ozone resides between 10 and 20 km. Moreover, this depletion is quite recent, becoming important after the 1970s. Ozone is known to be destroyed by reactive free radicals such as OH, O, Cl or ClO. The first two radicals are present in the upper stratosphere as a consequence of the photolysis of such molecules as H_2O or O_2. Their concentrations, however, have not varied for a long time and cannot consequently explain the quite recent appearance of this seasonal ozone depletion. It points to the responsibility of Cl and ClO radicals. No data indicate that their concentration suffered recent variations of importance, but the concentration of chlorinated molecules did increase in the atmosphere with the photolysis of gaseous chlorofluorocarbons (CFCs), which are now banned but have been widely used during the second half of the 20th century as refrigerants and propellants. In the atmospheric layer between 10 and 20 km where most of O_3 is found during the period August to October (winter and beginning of spring), chlorine does not exist in the form of Cl atoms, which are found in the layer around 40 km all around the Earth, as a result of the photolysis of CFCs. However, chlorine reservoirs are found in this 10–20 km layer in the form of HCl or $ClONO_2$ molecules. These molecules, also found all around the Earth, are inert towards O_3. How is it then that ozone is destroyed only during the Antarctic spring and only, or nearly so, over Antarctica? This is because the appearance of Cl radicals from inert chlorine reservoirs does not proceed from a homogeneous gas phase reaction, but is heterogeneously catalysed by ice of type-II PSCs. These are most common during the austral winter in the lower stratospheric layer, between 15 and 25 km, where most of the O_3 resides. They are formed at temperatures below 188 K (39), a temperature found in the higher troposphere and lower stratosphere over Antarctica, in winter, between July and September. It more recently appeared that such conditions may occasionally be also found over the Arctic.

In other latitudes where temperature is too high for PSCs to appear, ozone has a lifetime which extends over several years. It clearly points to the central role these PSCs play.

Mechanisms of ozone destruction

The role of these ice catalysts is to transform these inert chlorine reservoir gases into Cl_2 molecules, following the general reaction

$$ClONO_2 + HCl \rightarrow Cl_2 + HNO_3 \tag{8.3}$$

that may occur in several steps, such as

$$ClONO_2 + H_2O \rightarrow HOCl + HNO_3 \tag{8.4}$$

$$HOCl + HCl \rightarrow Cl_2 + H_2O \tag{8.5}$$

Cl_2 atoms are kept in/on these PSCs during winter, and are photolysed in spring, as soon as the sunlight reappears, around September. It produces Cl atoms, or ClO radicals originating from HOCl, which attack and destroy ozone. Let us note that HNO_3, which appears in these reactions, is found in an appreciable concentration in the atmosphere. It originates from fertilizers, formation of various NO_x nitrogen oxides in car engines or jet combustions, etc. In the Antarctic stratosphere it is found in relatively great quantities in the form of type-I PSCs, which are believed to be crystals of nitric acid hydrates, mainly trihydrates (40, 41).

Reactions described in eqs. (8.3)–(8.5) are now well-established general reactions for the ozone depletion mechanism. Their occurrence, favoured by the seasonal presence of PSC acting as catalysts, explains their seasonal appearance (spring) and localization (the Antarctic stratosphere). There remain, however, many obscure points that require a more detailed knowledge of these reactions. Thus most of them occur on the surface of ice crystals, as ice expels most molecules other than H_2O from its bulk. This is particularly true of the fixation of such molecules as HCl or HNO_3, which are strongly fixed on ice by establishing H-bonds with the somewhat disordered H_2O molecules present on the surface, most probably thanks to the easy ionization of these strong acids (37). This conclusion comes out of a comparison of the adsorptions of these strong acids with that of nonionizable molecules such as SO_2, which has a solubility in liquid water comparable to these acids, and which shows that the quantity of SO_2 adsorbed on the ice surface is three orders of magnitude lower than that of these acids (42). Much effort has been devoted to the study of this ionization process, which has been shown to occur on ice surface at temperatures higher than 50 K (43), most probably following a mechanism initially proposed by Gertner and Hynes (44), where HCl first establishes H-bonds on the first layer of H_2O molecules of the ice surface. Doing so, it strongly perturbs the H-bond network of this first layer and then finds itself in a favourable position to ionize: as already seen in Ch. 6, the surrounding H-bond network is a key actor in the ionization process of such species as HCl. It is then incorporated in a surface sublayer, but not in the bulk. This is in agreement with neutron diffraction and quasielastic scattering experiments (40), which clearly show that at 190 K (the temperature of PSCs in winter) the structure of the ice film remains that of ice Ih, whatever the HCl coverage is, strongly suggesting that HCl does not penetrate the film. It is also in agreement with another experiment where D_2O molecules adsorbed on the surface of ordinary ice made of H_2O molecules are shown by IR spectroscopy (45) to remain as D_2O molecules. It means that no H/D exchange occurs on the ice surface, in opposition to what occurs in the bulk where D_2O molecules are promptly transformed into HDO molecules when the D/H ratio is weak. On the other hand, this exchange readily appears with the presence of a small amount of HCl on the surface, which strongly suggests a modification of the H-bond network of surface H_2O molecules by a small amount of HCl molecules.

From all these results that extend in many directions and may appear somewhat confusing, we see that the uptake of such molecules as HCl or HNO_3 is important on ice surfaces. These molecules ionize, thus penetrating a surface sublayer, but not the ice bulk (46). This raises many questions. One of them is: how can Cl_2 molecules be formed and kept for a sufficiently long time to wait for a photolysis that can only occur some months later during the spring season? No answer to this question exists up to now. However, several considerations might indicate a direction. First, the ice surface of PSCs are not static, but constantly adsorb and desorb gaseous H_2O molecules, as shown by optical interferometry on ice surfaces (47). This may be a process by which such molecules as $ClONO_2$ may be encapsulated in an ice matrix and then hydrolysed into photolysable HOCl (eq. (8.4)) or react with HCl to give also photolysable Cl_2 (eq. (8.5)). Second, if HCl encounters difficulties penetrating ice bulk in crystals, this is apparently not so with smaller ice nanoclusters (48). It may then be possible that very small PSC ice crystals play a special role in the fixation of photolysable molecules. Furthermore, the ice particles of PSCs may be somewhat different from ice Ih crystals, because during formation of these particles, HCl and HNO_3 molecules are present. The consequence may well be that these particles are not crystalline, in which case their structure and reactivities may completely differ from those of ice Ih crystals. This is what IR spectra of H_2O/HCl mixtures of various relative concentrations, cocondensed in the laboratory at 190 K (49), has shown, putting into evidence incompletely crystallized phases at low concentration of HCl. All this explains why much effort is presently devoted to the understanding of the chemical reactions catalysed by ice clouds that lead to the ozone depletion. It also shows that we are far from completely understanding these atmospheric reactions.

CONCLUSION

The H_2O molecule exhibits an exceptional ability to establish numerous H-bonds around it. This specificity is of no consequence in the water vapour where nearly all these molecules are found as isolated molecules. In this state, the H_2O molecule takes on the form of a most stable molecule that is found at the end of many a chemical reaction. The only specific property this molecule displays then, is its strong electric dipole moment and small moment of inertia, which makes it a molecule that strongly absorbs in a broad IR region, and is consequently at the origin of two third of the "greenhouse effect" around the Earth. The exceptional ability of this molecule starts having consequences in ice where H_2O molecules are surrounded by similar molecules. It then develops an exceptionally dense H-bond network that makes the number of H-bonds equal to that of covalent bonds. Van der Waals intermolecular forces are then negligible, in opposition to nearly all other molecular species where their importance is much greater. It comes then as no surprise that ice displays singular properties due to specific properties of H-bonds: upon compression, H-bonds are first bent, their shortening appearing as a secondary effect only. This is at variance with most other solids. The consequence of this effect is the appearance in the phase diagram of ices of a line separating ordinary ice (Ih) and liquid water that has a negative slope (Figure 8.5). Another related consequence is the possibility of skating on ice, which is thus seen to owe much to the presence of this especially dense H-bond network. Also, when pressure increases, H_2O molecules become more and more strained, due to the 104.5° of the H–O–H angle of individual molecules having difficulties accommodating the 109.5° angle imposed by the tetrahedral organization of O-atoms at high pressures. Ice responds by adapting its H-bond network so as to minimize this strain, and this is at the origin of its particularly rich phase diagram displayed in Figure 8.5.

The presence of this hyperdense H-bond network also gives ice a particular chemical reactivity on its surface, which is characterized by the presence of numerous potential H-bond acceptors or donors that do not establish H-bonds. It thus has a central role in the seasonal ozone depletion that occurs mainly over the Antarctic, but also over the Arctic, where this type of PSCs that are made of ice Ih, called II PSC, play a fundamental catalytic role to transform inert chlorinated gas molecules into Cl or ClO radicals that attack and destroy O_3. Much work still remains to be done to fully understand this catalytic process. Singular properties of ice are thus a direct consequence of the presence of such a dense H-bond network. This network is rigid. Its transformation in liquid water into an as dense but flexible H-bond network has dramatic consequences, giving liquid water numerous exceptional properties. This is the object of Ch. 9.

REFERENCES

1. Y. Maréchal, J. Mol. Struct., 700 (2004) 207.
2. Y. Marcus, Ion Solvation, Wiley, Chichester, 1985.
3. T. Köddermann, F. Schulte, M. Huelsekopf and R. Ludwig, Angew. Chem. Int. Ed., 42 (2003) 4904.
4. A. J. Lock and H. J. Bakker, J. Chem. Phys., 117 (2002) 1708.
5. N. Goldman, R. S. Fellers, C. Leforestier and R. J. Saykally, J. Phys. Chem. A, 105 (2001) 515.
6. I. Ptashnik, K. M. Smith, K. P. Shine and D. A. Newnham, J. Roy. Meteor. Soc., 130 (2004) 2391.
7. M. Farnik, M. Weiman and M. A. Suhm, J. Chem. Phys., 118 (2003) 10120.
8. E. A. Raymond, T. L. Tarbuck, M. G. Brown and G. L. Richmond, J. Phys. Chem. B, 107 (2003) 546.
9. L. J. Larson, A. Largent and F.-M. Tao, J. Phys. Chem. A, 103 (1999) 6786.
10. C. E. Lund Myhre, D. H. Christensen, F. M. Nicolaisen and C. J. Nielsen, J. Phys. Chem. A, 107 (2003) 1979.
11. E. Whalley, in P. Schuster and G. Zundel (Eds.), The Hydrogen Bond: Recent Theory and Experiments, Vol. 3, Ch. 29, North Holland, Amsterdam, 1976.
12. E. R. Batista, S. S. Xantheas and H. Jonsson, J. Chem. Phys., 109 (1998) 4546.
13. J. W. Perram, in Schuster, G. Zundel and C. Sandorfy (Eds.), The Hydrogen Bond: Recent Developments in Theory and Experiments, Vol. I, Ch. 7, North Holland, Amsterdam, 1976.
14. G. P. Johari, J. Chem. Phys., 122 (2005) 194504.

15. H. Tanaka, J. Chem. Phys., 108 (1998) 4887.
16. J. E. Bertie, A. M. Khalique and H. H. Eysel, J. Phys. Chem., 93 (1989) 2210.
17. J. E. Bertie and K. H. Michaelian, J. Chem. Phys., 77 (1982) 5267.
18. J. P. Devlin, C. Joyce and V. Buch, J. Phys. Chem. A, 104 (2000) 1974.
19. M. W. Severson, J. P. Devlin and V. Buch, J. Chem. Phys., 119 (2003) 4449.
20. B. Rowland, M. Fisher and J. Paul Devlin, J. Phys. Chem., 97 (1993) 2485.
21. V. Sadtchenko and G. E. Ewing, J. Chem. Phys., 116 (2002) 4686.
22. H. Bluhm, D. F. Ogletree, C. S. Fadley, Z. Hussain and M. Salmeron, J. Phys. Condens. Matter, 14 (2002) 227.
23. S. Engemann, H. Reichert, H. Dosch, J. Bilgram, V. Honkimäki and A. Snigirev, Phys. Rev. Lett., 92 (2004) 205701.
24. S. Klotz, Th. Strässle, R. J. Nelmes, J. S. Loveday, G. Hamel, G. Rousse, B. Canny, J. C. Chervin and A. M. Saitta, Phys. Rev. Lett., 94 (2005) 25506.
25. O. Mishima, L. D. Calvert and E. Whalley, Nature, 310 (1984) 393.
26. C. Salzmann, I. Kohl, T. Lörting, E. Mayer and A. Hallbrucker, J. Phys. Chem. B, 106 (2002) 1.
27. W. B. Holzapfel, J. Chem. Phys., 56 (1972) 712.
28. P. Pruzan, J. C. Chervin, E. Wolanin, B. Canny, M. Gauthier and M. Hanfland, J. Raman Spectr., 34 (2003) 591.
29. M. C. Bellissent-Funel and L. Bosio, J. Chem. Phys., 102 (1995) 3727.
30. C. A. Angell, Ann. Rev. Phys. Chem., 55 (2004) 559.
31. E. Mayer, J. Microsc., 140 (1985) 3.
32. E. Mayer, in M.-C. Bellissent-Funel (Ed.), Hydrogen Bond Networks, Vol. 435, Kluwer Academic Publishers, Dordrecht, 1994.
33. B. Guillot and Y. Guissani, J. Chem. Phys., 119 (2003) 11740.
34. H. Kanno, K. Tomikawa and O. Mishima, Chem. Phys. Lett., 293 (1998) 412.
35. D. D. Klug, C. A. Tulk, E. C. Svensson and C.-K. Loong, Phys. Rev. Lett., 83 (1999) 2584.
36. J.-O. Baruch, M. Cassé and F. de Oliveira Santos, La Recherche, 382 (2005) 89.
37. O. Sokolov and J. P. D. Abbatt, J. Phys. Chem. A, 106 (2002) 775.
38. S. Solomon, R. R. Garcia, F. S. Rowland and D. J. Wuebbles, Nature, 321 (1986) 755.
39. J. Harnett, S. Haq and A. Hodgson, J. Phys. Chem. A, 106 (2002) 9226.
40. B. Demirdjian, D. Ferry, J. Suzanne, C. Toubin, S. Picaud, P. N. M. Hoang and C. Girardet, J. Chem. Phys., 116 (2002) 5143.
41. K. D. Beyer and A. R. Hansen, J. Phys. Chem. A, 106 (2002) 10275.
42. S. M. Clegg and J. P. D. Abbatt, J. Phys. Chem. A, 105 (2001) 6630.
43. L. Delzeit, K. Powell, N. Uras and J. P. Devlin, J. Phys. Chem. B, 101 (1997) 2327.
44. B. J. Gertner and J. T. Hynes, Science, 271 (1996) 1563.
45. M. A. Everest and C. J. Pursell, J. Chem. Phys., 115 (2001) 9843.
46. S.-C. Park and H. Kang, J. Phys. Chem. B, 109 (2005) 5124.
47. D. R. Haynes and S. M. George, J. Phys. Chem., 96 (1992) 8503.
48. N. Uras and J. P. Devlin, J. Phys. Chem. A, 104 (2000) 5770.
49. I. Xueref and F. Dominé, Atmos. Chem. Phys., 3 (2003) 1779.

– 9 –

The H_2O Molecule in Liquid Water

Liquid water is the same species as ice, but in a different physical state, that is with a different molecular organization due to changes of temperature and/or pressure. We have seen in the preceding chapter that a specific property of ice at the molecular level is the existence of an exceptionally dense H-bond network: as many H-bonds as covalent bonds, a nearly unique situation due to the unique property of the H_2O molecule, which is able to accept two H-bonds and at the same time establish (donate) two H-bonds. The rigidity of this dense H-bond network of ice is at the origin of a very rich phase diagram (Figure 8.5) that characterizes the various crystallographic stable structures obtained upon varying pressure or temperature. Another consequence of the rigidity of this dense H-bond network is the existence, in this phase diagram, of a negative slope for the line that corresponds to equilibrium between ice Ih and liquid water. It points to the exceptional possibility ice offers to skate upon it. Temperature and pressure are two quantities that strongly influence H-bonds, in opposition to stronger bonds such as covalent or ionic bonds, which are much less sensitive to these physical parameters. We therefore expect the H-bond network of liquid water to be different from that of ice. How different? The answer is not straightforward and is examined in the next section. Consequences of these differences are later examined in following sections.

H-BONDS IN LIQUID WATER

We may wonder what happens at molecular level when ice, a crystal, melts and becomes a liquid. In most liquids the correlations of positions and orientations of constituent molecules rapidly fall off when the distance between them increases. Furthermore, correlations between nearby molecules display rapid fluctuations that make the liquid a fluid. Most of these liquids, however, consist of molecules that mainly interact via weak Van der Waals forces, which are, at room temperature, much less directional than H-bonds, the only molecular interactions established by H_2O molecules in liquid water. The first idea that usually comes to mind is then that a relatively great proportion of H-bonds are broken in liquid water, so that H_2O molecules may gain some independence, making these correlations fall off rapidly with distance. Experiments tell us that the proportion of broken H-bonds is much too small in liquid water to be at the origin of its fluidity. We rapidly examine the first type of such experiments, thermodynamics, and describe in more detail a more informative type of experiment: IR spectroscopy, from which will emerge an image of the H-bond network of liquid water that will point to the origin of the loss of spatial correlations between distant H_2O molecules and to the rapid fluctuations of relative positions of nearly lying molecules.

Thermodynamics

As already seen in Ch. 8, Whalley (1) found that the energy of formation E_{HB} of H-bonds in ice is

$$E_{HB} \approx -23\,\text{kJ mol}^{-1} \tag{9.1}$$

This energy is that retrieved when transforming one H-bonded O–H···O group in ice into one free O–H group. It compares quite well with the H-bond energy of formation of a water dimer, which falls in the vicinity of $-21\,\text{kJ mol}^{-1}$ (2, 3). On the other hand, the heat of fusion of ice, ΔH_f is equal to

$$\Delta H_f \approx 80\,\text{cal g}^{-1} \approx 6\,\text{kJ mol}^{-1} \tag{9.2}$$

Supposing that the concentration of broken H-bonds (or free O–H groups) in liquid water is C, and that H-bonds have same energy in water and ice, we write:

$$\Delta H_f = -2CE_{HB} \tag{9.3}$$

where the presence of the factor 2 is due to the fact that in 1 mol of H_2O molecules, we have 2 mol of H-atoms, and consequently $2C$ mol of broken H-bonds. From these eqs. (9.2) and (9.3) we deduce:

$$C = 0.13 \tag{9.4}$$

This is too small a quantity for broken H-bonds to be at the origin of the fluidity of water, because with such a concentration of broken H-bonds, all H_2O molecules still remain strongly bound. In consequence, the correlations of positions and orientations of nearby H_2O molecules can hardly display rapid fluctuations. In order to do that, all H-bonds of ice should be transformed into somewhat different H-bonds of liquid water, instead of having only a few disrupted H-bonds. Note that we have supposed in eq. (9.4) that intact H-bonds have the same energy in liquid water and in ice. Relaxing this hypothesis would decrease the value of C, because part of ΔH_f would then be used to obtain weaker H-bonds, and the energy used to break H-bonds would consequently be smaller than ΔH_f. Increasing the value of C thus requires, in this simplified scheme, that H-bonds in liquid water are stronger than in ice, a supposition that is hard to justify and is supported by no experimental result.

IR spectroscopy

O–H free groups?

As already seen in Ch. 4, IR spectroscopy is by far the most precise method to provide information on H-bonds. It is therefore interesting to examine this particular point of the existence of free O–H groups, or of broken H-bonds in liquid water. This is the object of Figure 9.1 where the IR absorption spectrum of a 1 μm thick film of liquid water is displayed together

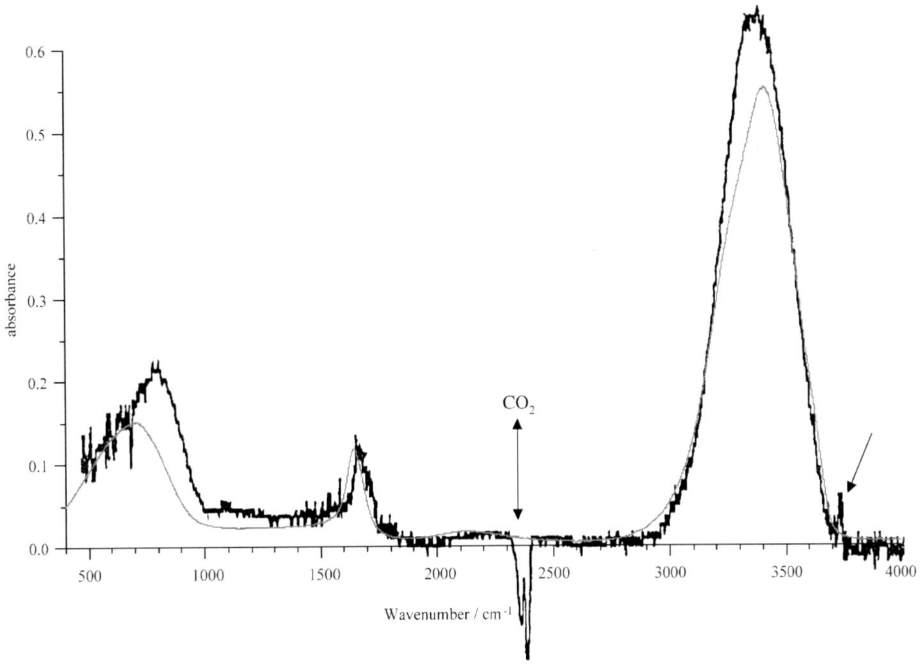

Figure 9.1 Infrared absorption spectra of a film of liquid water 1-μm thick (thin line) and of a cluster of H_2O molecules (coarse line; reproduced with permission from Ref. (4), copyright (2006) American Chemical Society). The intensity of this cluster spectrum has been magnified in such a way that its bending band around 1630 cm^{-1} has an integrated intensity comparable to that of the liquid water spectrum. A negative band due to residual gaseous CO_2 is indicated by a double-headed arrow. The inclined single-headed arrow points to the band due to free O–H groups in the spectrum of the cluster.

with that of a small cluster of H_2O molecules, magnified in view of an easy comparison. The somewhat unusual way in which the spectrum of liquid water has been obtained, due to the unusually strong absorption of liquid water in this mid-IR region, is described in Ch. 11 in detail. This spectrum of liquid water mainly consists of three bands that correspond to three types of vibrations of H_2O molecules: the most intense ν_s(O–H···) stretch band with maximum around 3400 cm^{-1}, the $\delta_{H\text{–}O\text{–}H}$ bending band (vibration of the H–O–H angle) with maximum around 1640 cm^{-1} and the "libration" band ρ_{H_2O} around 700 cm^{-1} (relative rotations of rigid H_2O molecules, also called hindered rotations). The intramonomer bands ν_s and $\delta_{H\text{–}O\text{–}H}$ correspond to those of water vapour (Figure 8.2). They are then called internal vibration bands. The integrated intensity of ν_s(O–H···O) in liquid water is about 17 times that of ν_s(O–H) in vapour (5), whereas that of $\delta_{H\text{–}O\text{–}H}$ has similar magnitude in both states. The shapes of these two bands are completely different due to the presence, in the spectrum of water vapour, of a rotational substructure, which is absent in the spectrum of liquid water. In water vapour, isolated H_2O molecules perform free rotations of great amplitudes. No force constant governs these free rotations. Transitions between these various rotational levels appear in the microwave region for molecular complexes with great moments of inertia (Ch. 3). For such molecules as H_2O with small moments of inertia, these rotational transitions also appear in the far infrared (FIR) region. In the case of water vapour, they thus appear in the region 0–200 cm^{-1}. In addition, rotational transitions accompany a (internal) vibrational transition, giving the corresponding bands of water vapour in Figure 8.2 their "rotational structure". In liquid water these rotations become relative rotations of individual neighbour H_2O molecules that are governed by a force constant due to the H-bonds established between these neighbour H_2O molecules. These rotations thus acquire a vibrational character. Hence their name: librations. They appear as well-isolated vibrational bands, ρ_{H_2O}. As a consequence, no vibrational band exhibits any rotational structure in the spectrum of liquid water. This is also true for the IR spectrum of ice. Librations in liquid water have smaller amplitudes than the free rotations of H_2O molecules in water vapour. Other differences furthermore appear between the internal vibration bands of water vapour and their corresponding intramonomer bands of liquid water. Thus, the ν_s band of water vapour exhibits two bands, one symmetric ν_1 at 3657 cm^{-1} and one antisymmetric ν_3 at 3756 cm^{-1}, which reflect the symmetry of the isolated H_2O molecule. No two such components appear in the ν_s band of liquid water, which is, in a first approximation, a band with a well-defined peak around 3400 cm^{-1}. This is because the coupling between the two ν_s vibrations of the two O–H groups on a same H_2O molecule, which is responsible for the $\nu_1 - \nu_3$ splitting in water vapour (3657–3756 cm^{-1}), is much smaller than the shifts of the frequencies of these ν_s bands due to H-bonds (from 3700 cm^{-1}, the average value for ν_1 and ν_3 in water vapour, to about 3400 cm^{-1} for O–H···O in liquid water). It means that the two H-bonds established by the two O–H groups of a single H_2O molecule destroy the symmetry of this H_2O molecule, because they display differences that, even if small, are visible in the IR spectra. This rupture of symmetry by H-bonds in water already appears in water oligomers made of clusters of more than two water molecules, where one distinguishes only ν_s of free O–H groups and ν_s of H-bonded O–H groups (6). The situation we find also at the surface of liquid water for which sum-frequency generation (SFG) spectroscopy evocated in Ch. 4 observes two kinds of O–H groups: "dangling" and H-bonded.

The spectrum of clusters of H_2O molecules (Figure 9.1) displays noise, but we should remember that this spectrum corresponds to a small number of H_2O molecules and has been magnified by a factor in the vicinity of 100, if we refer to the absorbance at the maximum of the $\delta_{H\text{-}O\text{-}H}$ band around 1640 cm^{-1}, which is of about 1.2×10^{-3} in the original publication (4) and of about 0.12 in Figure 9.1. The instructive point in this spectrum is the presence in this cluster of free O–H groups, which clearly appear in the form of a narrow band in the vicinity of 3700 cm^{-1} (slanted arrow), the wavenumber at which "dangling" O–H groups of amorphous ice, that is free O–H groups on the surface, display their ν_s bands (7). These dangling O–H groups are particularly numerous in small size clusters where the surface-to-volume (S/V) ratio is particularly great as compared to that of bigger samples of same shape (it varies as the inverse of the size of the sample for a given geometry). The small width of this band around 3700 cm^{-1} allows it to be easily detected, in spite of the relatively weak integrated intensities of ν_s bands of free O–H groups (Ch. 4). Vibrational SFG spectra of liquid water (8–11), also described in Ch. 4, display bands due to free O–H groups at the liquid–vapour interface that may have an amplitude greater than the band due to H-bonded O–H groups (their integrated intensity remain smaller), making them most visible. Such a band due to O–H free groups is conversely clearly absent in the spectrum of liquid water in Figure 9.1. It means that the concentration of such free O–H groups is low in liquid water, certainly much less than 5%. And indeed independent analyses of spectra of water, heavy water and mixtures of them (12, 13) gave a concentration of free O–H groups smaller than 1% in liquid water. Furthermore, recent nonlinear time-resolved IR spectra of HDO in heavy water, described in Chs. 4 and 7, unambiguously showed that broken H-bonds are intrinsically unstable and return to a H-bonding partner extremely rapidly, within 200 fsec (14). It implies that "a broken H-bond that is stabilized by liquid disorder appears to be more of a curiosity than a key player".

IR spectra of liquid water and of ice

These IR spectra therefore clearly point to a very low concentration of broken H-bonds in liquid water. It means that, as in ice, H_2O molecules in liquid water establish the maximum of H-bonds they can, which implies that each O-atom is surrounded by four nearest-neighbour O-atoms: two from which it accepts H-bonds and two on which it establishes H-bonds. In other words, the H-bond network of water is, as that of ice, a very dense H-bond network, with each H_2O molecule surrounded by four neighbour molecules on which it establishes or accepts H-bonds. There nevertheless remains that H-bonds of water and ice must display differences because, passing from the H-bond network of ice to the H-bond network of liquid water requires injecting thermal energy that is almost entirely absorbed by the H-bond network: individual H_2O molecules can hardly absorb such an energy, because their internal vibrations have all energy separations $h\nu$ much greater than kT at $T = 273$ K, not to speak of their electronic levels which are separated by even greater quantities. This thermal energy is consequently absorbed by intermonomer modes, which are relative vibrations of neighbour H_2O molecules of lower frequencies and may consequently be such that their $h\nu$ is of the order of kT at 273 K. These intermonomer modes, which include librations, are vibrations of the H-bond network. Furthermore, ice is a solid whereas water is a liquid, and in the same way, individual H_2O molecules are not expected to be different between these two states. This difference between crystal and liquid must consequently appear as a difference in the H-bond network. It appears in the X-ray scattering pattern of liquid water, where pair-correlation functions for O-atoms shown in Figure 3.4 display maxima that correlate quite well with Bragg peaks of ice at short distance, but disappear at greater distances and consequently show no correlation with Bragg peaks of ice of higher h, k and l indices. This was expected as, for any liquid, the ordering of atoms is well marked at short distances but disappears at longer distances. No matter how interesting these results on distances between O-atoms are, they give no precise information on H-bonds themselves. They can consequently hardly tell us what this loss of order at greater distances consists of in the H-bond network. We therefore have to turn towards other methods, and once more we turn towards IR spectroscopy. We then expect relatively precise conclusions concerning the H-bond network of liquid water, but we have to pay it at the price of a somewhat detailed analysis of IR spectra. It constitutes a good illustration of the power of IR spectroscopy in the study of H-bonds, and we perform it in the following. A reader who is not interested in the details of this analysis may skip it, directly reading the corresponding conclusion in the last paragraph of this subsection, before reading the following subsection "Structure of the H-bond network of liquid water".

Small differences between the spectra of ice and liquid water indeed appear in Figure 9.2 that displays absorption spectra of 1-μm thick films of both species. We have already identified in Figure 9.1 the three main bands ν_s, $\delta_{H\text{-}O\text{-}H}$ and ρ_{H_2O} that constitute the spectrum

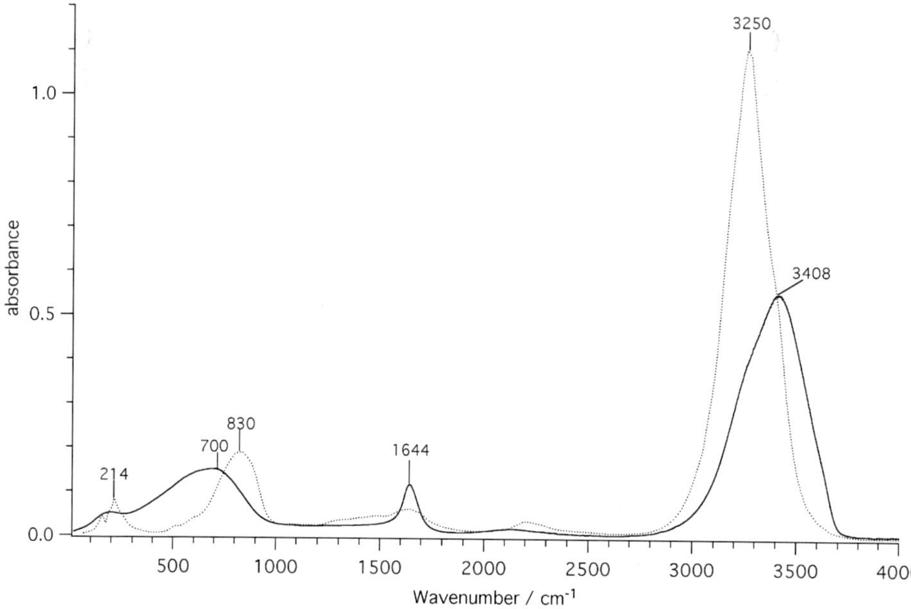

Figure 9.2 Infrared absorption spectra of 1-μm thick films of liquid water at room temperature (solid line) and of ice at −7 °C (dotted line). Absorbances $\log(I_0/I)$ for both species are calculated from the values of the absorption coefficient $k(\tilde{\nu})$, using eq. (5.A8) written in the form: $\log(I_0/I) = 4\pi\tilde{\nu}k(\tilde{\nu})l\log(e)$, with $l=10^{-4}$ cm and wavenumbers $\tilde{\nu}$ expressed in cm^{-1}. Values of k for ice are those given by Warren (16). Values of k for liquid water are calculated from ATR spectra (17) in the mid-IR region ($\tilde{\nu} > 550$ cm^{-1}) and are those given by Zelsmann (18), as calculated from absorption spectra, in the FIR region ($\tilde{\nu} < 600$ cm^{-1}).

of liquid water in the mid-IR region ($\tilde{\nu} > 400$ cm^{-1}). In Figure 9.2, we also include the FIR region ($\tilde{\nu} < 400$ cm^{-1}). It puts into evidence a supplementary hump around 200 cm^{-1}, the intermonomer stretching vibration band, which is the band due to relative stretching vibrations between two H-bonded H_2O molecules. In the spectrum of ice, ν_s, ρ_{H_2O} and this latter intermonomer stretching vibration are clearly visible with their maxima at 3250, 830 and 214 cm^{-1}, respectively. In opposition, the $\delta_{H\text{–}O\text{–}H}$ band is considerably broadened, and consequently less apparent. It spreads over two broad bands with maxima around 1650 and 2200 cm^{-1} due to a Fermi resonance of $\delta_{H\text{–}O\text{–}H}$ with the first overtone of ρ_{H_2O} (15). The bands of ice thus clearly display a close correspondence with those of liquid water, which is a mark of their H-bond network being closely related. They also display, however, differences that we analyse in the next subsections, in view of extracting information on the H-bond network of liquid water. These differences can in no way be explained by the relatively small temperature difference (–7 and 25 °C) of the spectra of ice and liquid water.

Intermonomer bands

Intermonomer bands, due to relative vibrations of H-bonded H_2O molecules, both of a translational type (the intermonomer stretching band) and of a rotational type (librations ρ_{H_2O}), display a dramatic increase of their widths when passing from ice to liquid water. This is particularly marked for ρ_{H_2O}, which, in the case of ice, is the band with maximum at 830 cm^{-1}. This band is not really narrow, but we have to remember that it indeed consists of three bands, corresponding to rotations of individual H_2O molecules around their three axes of rotations. The libration band ρ_{H_2O} in liquid water, with its broad maximum at 700 cm^{-1} is strikingly broader, as it extends towards very low wavenumbers and overlaps the stretching intermonomer band, the hump around 200 cm^{-1}. In ice this stretching intermonomer band, with its maximum at 214 cm^{-1}, is much sharper and is well separated from ρ_{H_2O}. It means that in ice, the amplitudes of intermonomer vibrations remain small and can be well described by harmonic potentials, which are at the origin of narrow bands in IR spectra. This is true even for libration bands, meaning that ρ_{H_2O} in ice can be described by three nearly resonant harmonic potentials at the origin of three bands with relatively small widths. This is no longer true for ρ_{H_2O} and intermonomer stretching vibrations in liquid water, which have *so great amplitudes* that they cannot be described by harmonic potentials, and should rather be called oscillations. The increase of the integrated intensity of ρ_{H_2O} in water, as compared to that of ice, also clearly indicates a significantly greater amplitude of librations in liquid water, rejecting the possibility that the broadening be due to the loss of order in liquid water, as compared to ice. It is compatible with a weakening of the force constant that is responsible for the shift of this band towards lower wavenumbers (700 cm^{-1}, as compared to 830 cm^{-1} in ice).

Intramonomer bands

The intramonomer bands exhibit less marked differences, if we do not take into account the effect of the Fermi resonance on $\delta_{H\text{–}O\text{–}H}$ in ice, which makes a direct comparison of $\delta_{H\text{–}O\text{–}H}$ in liquid water and ice difficult. However, these bands are not the same. Thus, the ν_s band of ice, with its sharp maximum around 3250 cm^{-1}, appears at lower wavenumbers than that of liquid water, which culminates at 3408 cm^{-1}. It is also markedly narrower and its integrated intensity is greater by a factor of 1.4 (5). This is surprising in view of what we have seen in Ch. 4 on the IR spectra of H-bonded systems: a shift of ν_s towards lower wavenumbers, due to a strengthening of an H-bond and accompanied by an enhancement of its integrated intensity, is also accompanied by a broadening of this band, not by a narrowing. Let us examine this point in more details. We have seen in Ch. 4 that the predominant factor that determines the position of the centre of ν_s is the amplitude of the anharmonic coupling of ν_s with low frequency intermonomer vibrations, which characterizes its strength. This is the only factor in the case of an isolated H bond. In the case of identical interacting H-bonds, such as those found in ice and liquid water, harmonic couplings between ν_s vibrations of neighbour O–H groups also contribute (Eq. (5.A51)) to the position of the centre of ν_s with, however, a much smaller effect. For H_2O molecules we have two kinds of such couplings: each O–H stretching vibration interacts with the stretching vibration of the other O–H group situated on the same H_2O molecule, and also with the two O–H groups of the neighbour molecule on which it establishes an H-bond. The first interaction is not expected to vary in a significant way between ice and water, because it is a property of the H_2O molecule, which can only very slightly be modified by a change in the H-bonds that particular H_2O molecule establishes. The second interaction may be more affected by a change in the H-bond between the two H_2O molecules. It is, however, expected to be at the origin of a separation of the centres of the ν_s bands of ice and liquid water of the order of some tens of cm^{-1}, much smaller than the $\approx 160\,cm^{-1}$ separation that appears in Figure 9.2 between the two ν_s bands at 3250 and $3408\,cm^{-1}$. In cubic ice Ic, the magnitude of this interaction has been found equal (19) to $61\,cm^{-1}$. As this interaction is expected to be of the same magnitude between ice Ih and ice Ic, a variation of at most some tens of cm^{-1} is consequently expected between ice Ih and liquid water. This shift of $\approx 160\,cm^{-1}$ is consequently a consequence of a different anharmonic modulation of ν_s by low-frequency intermonomer modes of the H-bond network, the preponderant factor that determines the positions of the centres of ν_s bands in isolated H-bonds.

If one admits that ν_s is anharmonically coupled to stretching intermonomer vibrations Q_s only, which is sufficient to precisely reproduce the lineshapes (20) of ν_s of most H-bonds, one nevertheless encounters a difficulty. We have seen in Ch. 4 that the linear dependence on Q_s of the frequency ν of ν_s, or more precisely of $\omega = 2\pi\nu$, is the key parameter that defines this lineshape, at least in the case of weak and intermediate H-bonds, which is the case of liquid water and ice. The centre $\bar{\omega}$ of the ν_s band is then equal (eq. (5.A43)) to $\omega(\bar{Q})$, where $\bar{Q}$ is the experimental equilibrium distance of the H-bond. It implies that the average value $\bar{Q}$ of Q_s is appreciably shorter in ice than in liquid water, so as to have ν_s in ice shift of $-160\,cm^{-1}$ with respect to water. It contradicts the well-known result that the density of ice is smaller than that of liquid water, and also X-ray diffraction (1, 21) data, which indicate that both $\bar{Q}$ are nearly the same in liquid water and ice, around 2.75 Å (see Figure 3.4 for liquid (heavy) water). Introducing quadratic variations of ω with Q_s does not solve the problem: it is at the origin (eq. (5.A47)) of a supplementary shift proportional to the average square amplitude of vibration $\overline{\langle (Q_s - \bar{Q})^2 \rangle_Q}$. This shift occurs towards lower wavenumbers, because the coefficient for that supplementary shift is $d^2\omega/dQ_s^2$ and is negative, as indicated by the ω dependence on Q_s, illustrated for instance in Novak's (22) curve (Figure 4.5) or as given by Mikenda for hydrates (23). As $\overline{\langle (Q_s - \bar{Q})^2 \rangle_Q}$ is expected to be greater in liquid water than in ice, this quadratic term consequently enhances the contradiction. Consideration of the widths of both ν_s in liquid water and ice also leads to the same contradiction, as with this coupling alone the width of ν_s in ice should be greater than that in liquid water. This is not what Figure 9.2 shows.

It means that this modulation is not sufficient to solve this contradiction between spectra of ice and liquid water. In order to give a consistent explanation of the $160\,cm^{-1}$ shift of ν_s, we have to suppose that *librations*, which are of the same nature as bending intermonomer vibrations for isolated H-bonds, *strongly modulate* ν_s. Such a modulation takes on the form of eq. (5.A48). The centre and width of ν_s are then given by eq. (5.A49), with $\partial^2\omega/\partial Q_\delta^2$ now positive, at least for not too great amplitudes of Q_δ. It ensures that, with the square amplitude of vibrations $\overline{\langle Q_\delta^2 \rangle_{Q\delta}}$ and terms of order 4 of the form $\overline{\langle (Q_\delta - \langle Q_\delta^2 \rangle)^2 \rangle_{Q\delta}}$ greater in liquid water than in ice, the centre of ν_s appears at higher wavenumbers in liquid water than in ice, and its width is greater in liquid water than in ice. This is what we observe in Figure 9.2. We conclude that in liquid water, librations strongly modulate ν_s, and the anharmonic coupling responsible for this modulation is preponderant, greater than that established by translational intermonomer modes (stretching type), the predominant anharmonic coupling in nearly all other H-bonded systems. *It stresses the central role librations play in liquid water.* It also shows the *singularity of the ν_s band of liquid water, which is predominantly* *modulated by librations*. This result comes out of the comparison of the ν_s bands of liquid water and ice: that of liquid water appears at higher wavenumbers, but is at the same time markedly broader.

The IR spectra of ice and liquid water thus clearly show that the main difference between the H-bond networks of liquid water and ice is an appreciable increase, in liquid water, of the amplitude of librations. Their especially important role has also been recently emphasized in analyses of femtosecond IR spectra (24). Let us note that this conclusion does not contradict the usual simplified picture that H-bonds in liquid water are weaker than H-bonds in ice, as great rotational amplitudes of individual H_2O molecules imply weakening of the H-bonds. It is more precise, as it points to the mechanism at the origin of this weakening, and will consequently allow a more precise description of the H-bond network of liquid water. It implies that the strengths of H-bonds in liquid water extend over a wide range and rapidly vary with time, at the frequency of librations.

Structure of the H-bond network of liquid water

The great amplitudes of the rapid librations are at the origin of the fluidity of liquid water as they favour a decrease of the orientational correlations between neighbour H_2O molecules. We then rejoin other liquids, which we have seen characterized by a rapid decrease of their spatial correlation functions with distances. The origin of this mechanism in liquid water, where H_2O molecules interact nearly only by establishing directional H-bonds, is, however, not the same as in most other liquids, ruled by Van der Waals intermolecular interactions that are weakly directional at room temperature. As for the loss of positional correlations with distance, that is correlations of the centres of gravity of molecules, it may be qualitatively understood by the weakness of these Van der Waals interactions in most liquids. Which precise mechanism gives liquid water this same loss of positional correlation? The answer to such a question is not clear (25). It has certainly much to do with the weakening of the H-bonds induced by these librations of great amplitudes, which may allow diffusion of H_2O molecules or embedding of other molecules, as in other liquids. Thus, liquid water is a liquid, but a singular one that keeps the same average local tetrahedral symmetry for O-atoms as in ice, with four nearest O neighbours, as indicated by X-ray scattering (Figure 3.4), by the absence of bands due to free O–H groups in IR spectra, and also by the very fast return to H-bonding transient free O–H groups display in time-resolved nonlinear IR spectra (14). It means that at a given time, individual H-bonds are significantly bent. This bending is, however, dynamic, not static, as it oscillates at a frequency around 2×10^{13} Hz, which corresponds to $\tilde{\nu}$ around 700 cm^{-1}, the characteristic wavenumber of hindered rotations (librations) ρ_{H_2O} in liquid water.

From this analysis we deduce that *the H-bond network of liquid water is, in the average, the same as that of ice*, made of H_2O molecules that establish two H-bonds on two neighbour molecules and accept two H-bonds from two other neighbour molecules. We do not pass from ice to liquid water by disrupting H-bonds, as often thought, but by inducing, *within the H-bond network of ice*, rotations of individual H_2O molecules of great amplitudes. The H-bond network of liquid water consequently suffers important dynamic distortions, due to these librations (hindered rotations of the whole H_2O molecule) of great amplitudes, which cannot be fully accounted for in the harmonic approximation and are at the origin of the important widths of intermonomer and ν_s bands. They are also at the origin of the fluidity of liquid water and at the origin of the flexibility of this H-bond network. In the following, we shall call the H-bond network of liquid water *a hyperdense flexible H-bond network*. In ice, librations have relatively small amplitudes, which can be well described within a harmonic approximation, and keep ice a crystal. The structural consequence is that a well tetrahedral order appears in ice, which extends towards infinity as shown by X-ray diffraction. In liquid water, the distortions of H-bonds has but a small influence on this ordering at short distances, where we retrieve a conformation which is, on an average, the same as in ice, that is nearly tetrahedral, but with nevertheless a greater dispersion, meaning that fluctuations around this same average configuration are greater in liquid water than in ice. At longer distances, beyond 15–20 Å as shown by both X-ray scattering (Figure 3.4) and neutron diffraction (26, 27), no memory is kept of this tetrahedral ordering, which is typical of a liquid. This distance, 15–20 Å, is nevertheless great for such a small molecule as H_2O. Years ago Pople (28) was the first one to explain these structural features with the supposition of the existence of a very dense H-bond network with bent H-bonds. It is not so far from our present view of liquid water. It differs by seeing the distortions introduced by bent H-bonds as static, which may be more adapted to amorphous solid ice than to liquid water itself. Let us note that pump-probe IR spectroscopy, a time-resolved nonlinear technique evocated in Ch. 4, has shown that librations of confined H_2O molecules have standard amplitudes (29) that are much smaller than in liquid water, even if their frequencies are similar. It means that the H-bond network of liquid water is really unique and that many of its properties are due to H_2O molecules performing within it librations of great amplitudes.

THE EXCEPTIONAL PROPERTIES OF LIQUID WATER

The property of liquid water and ice to have a hyperdense H-bond network is, as already mentioned, a unique property due to the unique structure of the H_2O molecule, which is able to establish two H-bonds with neighbour molecules and at the same time accept two other H-bonds from two other neighbour molecules. It is at the origin of an exceptional macroscopic property of ice, which exhibits a complicated and rich phase diagram, and within this diagram the line that separates ice Ih from liquid water exhibits a negative slope (Figure 8.5). This negative slope has much to do with the possibility of skating on ice. The presence in liquid water of such a dense H-bond network, which is now flexible, gives water even more exceptional properties. Most "ordinary liquids" are made of big molecules (big as compared to H_2O), which interact through Van der Waals forces which are weakly directional at room temperature. In opposition, the H_2O molecules in liquid water are held by H-bonds that even if dynamically distorted are still directional on an average. It is consequently not so much surprising that liquid water, when considered at our macroscopic scale, exhibits exceptional properties and, in opposition to an intuitive thinking, liquid water, by far the most common liquid on Earth, is not at all a good representative of liquids. In this section we examine the really exceptional physical and chemical properties of liquid water. We start with the chemical properties, which are relatively well understood in terms of chemical mechanisms. We shall see this is not so of the exceptional physical properties of liquid water, which we examine in a subsequent section. We shall not examine biological properties because, even if it is common knowledge that life occurs in water, and only in water, it is not evident that liquid water displays exceptional biological properties. The properties of liquid water may be widely used in many biological processes, but the presence of liquid water is not a prerequisite for life to occur, even if life slowly appeared, around 3.5 billions years ago, in the ocean and remained there for still a few billion years. The existence of some plants such as *Panicum turgidum*, found in the Namibian desert, which not only survive during drought but become green again when the air is humid but with no rain, suggests that some form of life occurs without liquid water. It nevertheless requires the presence of H_2O molecules that do not appear in the form of liquid water. Also, many enzymes are active as soon as a minimum amount of water is present (30). We shall see in Ch. 10 that the prerequisite for life is consequently the presence of H_2O molecules with their exceptional possibilities to develop around them an extended and dense H-bond network, forming nanodroplets associated with other molecules inside which bioreactivity may proceed. Water nanodroplets do not behave the same way as liquid water, and the special properties of liquid water itself are consequently of a minor relevance in bioprocesses.

Exceptional chemical properties

Water is so familiar a liquid that we do not always realize that a number of casual looking chemical properties it displays are indeed exceptional. Liquid water is thus unique to ionize acids and bases, to dissolve such ions as H_3O^+, OH^-, or other cations and anions. It is also an exceptional solvent for organic molecules, and in some cases, the asymmetric solubility of amphiphile molecules is at the origin of well-defined structural entities, which play important roles in both chemical and biomedia. As for any other liquid, the fluidity of liquid water allows it to have molecules different from H_2O embedded in it. The presence of a hyperdense H-bond network furthermore provides liquid water with possibilities to accommodate embedded molecules other than H_2O, D_2O or HDO. Other liquids are far to display these possibilities to the same extent. We examine these different points successively.

Ionization and solvation of acids and bases

In Ch. 6 we considered transfers of protons that are at the origin of the ionization of acids and bases and of the solvation of the resulting ions $H_3O^+$and OH^-. We have seen they are more complicated than simple tunnellings of protons through double-well potentials and we are far from really understanding them. We nevertheless know they strongly depend on the structure of the H-bond network that surrounds the acidic or basic molecules or ions, and the exceptionally dense and flexible tridimensional H-bond network of liquid water strongly favours ionization of X−H (X a halogen atom) into X^- and H_3O^+, as compared to ionization by a small number of H_2O molecules. This is because it offers more possibilities to optimize this ionization than the reduced H-bond network developed by a cluster of a few H_2O molecules. This is not so for the as dense, but rigid, H-bond network of ice, where H_3O^+, with its planar structure, cannot take the place of an H_2O molecule in the tetrahedral ice lattice. The consequence is that bulk ice does not ionize acids, even strong ones, and on freezing, water that contains ions expels most of them, because they cannot take the place of an

H_2O molecule without distorting the ice crystal. The same is true for bases. It illustrates the crucial importance of the flexibility of the H-bond network of liquid water. In liquid water this exceptionally dense H-bond network offers H_3O^+or OH^- ions the possibility to occupy many different sites with equal energies: they may be located on any sites occupied by an H_2O molecule. The corresponding decrease of free energy, due to a gain in entropy, is consequently important, and we have seen in Ch. 6, that for the dissociation of HCl in water, this gain in entropy, mainly due to this delocalization of the H_3O^+ions, is of the order of $-45\,\mathrm{kJ\,mol^{-1}}$, which represents an appreciable part of the free energy of solvation, $-75\,\mathrm{kJ\,mol^{-1}}$. This important entropic gain allows weaker acids to be nevertheless dissociated in water, despite their endothermic ionization step that occurs with an increase of enthalpy ΔH. We have also seen that this exceptionally dense and flexible H-bond network of water, which optimizes proton transfer, is at the origin of the great proton conductivity of liquid water.

The role of this hyperdense and flexible tridimensional H-bond network is thus central for liquid water to dissolve acids and bases. This is somewhat different from the usual picture that attributes this exceptional possibility of liquid water to its particularly great dielectric constant $\varepsilon = 80$, which is supposed to be at the origin of an efficient screening of the ions. This usual image, though not in contradiction with the mechanism we have given above, is only part of the story. It is a condition that is necessary but largely insufficient. It does not explain why strong acids or bases can be found ionized when H-bonded to a small cluster of H_2O molecules, where screening has no meaning. More important, it does not explain why liquids with greater dielectric constants, such as cyanhydric acid HCN, with its $\varepsilon = 115$, or *N*-methylformamide CH_3NHCOH, with its $\varepsilon = 189$, and others (31), are much less efficient than water to ionize acids and bases. It means that molecules with a big electric dipole moment, such as those which constitute liquids with high dielectric constants, do not efficiently ionize acids and bases if they are not in addition held together by relatively strong intermolecular forces. Their electric dipole moment may offer an efficient screening if clusters of these molecules can be oriented in a proper direction to oppose the electric field of an ion. However, this particular configuration must be mechanically stable to hinder recombination of ions. This is what the H-bond network of liquid water ensures. It does not occur in liquids made of molecules with a great electric dipole moment, which do not establish H-bonds. It neither occurs in H-bonded liquids such as fluorhydric (HF) or cyanhydric acids (HCN), which also display a great density of H-bonds, but are far from being as efficient as water to ionize acids and bases, because the H-bond network they develop is unidimensional, with the H-bonded molecules forming zigzag chains. Different chains are only loosely bound by Van der Waals forces. It shows that, in addition to an important electrical dipole moment, which allows screening of the ions, the constituting molecules of the liquid should also develop interactions that offer screening configurations they form a relatively great mechanical stability. This is what the 3D hyperdense H-bond network of liquid water does in liquid water, in addition to providing an efficient electrical screening. We conclude that reducing the origin of the exceptional properties of liquid water to its great dielectric constant is a poor and largely insufficient simplification, as it neglects the stability of the screening configuration.

Ionization and solvation of salts

Liquid water is also very efficient to solvate salts. It also occurs in two steps, which are ionization of the salt MX itself into a cation $M^+$and an anion X^-, followed by solvation of these ions. In opposition to the solvation of acids or bases, these mechanisms do not involve proton transfers. Solvation occurs with a gain in entropy, similar to that encountered for solvation of acids. It seems to require a minimum number of H_2O molecules, appreciably more than 10 (32) for NaCl. The structure of completely solvated sodium chloride is drawn in Figure 9.3. It is representative of other ions, with the first layer of water molecules around the anions pointing their O–H groups towards these anions and establishing H-bonds on them and the first layer of other water molecules presenting lone pair electrons of their own towards the cations. This kind of structure does not perturb very much the H-bond network of liquid water, when cations and anions have sizes similar to those of H_2O molecules, that is when they are neither too small nor too big. This is the case of Cl^-, which allows the establishment of four H-bonds around it, the same number as that of H-bonds around a single H_2O molecule in pure liquid water. This is also the case of Na^+ that also accepts four to five H_2O molecules around it (33). This structure provides these ions an efficient electrical screening. Their presence only slightly disturbs the H-bond network of water. They affect slightly its flexibility, as shown by femtosecond pump-probe spectroscopy (34) described in Ch. 4, even if the first H_2O layer around each ion constitutes a relatively rigid sphere (35), which increases the viscosity of liquid water. In addition to these methods, near edge X-ray absorption spectroscopy, evocated in Ch. 3, has also shown that only the water molecules in the first solvation sphere of ions are affected by the presence of the ions (36), leaving the H-bond network out of this first sphere unaffected. It is interesting to note that a solution of NaCl in water keeps this configuration up to its limit of solubility, 357 g l^{-1}. Smaller ions, such as F^-, encounter greater difficulties in being solvated. This is because, when embedded in water, they disturb the H-bond network of bulk liquid water more than Na^+ or Cl^-. This is also the case of bigger

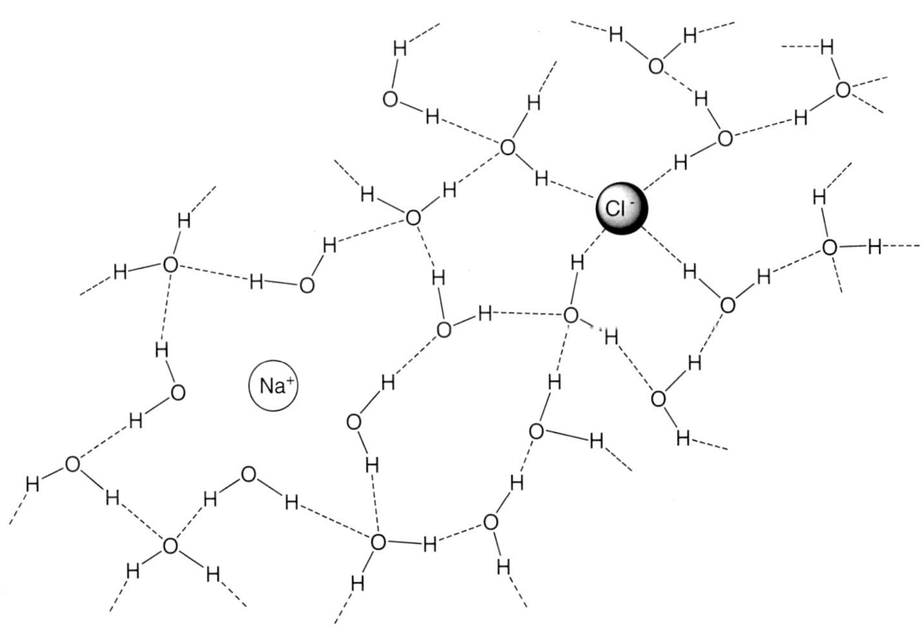

Figure 9.3 Na^+ and Cl^- ions in liquid water.

ions such as I^- or Cs^+, which are also less easily solvated. As in the case of acids and bases, solvation of salts by liquid water cannot be understood only on the basis of its great dielectric constant. It is due to the existence of this dense 3D H-bond network that, in addition to providing an efficient electrical screening of the ions, also ensures the necessary stability of the screening configuration. Most theories that attempt to calculate solvation energies of ions, such as the "dielectric continuum theory" encounter great difficulties taking this central point into account. They often work quite well (37) when the solvent is not liquid water but is made of polar molecules such as acetonitrile or dimethylsulfoxide that do not establish H-bonds. In the case of liquid water, neglecting the crucial effect of the H-bond network leads to having to introduce unreasonable cavity sizes to retrieve solvation energies in agreement with experimental ones.

Figure 9.3 clearly shows that solvation of an anion is different from that of a cation: an anion accepts H-bonds from H_2O molecules, whereas a cation attracts the lone-pair electrons of H_2O molecules, thus having the same structural effect as an H-bond donor. This dissymmetry of solvation makes anions and cations behave differently at water surfaces or interfaces, a result that has been experimentally put into evidence by sum-frequency vibrational spectroscopy (38), a surface-specific method described in Ch. 4. Cations thus display tendencies to stay in the bulk, because on arriving at the surface, some of the four H_2O molecules bound to them cannot establish H-bonds when arriving on the surface. This is not so for anions, as the H_2O bound to them already establish H-bonds on them. The surface therefore favours the presence of anions. This effect is at the origin of the oxidative power of sea water, which is ascribed to Cl^- anions positioned on surfaces of liquid water droplets.

Solvation of organic molecules

The power to solvate organic molecule is another casual looking but nevertheless exceptional property of liquid water. In this case, the presence of the electric dipole moment of H_2O is of a secondary importance, whereas the presence of the H-bond network is central. In order to have an idea of what goes on when we introduce an organic molecule in liquid water, let us first have a rapid evaluation of the thermodynamic balance of this mechanism. A more precise description of the thermodynamics of this solution is given in textbooks (25). The gain in free energy that corresponds to solvation of small organic molecules into liquid water, starting from neat liquid water separated from these organic molecules in the form of a liquid or a solid, can be written in a first approximation as:

$$\Delta G = E_{\mathrm{HB}} \Delta n_{\mathrm{HB}} + \Delta H_{\mathrm{NHB}} - T\Delta S \qquad (9.5)$$

where E_{HB} is the negative internal energy of an H-bond in liquid water, Δn_{HB} the change of the H-bond number upon the introduction of organic molecules in water, ΔH_{NHB} the gain in enthalpy due to other intermolecular interactions than H-bonds and ΔS the gain in entropy. In this first approximation we furthermore suppose, in view of simplification, that H-bonds established or accepted by organic molecules have same internal energy E_{HB} as H-bonds of liquid water. The term ΔH_{NHB}, which incorporates Van der Waals interaction of these organic molecules prior to their introduction in liquid water, is positive, because Van der Waals interactions between organic molecules and water are very weak, whereas they may not be weak between organic molecules before their introduction into water. They may nevertheless be neglected within the present approximation, because Van der Waals interactions are significantly weaker than H-bonds. We are thus left, in this approximation, with two terms, the enthalpic term in Δn_{HB}, which describes the change of the H-bond number and the entropic one in ΔS. The first term, in Δn_{HB}, is positive with our assumption of all H-bonds having same energy, because the H-bond network of liquid water is the optimized structure that allows the establishment of the maximum of H-bonds in an assembly of H_2O molecules. Disturbing it by the introduction of other molecules causes Δn_{HB} to be negative as it can but disrupt this optimized structure, consequently diminishing the number of established H-bonds. As E_{HB} is also negative, this enthalpy term is positive, that is unfavourable to solvation. The other term, the entropy term in ΔS, is conversely favourable, that is negative, as it always corresponds to an increase in entropy, by dispersion of the organic molecules among the water molecules, at least as long as the number of water molecules is greater than that of solvated molecules.

The solubility depends on the balance between these two terms. Thus, for small molecules which have hydrophilic groups and a relatively small number of hydrophobic groups, such as acetone, methanol, ethanol, etc., the balance is largely negative: Δn_{HB} may be slightly negative because of the introduction in liquid water of hydrophobic groups which, even if in a small number, replace some O–H hydrophilic groups. It is however small, due to the possibility for these molecules to establish new H-bonds with surrounding H_2O molecules, after they have been introduced in water. The entropic term is consequently largely predominant and, when these organic molecules are found in the form of a liquid before introduction in water, such a liquid is miscible in all proportion with liquid water. This is still true for tertiobutanol $(CH_3)_3$–C–OH, abbreviated as *t*-butanol. Amazingly, this is not the case of its isomer molecule, *n*-butanol CH_3–$(CH_2)_3$–OH, which is solvated in water only at lower concentration. The reason for this difference is to be found in the approximately spherical form of *t*-butanol. The sphere is the form that displays a minimum for its S/V surface to volume ratio. It consequently minimizes the absolute value of Δn_{HB}, which nevertheless remains negative. When concentration of *t*-butanol increases in liquid water, this enthalpy term Δn_{HB} never overcomes the entropy term, even when molecules assemble in small aggregates. *n*-Butanol, with its zigzag form and consequent greater S/V ratio, requires an appreciably greater number of H-bonds of H_2O molecules to be broken for accommodation in water, that is a greater Δn_{HB}. The enthalpy term in Δn_{HB}, which monotonously increases with its concentration C (it is proportional to C at low concentrations), becomes equal to the entropy term when C reaches a value C_m, the limit of solubility. At higher C it becomes preponderant and solvation unfavourable, leading to precipitation of *n*-butanol.

When the ratio of hydrophobic to hydrophilic groups of the organic molecule increases, liquid water may disrupt a small quantity of H-bonds in the vicinity of organic molecules so as to form a cage-like structure, inside which these organic molecules may be solvated. Such a cage is shown in Figure 9.4, on the left part of the solvated molecule. It makes Δn_{HB} take an appreciable negative value. Solvation is still possible at low concentration C, where Δn_{HB} may be predicted proportional to C as long as organic molecules are dispersed in water, making no close contact. At higher C, solvation may still occur by grouping clusters of organic molecules in bigger cages. Inside this cage the organic molecules are organized in such a way that their few hydrophilic groups are oriented towards the exterior of the cluster, so as to still establish the maximum of H-bonds with the surrounding H_2O molecules, which has the effect of decreasing the absolute value of Δn_{HB}. In such a configuration, we can however no longer

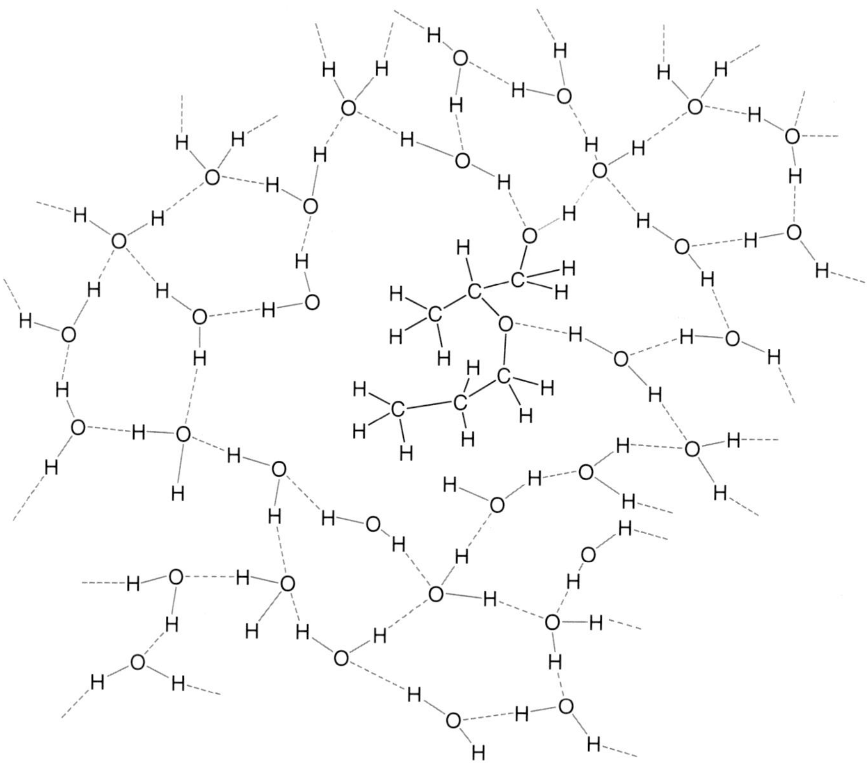

Figure 9.4 An organic molecule solvated in a cage inside liquid water.

neglect, as in eq. (9.5), the decrease of entropy due to the partial ordering of these organic molecules, and to that of H_2O molecules around such a solvated cluster. This decrease of entropy reflects itself in the decrease of the solubility when temperature increases (39). It illustrates the delicate thermodynamic balance of the solvation process of such organic molecules. In biochemistry this latter kind of structure, grouping of hydrophobic parts of organic molecules inside clusters, is often attributed to a "hydrophobic interaction". This term should be avoided, because this is not interaction, but a special arrangement of the H_2O molecules around the cluster of solvated molecules that expose the maximum of their hydrophilic groups towards H_2O molecules and consequently groups hydrophobic parts in a confined space. The denomination "hydrophobic interaction" is consequently misleading as it has more to do with confinement than with an interaction.

Finally, when the organic molecule has no hydrophilic group, such as in alcane, alcenes, benzenics, etc., the positive enthalpic term always overcomes the entropic one, except, may be, at high dilution for small molecules such as methane, when Δn_{HB} is still very small. The organic molecule of a sufficient size does not mix with liquid water. Water and such organic molecules separate into two phases, and oil, for instance, spreads on the surface of water, even forming a very thin molecular monolayer at high dilution. Almost no molecule of oil is found inside water.

Let us point out that the above analysis we have made is qualitative. It gives a simple picture that emphasizes the role of the H-bond network of liquid water and is only an Ariane thread to understand how solvation occurs in water. Quantitative descriptions require a finer analysis and are the objects of an intense research field. They accept the above description as a coarse starting scheme, but refining it requires to release such approximation as an equal energy of formation for all H-bonds (the H-bonds established by H_2O molecules at the surface of a cage generally appear stronger than in the bulk (40), but may be weaker (41) when this surface is great, as seen by sum-frequency vibrational spectroscopy described in Ch. 4), to introduce Van der Waals interaction, etc.

Organized structures of amphiphile molecules

These are bigger organic molecules which have two well-identified and separated parts, one hydrophobic and one hydrophilic. Hence their names: amphiphile = that loves on both sides. In contact with liquid water they are at the origin of various original organized molecular structures. A set of the most current amphiphile molecules is shown in Figure 9.5. They are at the origin of organized structures we describe in the following. These organized structures appear in a restricted and well-defined region of concentration of these molecules in liquid water. Special experimental methods to observe them are evocated in Ch. 4 (IR and sum-frequency vibrational spectroscopy) and also in Ch. 11.

Micelles

The simplest amphiphile molecules are molecules of the type sodium or potassium stearate $CH_3–(CH_2)_{14}–COO^-$, which constitute ordinary soap. In water these molecules adopt the form of micelles, drawn in Figure 9.6, that are roughly spheres where hydrophilic parts made of carboxylate ions with loosely bound cations are all at the outer part of the sphere, so as to establish H-bonds with the surrounding H_2O molecules. Hydrophobic parts are confined within the interior of the sphere where it can accommodate smaller hydrophobic molecules, drawn in green in Figure 9.6. Such amphiphile molecules thus act as detergents, allowing strongly hydrophobic molecules to be indirectly solvated into water, that is incorporated in water but with no direct contact with H_2O molecules. Soaps are detergents mainly made of K^+stearates. Shampoos are preferably made of sodium dodecylsulfate molecules $CH_3–(CH_2)_{11}–SO_4^-$, another amphiphile molecule drawn in Figure 9.5. The molecular constituent of cleansing agents in general are often other amphiphiles such as polyoxyethylene glycols, $CH_3–(CH_2)_n–(CH_2–O–CH_2)_m–OH$, which, in opposition to the preceding detergents, are not ionic. They are consequently more insensitive to the presence of other salts in water than stearates or dodecylsulfates. Another detergent of great interest is sodium hexadecylbenzene-sulfonate $(CH_3–(CH_2)_5)_2–C=C_6H_4–SO_3^-$ (Figure 9.5) used to extract oil from porous rocks.

When dissolved in a liquid made of completely hydrophobic molecules, these amphiphile molecules may adopt the structure of a "reverse micelle", drawn in the lower part of Figure 9.6. In this reverse micelle, all hydrophilic groups are directed inwards the centre of the structure, where they can establish H-bonds with H_2O molecules of a nanodroplet of water (42), which can thus be also indirectly solvated in oil.

Such amphiphile molecules are surface active agents, abbreviated as "surfactants". This means they decrease the surface tension at the interface between liquid water and a solid, liquid or gas made of other molecules that have a low affinity for H_2O. The surface of liquid

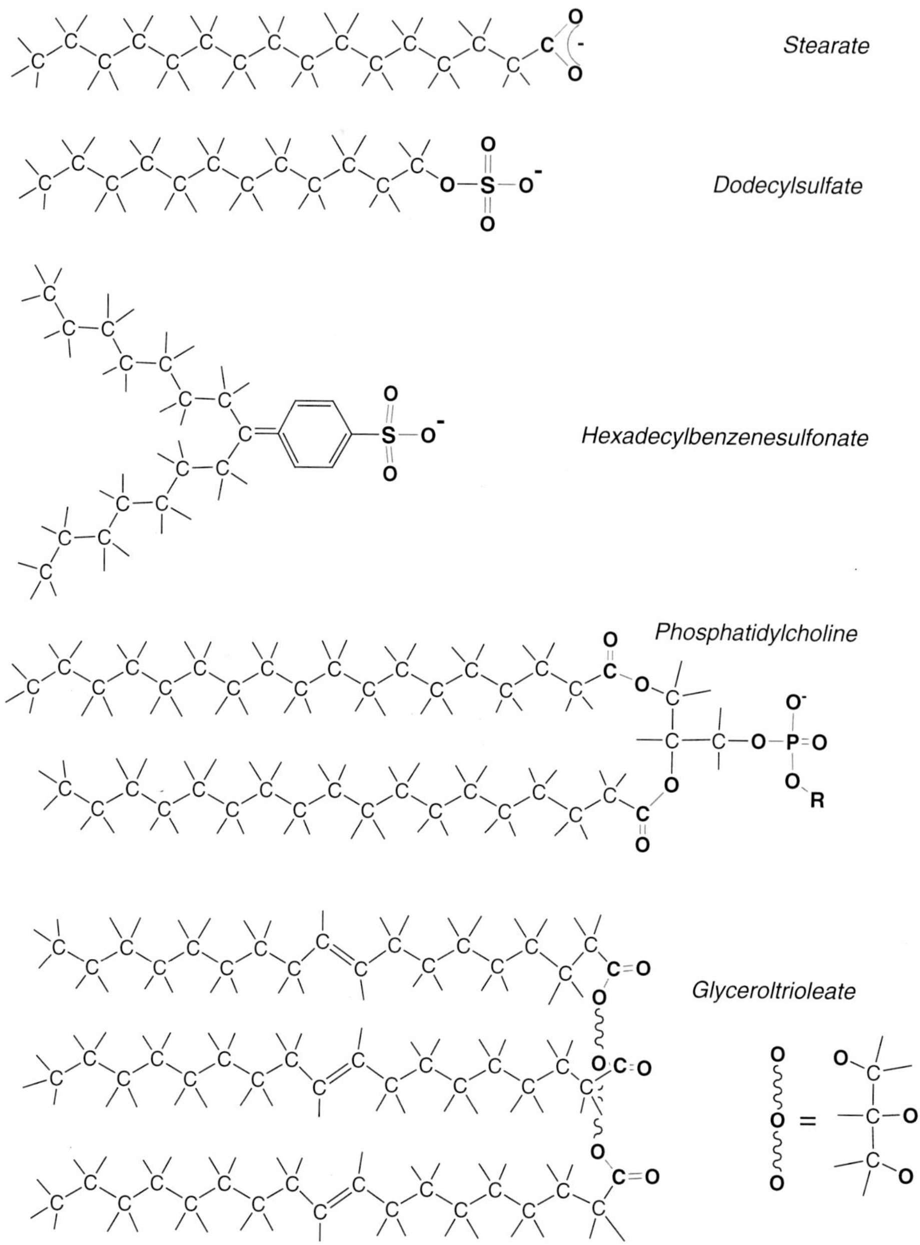

Figure 9.5 Some amphiphile molecules. In glyceroltrioleate, the glycerol part, not positioned in the plane of the figure, is symbolized in the form of an ondulation.

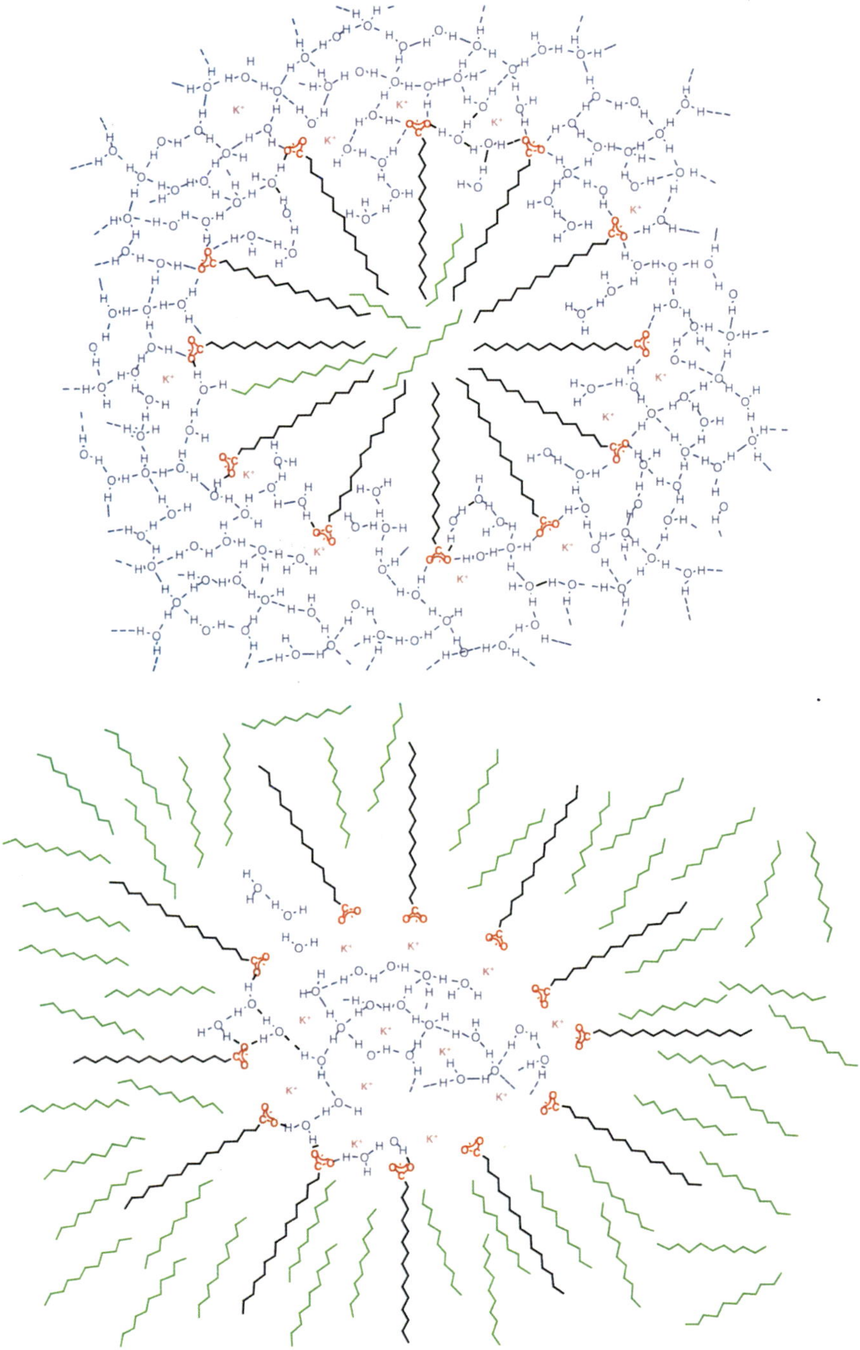

Figure 9.6 Structures adopted by sodium or potassium stearate (see Figure 9.5): a micelle in water (upper drawing) and a reverse micelle in oil (lower drawing). Inside the micelle hydrophobe molecules, drawn in green, can be solvated.

water, for instance, contains H_2O molecules that are frustrated, as compared to H_2O molecules in the bulk. These molecules of the bulk have four H-bonds established around each O-atom. The molecule on the surface can establish H-bonds towards the bulk, but none towards air. As a consequence, free OH groups, often called "dangling OH groups", appear at the surface of liquid water, which has the tendency to minimize its surface where it cannot establish H-bonds, so as to reduce the number of these free OH groups. Thus, water droplets are spherical because, as already seen, the sphere is the volume that displays a minimum S/V surface to volume ratio. As the number of frustrated OH groups that cannot establish H-bonds is proportional to this surface, when the droplet is made of a sufficiently great number of H_2O molecules, such a spherical form minimizes the number of frustrated groups. It implies that two colliding water droplets will form a more stable bigger single droplet, with a S/V ratio smaller than that of the two original droplets. We have seen the implication of this mechanism in the formation of raindrops discussed in Ch. 8. This trend to minimize the interface points to the origin of capillarity. In order to increase the surface of liquid water, energy should be provided because the number of free O–H groups increases. In physics this is described by the surface tension coefficient γ, which is such that the (free) energy necessary to increase a liquid surface of δS is equal to $\gamma\delta S$. This energy is equal to $N_F|E_{HB}|$ where N_F is the number of frustrated H_2O molecules found on this surface δS and E_{HB} the (negative) internal energy of an H-bond in liquid water. As E_{HB} is greater than Van der Waals energies and as the density of O–H groups is particularly high in liquid water, it makes γ have an especially great value in the case of liquid water, as compared to liquids of organic molecules (25). Along the surface between liquid water and a group of amphiphile molecules, no such frustration appears: the H_2O molecules in contact with the amphiphile molecules are not frustrated because they establish H-bonds with them. This is equivalent to having N_F equal to 0. The surface tension between water and a surfactant then becomes small, and only due to the decrease in entropy an ordered structure such as the micelle of Figure 9.6 requires when it is established. Hence the name "surface active agent". Incidentally, it allows, at some defined concentration of soap in water, to have bubbles, or foams. In these structures a very thin layer of H_2O molecules is placed between two layers of Na^+or K^+stearate, which all present their carboxylate hydrophilic heads towards this water layer, exposing their hydrophobic tail towards air. This is a configuration known as a "Langmuir–Blodgett" layer described in a subsequent subsection. As compared to a spherical droplet of liquid water alone, the S/V ratio for this configuration is increased by several orders of magnitude. This is made possible, because addition to water of such amphiphile molecules considerably decreases the surface tension. These properties of micelles to reduce surface tension are widely used in the industry. The problem, here, is their stability. In order to improve it, more than one surfactant is often used. It favours the appearance of other original but more complicated organized structures such as the interpenetration of two nonmiscible liquids, water and oil for instance, in the form of bicontinuous miscible liquids that both extent throughout the offered space and allow to travel between any two points in the same liquid without leaving this liquid (43). Micelles in general are the object of an intense research work in a domain of physics (25) known as that of "soft matter".

Biological membranes

Phosphatidylcholine, a glycerolphosphatediester drawn in Figure 9.5, is the constituent molecule of "phospholipidic membranes", which form the envelopes of biological cells, cell nuclei, etc. The esterifying acids of this molecule are fatty acids. The ratio of the volume of the polar part to that of the apolar part of this molecule is smaller than in the preceding case of micelle forming molecules. When immersed in water this molecule then forms bilayers made of phosphatidylcholine molecules that present all their hydrophilic parts, made of phosphate and carboxylate groups, towards liquid water (Figure 9.7). The inner liquid water is thus isolated from outer water. This structure may nevertheless accommodate proteins, which have sufficiently large hydrophobic parts that can be easily inserted inside the hydrophobic part of the membrane. These transmembrane proteins allow the control of all exchanges between inner and outer liquid water of the cell or nucleus, etc. They are highly selective for what concerns exchanges of H_2O molecules, ions, small organic molecules, etc. They can easily migrate within the membrane.

This particular structure of a bilayer is stable in liquid water at room temperature, that is in biological conditions. In different conditions, such as at high temperature, at pH different from neutral or in presence of an appreciable concentration of dissolved salts, which modify the H-bond network of liquid water around the membrane, they are no longer stable and collapse. The inner water then comes in contact with the outer liquid water, an often lethal phenomenon for the medium inside the membrane. Some bacteria living in extreme conditions of high temperature, such as archeobacteria of geysers which develop in water up to 80 °C, or thermophile bacteria found in hydrothermal springs at the bottom of oceans and are able to live in water at 110 °C, find a way to accommodate these conditions: instead of using phosphatidylcholine molecules, they use a diphospholipid, made of two phosphate groups covalently linked by two hydrophobic chains. It allows a configuration similar to that drawn in Figure 9.7, with the bilayer being transformed into a monolayer. Such a membrane can resist "extreme conditions" defined above. It is, however, much stiffer than usual membranes and does not allow close contact between various cells. It consequently only appears in unicellular organisms.

Langmuir–Blodgett layers

These are monomolecular layers that appear on the surface of liquid water, when a small quantity of amphiphile molecules, such as fatty acids, are spread there on a limited water surface. Fatty acids are molecules with a carboxylate COO^- or COOH hydrophilic head, which is covalently linked to a hydrophobic tail of the form CH_3–$(CH_2)_n$ that may intercalate an unsaturated CH=CH group within its CH_2 chain. Stearate, drawn in Figure 9.5, is a typical fatty acid. The generic name biologists give these kinds of molecules is lipids. A small number of such molecules, spread over the surface of liquid water, organize themselves with their hydrophilic head accepting H-bonds from H_2O molecules on the surface of liquid water. With a sufficient concentration, the hydrophobic chain stands away from the surface, taking a direction roughly perpendicular to it, as drawn in Figure 9.8. We have already evocated such a structure found in soap bubbles. On increasing the concentration of amphiphile molecules, the whole surface of liquid water may become occupied. A second layer of amphiphile molecules is then thought to stack on this first layer. The molecules of this second layer keep the same orientation as the first layer, perpendicular to the surface of liquid water, but they have been rotated of 180°, so as to have their hydrophobic tails facing the hydrophobic tails of the first layer. Successive layers are thought to form, on increasing concentration, keeping the same direction but each with a 180° rotation (44).

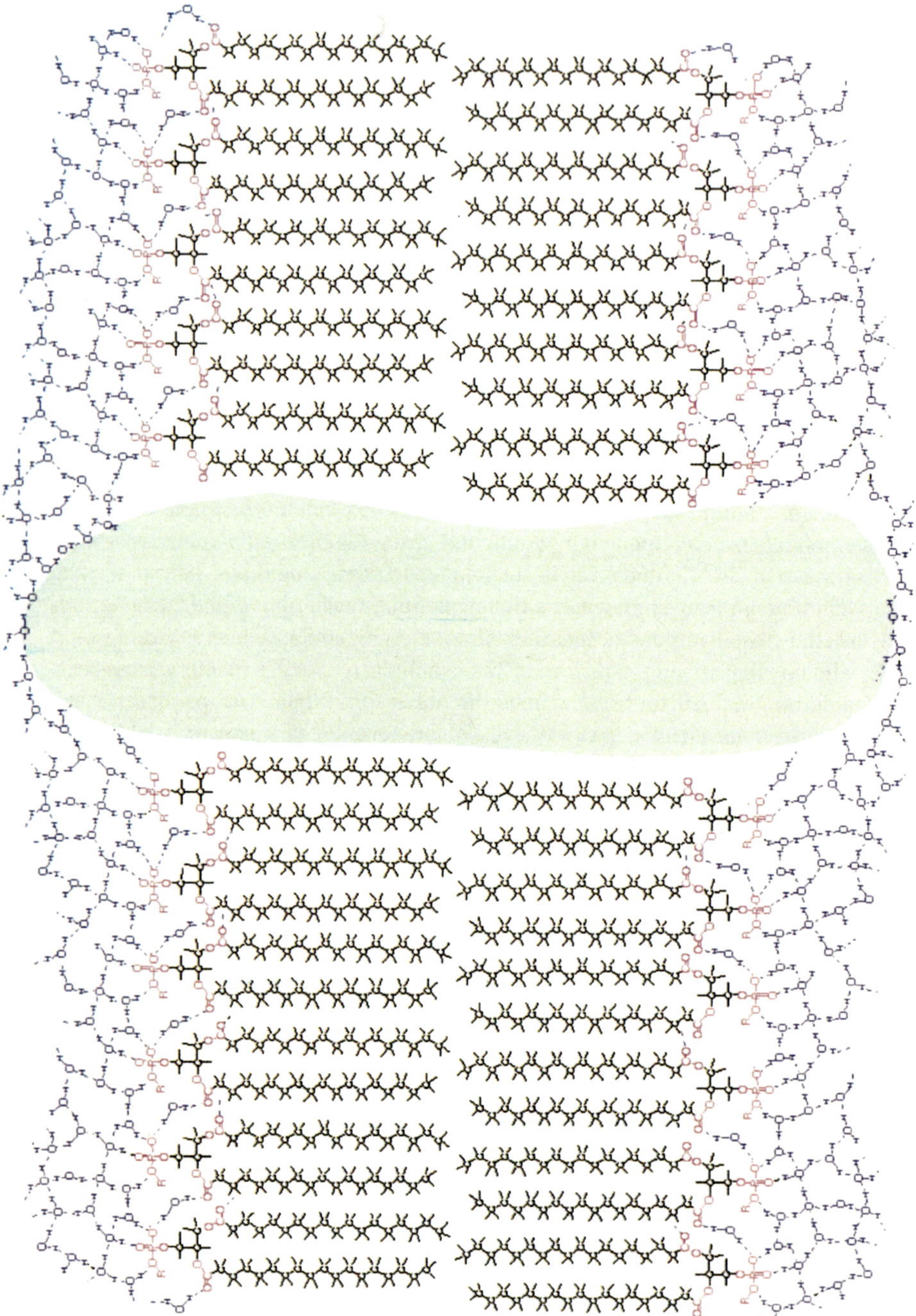

Figure 9.7 Section of a phospholipidic membrane in liquid water: a bilayer of organized phosphatidylcholine molecules (Figure 9.5) with their hydrophilic phosphate and carboxyl groups (purple) H-bonded to outer H_2O molecules (blue) and accommodating a transmembrane protein (green).

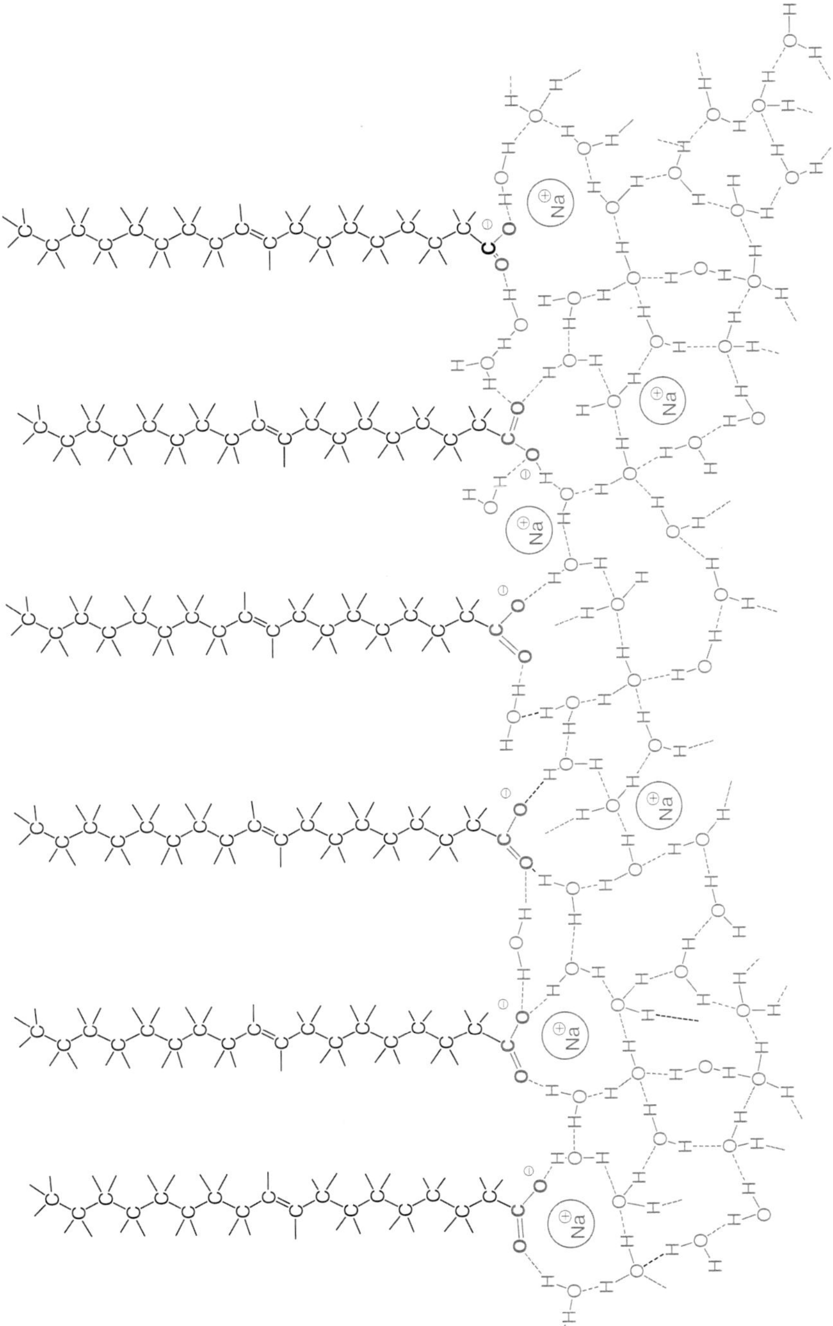

Figure 9.8 A Langmuir–Blodgett layer of sodium oleate at the surface of liquid water.

These strongly orientated layers are used in the design of organic electroluminescent materials and devices.

As already seen in the case of micelles, the absence of frustration of the H_2O molecules at the surface of liquid water with Langmuir–Blodgett layers has the effect of significantly decreasing the surface tension of water. This property was known to ancient Greek sailors who used to spread olive oil around their ship when surprised by gales. It hindered dangerous unfurling waves to form, keeping the sea surface smoother. Olive oil is a typical lipid made of glyceroltrioleate, an ester formed by condensation of three molecules of oleic acid and one molecule of glycerol, an amphiphile molecule shown in Figure 9.5.

Exceptional physical properties

Figure 9.9 displays some exceptional properties of liquid water. The dependence on temperature of the density *d* may be the best known one. The singularity of water appears with the presence of a maximum at 4 °C. It extends with a pronounced effect, that is a positive slope which becomes more and more so when temperature decreases, in the supercooled region, where water is a metastable liquid at temperatures below 0 °C. By comparison, most other liquids display a monotonous variation, schematically drawn as a dashed line, with a negative slope in the whole temperature range. This singularity of the density of liquid water is also connected to the exceptional change in the density upon freezing, which leads to a crystal, ice, with a markedly lower density. By comparison, on freezing almost any other liquid gives a solid with a higher density, schematically shown by the discontinuity in the dashed line. This exceptional behaviour of liquid water and of the water/ice transition has important macroscopic consequences. Thus, the temperature at the bottoms of the oceans and of many lakes that freeze in winter is 4 °C, not 0 °C. Another well-known consequence is that ice stays on the surface of the Arctic ocean, providing an efficient thermal screen that keeps heat in the oceanic waters. It hinders too rapid a dissipation of heat in the atmosphere and consequently in the space surrounding the Earth. It is an important climatic consequence of the special structure of liquid water at molecular level. Another consequence is that life is preserved under the ice of freezing lakes: water, with no contact with the cold atmosphere, remains liquid under the ice surface. If water had not a density markedly greater than ice, the solid that would form on freezing would fall in the bottom of lakes, water would remain on the surface and would then be in a direct contact with the colder atmosphere. It would most rapidly be entirely transformed into its solid counterpart, leaving no hope to all animals or plants in the lake to survive.

We may qualitatively understand the origin of this exceptional property. As already seen, the H_2O molecules of ice are held by rigid H-bonds, which perform small vibrations only around their equilibrium positions. It results in a tetrahedral structure for the "heavy" O-atoms, which have each four nearest O-atoms. Such a structure is noncompact, that is at the origin of a somewhat low density solid. For a comparison, in a face centred cubic structure of identical atoms, each atom has 12 nearest neighbours. It is thus not so much surprising that on increasing temperature which, as already seen, has the effect of increasing the amplitudes of librations, that is increasing the dynamic distortions of H-bonds, we destroy local order, especially at the level of second neighbours. It gives a more compact structure the possibility to appear.

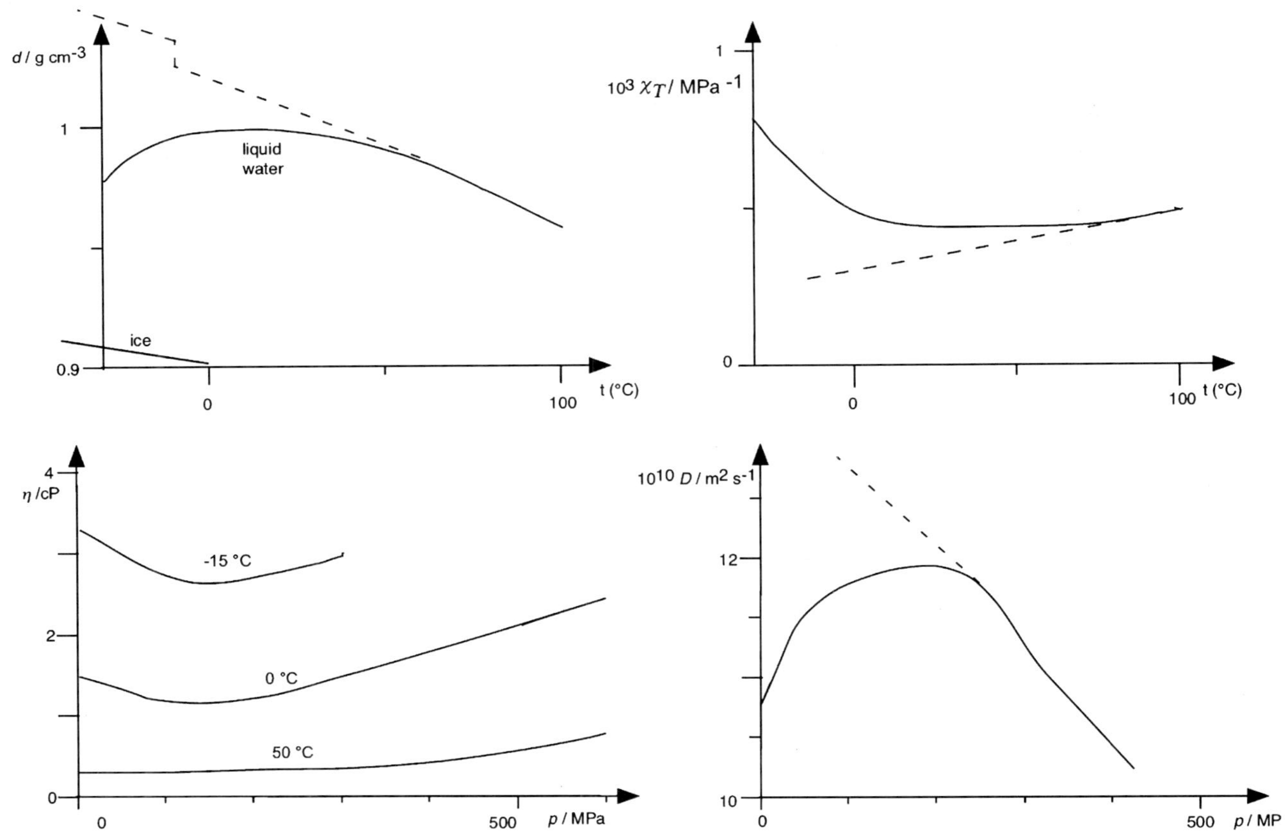

Figure 9.9 Exceptional physical properties of liquid water (solid lines): temperature dependences (upper diagrams) of the density d (45) and isothermal compressibility χ_T (adapted from Refs. (45–47)); pressure dependences (lower drawings) of the shear viscosity η at various temperatures (adapted from Ref. (48)) and of the isothermal diffusion coefficient D at 0 °C (adapted from Ref. (49)). Dashed lines sketch typical dependences displayed by almost all other liquids. Note that at -15 °C no value is given for η at $p > 300$ MPa, because of a phase transition towards ice V (Figure 8.5).

When ice melts, we consequently replace a noncompact structure by a slightly more compact one. In liquid water, the same mechanism proceeds and is at the origin of an enhancement of the density, even if this enhancement is much less marked than on ice melting. Furthermore, the effect of this dynamic distortion of H-bonds is somewhat counterbalanced by the usual anharmonicity at the origin of the dilatation of all solids and liquids. More precisely, this usual anharmonicity takes on the form of a Morse potential (Figure 4.1 and eq. (4.6)) for the intermonomer O···O stretching vibrations. The lower levels of such vibrations are separated by about 200 cm^{-1}, which is equivalent (Ch. 1) to a temperature of about 300 K. It means that around this temperature, which is typically the temperature of existence of liquid water, excited states of this vibration become populated, and when temperature still increases the population of these excited states also still increases. As the equilibrium O···O distance also increases with the degree N of excitation (N stands for the Nth excited level—see eq. (4.4) where it is written as N_s), the thermally populated excited states of the intermonomer stretching vibrations induce an increase of average O···O distances, that is a dilatation. This latter mechanism becomes preponderant at temperatures above 4 °C, where liquid water becomes a "normal" liquid, at least for what concerns its density which then exhibits a negative slope for its variation with temperature. This mechanism is already present in the supercooled region, at temperatures below 0 °C, but the former gain in compactedness due to distortions of H-bonds is predominant, all the more so that we are in the deep supercooled region, around −35 °C, for instance, where the slope dd/dT is largely positive, and water consequently largely anomalous.

The temperature dependence of the isothermal compressibility $\chi_T = -(1/V)(\partial V/\partial p)_T$, where V is the volume of 1 mol of liquid water and p is pressure, also exhibits an exceptional behaviour, as displayed in Figure 9.9. As shown by the dashed line, almost all liquids exhibit a variation with a positive slope, which corresponds to a softening of the interactions between molecules when temperature is raised. This softening makes the liquid easier to compress. Liquid water conversely exhibits a negative slope at low temperatures, which becomes positive at higher temperatures, with a corresponding minimum around 45 °C. The negative slope, an exceptional property that is magnified in the supercooled region, means that on increasing temperature in this region, water becomes harder to compress. This is due to the increase of the density when temperature is raised in this region, another exceptional property we have just discussed above.

The two preceding exceptional properties are usually considered as related to static properties of liquids. Dynamic properties of liquid water may also be exceptional when compared to almost any other liquids. Thus, the viscosity, η, of an ordinary liquid increases when pressure is applied at a fixed temperature. Upon pressure the average distances between molecules shorten, restricting the amplitudes of their relative rotations and translations, thus making this liquid more viscous. Liquid water behaves differently: as displayed in Figure 9.9, the viscosity at −15 °C starts decreasing when pressure is applied, before reaching a minimum and then increasing. This behaviour is also found at 0 °C. Only at a temperature above 50 °C does liquid water display the behaviour of other liquids, with η monotonously increasing with pressure. Once more this effect is more marked in the supercooled region than at temperatures above 0 °C. It means that, in this region, liquid water starts becoming more fluid when pressure is applied. And indeed pressure partly destroys the tetrahedral symmetry adopted by O-atoms, a subtle effect detected by neutron scattering (27). It means that pressure first bends H-bonds, changing the equilibrium positions of librations of individual H_2O molecules, an effect we have evocated in that part of Ch. 4 that is related to anharmonicities of intermonomer vibrations. This bending of H-bonds is accompanied by a small decrease of the O···O distance. This decrease is preponderant in most liquids, where interactions between molecules are not H-bonds. In liquid water, the main effect of pressure, bending of H-bonds makes H_2O molecules more loosely bound to their neighbours, enabling a greater relative motion of all H_2O molecules, making water in the whole more fluid. At higher pressures the decrease of the O···O distance overcomes that exceptional mechanism, and then liquid water starts behaving as other liquids: it becomes more viscous when pressure is applied. This exceptional enhancement of the fluidity of water with pressure at low temperatures is due to the same molecular mechanism that is at the origin of the negative slope of the line that separates liquid water and ice Ih in the phase diagram of ice (Figure 8.5), and which leads to transforming rigid ice into fluid water when pressure is applied. It is a consequence of the fact that the primary effect of pressure is to bend H-bonds. Let us point that this bending of H-bonds by pressure has a static character: it results in changing the equilibrium positions of librations. It is different from the effect of temperature, which has a dynamic character, changing in a first approximation the amplitudes of librations, but not their equilibrium positions.

The pressure dependence of the self-diffusion coefficient D, which describes the motion of the centre of gravity of a particular H_2O molecule in liquid water, also displays an exceptional behaviour with pressure: it increases when pressure increases, which means that it becomes easier for a particular H_2O molecule to diffuse inside compressed liquid water, at least in the region where pressure is not too high. Above 200–300 MPa, diffusion becomes more difficult when pressure is increased, and liquid water then behaves as almost all other liquids. This anomalous behaviour occurs at 0 °C, as represented in Figure 9.9. It becomes less marked at higher temperatures, but more marked in the supercooled region. The reason for this behaviour is the same as that encountered for viscosity: because in this region pressure bends H-bonds between water molecules, it makes translation of a particular H_2O molecule easier, at least as long as water molecules are not too close to hinder such a translation as in almost all other liquids.

These are the main exceptional physical properties of liquid water. There are some other ones, which may be less spectacular. Thus, the lifetime of the first excited state of ν_s, of the order of 1 psec (50), has recently been shown, using femtosecond IR pump-probe spectroscopy (51) (see Ch. 11), to increase with temperature. Lifetimes of excited vibrational states usually decrease when temperature increases, and long lifetimes are obtained at low temperatures. This is because relaxation of these excited states towards ground state occurs via a direct transfer of the excitation towards highly excited states of intermonomer modes (phonons in a solid), a transfer that is made possible by an anharmonic interaction between this vibration and intermonomer modes (51). Raising temperature increases the populations of the excited states of the intermonomer modes, increasing the average transfer rates of the energy of the excited state towards intermonomer modes. Relaxation itself occurs within various states of these intermonomer modes towards a thermal equilibrium of their populations. In the case of ν_s of liquid water, the relaxation path is different and occurs via a succession of different steps. The first and rate-determining one is the transfer of excitation of ν_s towards the overtone $2\delta_{H\text{-}O\text{-}H}$ of the bending vibration of H_2O via Fermi resonance (see Ch. 4), and is followed by relaxation of these bending vibrations, which have been shown to have themselves a lifetime (52) of 1.4 psec. As already seen, raising temperature makes ν_s shift towards higher wavenumbers and δ_{H-O-H} towards lower wavenumbers. It decreases the overlap between the first excited state of ν_s and the first overtone of δ_{H-O-H}, thus making this $\nu_s \rightarrow 2\delta_{H-O-H}$, transfer less efficient. The lifetime of the first excited state of ν_s is therefore longer (50).

Another property of liquid water, which is more singular than exceptional, but has more important consequences, is the behaviour of its heat capacity C_p: nearly a constant for temperatures above -10 °C and extending up to 100 °C, but with a value more than twice that of ice, it considerably increases in the supercooled region below -10 °C (45, 53). It is on the limits of an exceptional property, because the nearly constant value above -10 °C has been shown to really be a slightly decreasing value when temperature increases at low temperature, followed by an also slightly increasing value at higher temperature, the whole thing displaying a slightly marked minimum at 35 °C (31, 45, 48). The exception of C_p is thus not so much in this singular behaviour with temperature, but more in the unusually great value it takes on above 0 °C: $1\,\text{cal}\,\text{g}^{-1}\,\text{T}^{-1} \approx 75\,\text{J}\,\text{mol}^{-1}\,\text{T}^{-1}$. It has important well-known consequences at our level, making for instance oceans a highly efficient regulator of the Earth surface temperature. It is also at the origin of other climatic effects, such as a great amplitude of thermal variations where water is absent, as in deserts, etc.

OUR UNDERSTANDING OF LIQUID WATER

From the preceding sections, it clearly appears that water, the most familiar liquid, is indeed an exceptional liquid. Its exceptional properties are due to the presence of this flexible hyperdense H-bond network that links all H_2O molecules. Ice, which has an as dense H-bond network, does not exhibit so many exceptional properties. This is because the H-bond network of ice is rigid. In opposition that of liquid water is flexible, due to the possibilities for each H_2O to perform rotations of great amplitudes. Water is a liquid and as such may accept inside it different molecules. The exceptional presence of this hyperdense H-bond network gives it unique possibilities to accommodate embedded molecules, making water an exceptional liquid to solvate ions, organic molecules, or to give with them original molecular structures, such as micelles of detergents, biological membranes made of phospholipid bilayers, Langmuir–Blodgett films, etc. All these properties are reasonably well understood, in terms of chemical processes, even if they often require a more accurate thermodynamic assessment than the illustrative one we have discussed on the basis of eq. (9.5) (54). This is also necessary when one has to do with more complicated systems, such as when adding cosolvents to surfactant molecules in order to give micelle structures a greater stability (43), for instance. Also, such complex systems as biosystems also require a more precise knowledge than the simplified description given above.

The situation of the exceptional physical properties, which are described in the preceding section, is less favourable. The hyperdense H-bond network makes such physical quantities as those shown in Figure 9.9 vary in a direction opposite to that of most other liquids when temperature or pressure are changed in the low temperature region. We gave qualitative explanations for these variations. They absolutely need receiving a more precise quantitative treatment before they can be considered as satisfactorily understood (55, 56). In other words, we still have no precise description of liquid water at molecular level. It sounds as a paradox, when one realizes that the situation is much better for most other liquids, which are, for all of them, much less familiar than liquid water. On the other hand, they do not exhibit exceptional physical properties, and this makes them easier to describe. The familiarity of liquid water, which we everyday drink and use in various ways, makes us erroneously think that it is a simple and ordinary liquid. At molecular level this is far from true, as the presence of this hyperdense H-bond network gives it original properties that are not easy to precisely describe. Thus, molecular dynamics (MD) methods, which have displayed success describing many properties of other liquids, up to now failed to reproduce the exceptional properties of liquid water and to give a description of liquid water at molecular level. MD consists of calculating the evolution with time of the molecular structure of an assembly of several hundredths of composing molecules, subject to defined intermolecular forces, and which have been given an initial structure at time $t = 0$. For most liquids these forces can be reduced to Coulomb interactions between point charges that are optimally spread along the molecule, an approximation that is necessary to handle a sufficiently great number of molecules. It works usually well for ordinary organic liquids, which are made of relatively big molecules. Such a treatment is inappropriate for mimicking the highly directional H-bonds that constitute the interactions of small H_2O molecules inside the hyperdense H-bond network they build, and for describing the well-defined librations of great amplitudes IR spectroscopy has shown to occur and are responsible for its fluidity. MD simulations could consequently only fail to explain how translational and rotational diffusion are comparable to those of other liquids, despite the presence of this highly directional but nevertheless flexible H-bond network. Thus, despite a throng of MD models devoted to liquid water, none of them can claim any success to give a molecular description of liquid water. Most of all find geometrical arrangements for H_2O molecules in agreement with X-ray and neutron scattering experiments. This is considered a test for a good definition of their potentials, which is then used to predict other properties. Unfortunately, these scattering experiments concern only distances and are not easily related to the dynamic structure of the H-bond network itself, which is therefore ignored. As we have seen that dynamics of H_2O molecules is central for most properties of the H-bond network of liquid water, it is not so much surprising that no definite description of liquid water at molecular level exists.

More original attempts by Robinson and coworkers to fully take into account these librations (57), which we suspect to be at the origin of the fluidity of water, have not been satisfactory. Even if the authors of such an approach claim success in precisely reproducing such anomalies as the change in viscosity with pressure (58), or the variation of the density d with temperature (59), which may be considered a key-step to arrive at a satisfactory description of liquid water at molecular level, the potential they use is too "ad hoc" and lacks generality to give a satisfactory description of liquid water. A former original analysis by Stanley and Teixeira (60) of the connectivity of the H-bond network has also given hope that the specificity of liquid water could be understood, as it gave evidence that H_2O molecules with established H-bonds were not randomly distributed inside the liquid bulk, but formed tiny patches of four-bonded molecules. This is a picture that was proposed long time ago, by Röntgen himself (that one who discovered X-rays). Such patches are known as "ice-like aggregates". Anomalous properties were then predicted to originate from the existence of these patches. However interesting the Stanley–Teixeira's analysis could be, it was hampered by the starting hypothesis that O–H groups could be divided into two groups, those which were H-bonded and those which were not. This contradicts IR spectroscopy that we have seen to exclude the presence of broken H-bonds in a significant proportion (Figure 9.1), but points to a continuous distribution of H-bonds of various energies that rapidly interchange and are due to great amplitude librations of H_2O molecules. Furthermore, these patches have never been put into evidence.

These numerous theoretical attempts to give a fair molecular description of liquid water have been accompanied by a multitude of new experiments, testing liquid water at various temperatures and pressures by X-ray and neutron scattering, NMR, Raman scattering and IR absorption spectroscopy, etc. It makes liquid water certainly the species for which the greatest number of data have been collected. Up to now with no corresponding satisfying description emerging from all these studies. The most precise results, those that come from IR and Raman spectroscopy, have, in an astonishing way, never been considered as starting points for a molecular description of liquid water. As previously seen, IR spectra stress the importance of librations for the properties of the H-bond network. Raman spectra of liquid water suggest, with the presence of isosbectic points (61, 62) also visible in IR spectra (17, 63), the presence of two distinct types of H_2O molecules. Recent ultrafast spectroscopy has recently put into evidence (64) the existence of H_2O molecules having two different relaxation rates, strongly supporting the existence of these two types of H_2O molecules. Interestingly, these distinct molecules have enthalpy differences that have been estimated to be equal to 2.6 kcal mol^{-1} (65) by Raman spectroscopy, or more recently to 1.5 kcal mol^{-1} by X-ray absorption spectroscopy (66) (see Ch. 3). Beyond the differences of these values, which is not so much surprising in view of the accuracy provided by such measurements, we may note that these two values correspond to wavenumbers $\tilde{\nu}$ equal to 900 and 500 cm^{-1}, respectively, if one applies the relation $E = h\tilde{\nu}/c$ (eq. (1.A3) and c, the velocity of light, expressed in cm sec^{-1}). These wavenumbers, 900 and 500 cm^{-1}, are those of librations, strongly suggesting that in liquid water, molecules in both their fundamental and first excited librational states coexist, and that in these excited states they display great rotational amplitudes. It implies that treating librations by quantum mechanics cannot be avoided. MD methods are poorly adapted to fully incorporate such quantum effects. A realistic description should thus preferably take into account the properties of librations from the start, establishing a potential for both librational and stretching intermonomer modes. It should lead to the possibility of having librations with great amplitudes, where couplings with stretching intermonomer modes should be included so as to explain why both ice and liquid water become more fluid when pressure is applied, ice being even transformed into liquid water (Figure 8.5). With such a coupling, pressure first bends H-bonds before compressing O···O distances. It highlights the fundamental role of the hyperdense H-bond network. It should also explain the different effects of temperature: ice exhibits a slight decrease in its density when temperature increases (45), as any other species, whereas liquid water exhibits an exceptional increase in its density. The effect of temperature on ice is that expected from a rather rigid H-bond network where all vibrations obey a harmonic potential, whereas on water temperature still increases the amplitudes of librations, which are already great, making liquid water denser because of the increase of the (dynamic) bending of H-bonds. It might explain another astonishing result of IR spectroscopy: the *intramonomer* bending band δ_{H-O-H} becomes narrower when temperature is raised (67) and even takes on a Lorentzian shape (63), characteristic of a rapid loss of coherence. This effect, recently acknowledged by pump-probe time-resolved IR experiments (68), has up to now received no definitive explanation, despite MD attempts (69). It is likely to be due to an anharmonic coupling of δ_{H-O-H} of H_2O with intermonomer bending modes. Such a modulation is certainly at the origin of the small relaxation time of δ_{H-O-H}, 170 fsec, as recently measured by pump-probe IR spectroscopy (70).

CONCLUSION

Liquid water, one of the most familiar species, which we consider a most simple liquid in ordinary life, reveals so shrewd a species at molecular level that, in opposition to most other liquids, we still lack any fair molecular description of it. Its peculiarity: it possesses a hyperdense H-bond network that keeps on an average a tetrahedral symmetry for O-atoms of nearest neighbour molecules, with, however, great and rapid fluctuations around this symmetry due to librations (hindered rotations) of individual H_2O molecules having great amplitudes. This H-bond network is as dense as that of ice. In opposition to that of ice, it is most flexible, due to the possibility all H_2O molecules have to perform within it rotations of great amplitudes, as shown by IR spectroscopy. In ice such rotations keep small amplitudes, which makes the H-bond network rigid. With the exception of some elusive minor points, such as the structure of its common form Ice Ih we mentioned in Ch. 8, the structure and dynamics of ice are understood. The really exceptional property displayed by ice is the existence of a line separating ice Ih from liquid water that displays a negative slope. Its macroscopic consequence is the possibility of skating. It is due to the presence of this hyperdense H-bond network which exhibits, when pressure is applied, a reaction typical of H-bonds, but which manifests only when H-bonds are as numerous as in ice, that is as numerous as covalent bonds: pressure first bends H-bonds, making compression appear as a secondary effect. In nearly all other species where intermolecular interactions are not exclusively H-bonds, compression is the primary reaction to pressure. As a consequence ice is exceptional for becoming a liquid when pressure is applied. Liquid water having an as dense H-bond network as ice exhibits the same behaviour: when pressure is applied, it becomes a more fluid liquid, at least at not too high a temperature, in opposition to all other liquids which become more viscous.

In ice the rigidity of this H-bond network hinders the appearance of other exceptional properties. Thus, temperature has the same effect as in other species where intermolecular vibrational modes, the only temperature sensitive modes at 300 K, exhibit no anharmonic couplings. Only a 1D Morse-type anharmonicity is present and responsible for the thermal dilatation, a general phenomenon. Rigidity also hinders embedding of organic or inorganic molecules, and ice consequently exhibits the weak reactivity of a common solid, if we disregard its surface reactivity. This is not so of liquid water, where introduction of flexibility within this hyperdense network has tremendous consequences. The first one is the great value of the surface tension coefficient γ of liquid water. It is a direct consequence of the presence of this hyperdense H-bond network, as increasing the surface of liquid water requires breaking a very large number of H-bonds. We have seen in Ch. 8 that the formation of raindrops owes much to this important capillarity effect. Furthermore, the flexibility of this H-bond network makes liquid water react differently than other liquids when temperature, for instance, is varied. It also makes it react differently than ice. Its contraction when temperature increases around 0 °C is a typical example of these exceptional physical properties. In opposition to other liquids, rotations of individual H_2O molecules in water are rapid and exhibit great amplitudes, which makes water a liquid, but this amplitude also increases with temperature, making the amplitudes of H-bond distortions (bending) increase with temperature. As we are far from a harmonic potential and of its complete independence of intermonomer modes, this increase of the amplitudes of the distortions leads to a slight increase of the density due to a shortening of the equilibrium distances of the intermonomer stretching modes. In ice this increase of the amplitude of bending with temperature remains within the limits of the harmonic approximation, a situation found in other liquids or solids: it has no effect on intermonomer stretching modes. The effect of the 1D anharmonicity of these stretching modes can then appear in the usual form of a dilatation. This singular reaction to temperature changes is itself at the origin of many exceptional physical properties of liquid water. Even if we understand qualitatively their origins, we do not yet know how to precisely describe them. This is some paradox of sort, as we "know how to workout the properties of ice and steam, but we have no clear idea why there is such a thing as ordinary liquid water" (71). The flexibility of the H-bond network of water is also at the origin of exceptional chemical properties. It is thus at the origin of exceptional capacities of embedding of organic and inorganic molecules, which make liquid water one of the most powerful solvent. These exceptional chemical properties are relatively well described in terms of chemical mechanisms, even if more accurate descriptions are needed to more precisely understand them.

In this chapter we did not investigate the exceptional biological properties of liquid water. The reason is that if it is well known that water is absolutely necessary for life to proceed, at least life as we know it on Earth, it is not evident that the presence of liquid water is a fundamental prerequisite for life to occur. The necessary condition for life to occur seems to be the presence of H_2O molecules with their specific ability to build around them an exceptionally dense H-bond network. Even if liquid water is the basic and most simple system that provides such a dense H-bond network, it is not the only one, as dense but slightly different networks may also exist in more locally defined structures with H_2O molecules. We examine this point in more details in Ch. 10.

Up to now, we have examined water in its three states: vapour, liquid and solid. Another state has relatively recently revealed to be of interest: the supercritical state, where an assembly of water molecules is under a pressure greater than 221 bar (1 bar = 10^5 Pa) and a temperature higher than 374 °C. In these conditions H_2O molecules still establish (weak) H-bonds. They are surprisingly most reactive and are thus candidates to be used in industry to decompose numerous polymers and hazardous chemicals (72). It may for instance rapidly hydrolyse organic molecules without any acidification (73). Supercritical water can thus act as an effective decontamination agent, itself devoid of any toxical effect. Also, its dielectric constant can be continuously varied, as may also its density or the density of H-bonds it contains, etc. How is it so? Another open question for that exceptional molecule, and certainly yet a lot of work for scientists.

REFERENCES

1. E. Whalley, in Schuster, G. Zundel and C. Sandorfy (Eds.), The Hydrogen Bond: Recent Theory and Experiments, Vol. 3, Ch. 29, North Holland, Amsterdam, 1976.
2. S. S. Xantheas, J. Chem. Phys., 104 (1996) 8821.
3. R. S. Fellers, C. Leforestier, L. B. Braly, M. G. Brown and R. J. Saykally, Science, 284 (1999) 945.
4. L. M. Goss, S. W. Sharpe, T. A. Blake, V. Vaida and J. W. Brault, J. Phys. Chem. A, 103 (1999) 8620.
5. J. E. Bertie, A. M. Khalique and H. H. Eysel, J. Phys. Chem., 93 (1989) 2210.
6. M. Farnik, M. Weiman and M. A. Suhm, J. Chem. Phys., 118 (2003) 10120.
7. C. A. Angell, Annu. Rev. Phys. Chem., 55 (2004) 559.
8. E. A. Raymond, T. L. Tarbuck, M. G. Brown and G. L. Richmond, J. Phys. Chem. B, 107 (2003) 546.
9. X. Wei and Y. R. Shen, Phys. Rev. Lett., 86 (2001) 4799.
10. C. M. Johnson, E. Tyrode, S. Baldelli, M. W. Rutland and C. Leygraf, J. Phys. Chem. B, 109 (2005) 321.
11. Q. Du, R. Superfine, E. Freysz and Y. R. Shen, Phys. Rev. Lett., 70 (1993) 2313.
12. J.-J. Max and C. Chapados, J. Chem. Phys., 116 (2002) 4626.
13. J.-J. Max and C. Chapados, J. Chem. Phys., 122 (2005) 014504.
14. J. D. Eaves, J. J. Loparo, C. J. Fecko, S. T. Roberts, A. Tokmakoff and P. L. Geissler, Proc. Natl. Acad. Sci., 102 (2005) 13019.
15. J. P. Devlin, J. Sadlej and V. Buch, J. Phys. Chem. A, 105 (2001) 974.
16. S. D. Warren, Appl. Opt., 23 (1984) 1206.
17. Y. Maréchal, J. Chem. Phys., 95 (1991) 5565.
18. H. R. Zelsmann, J. Mol. Struct., 350 (1995) 95.
19. J. P. Devlin, P. J. Wooldridge and G. Ritzhaupt, J. Chem. Phys., 84 (1986) 6095.
20. P. Blaise, M. J. Wojcik and O. Henri-Rousseau, J. Chem. Phys., 122 (2005) 64306.
21. S.-H. Chen and J. Teixeira, in I. Prigogine (Ed.), Advances in Chemical Physics, Vol. LXIV, Wiley, 1986.
22. A. Novak, in J. D. Dunitz, P. Hemmerich, R. H. Holm, J. A. Ibers, C. K. Jorgensen, J. B. Neilands and D. Reinen (Eds.), Structure and Bonding, Vol. 18, Springer-Verlag, Berlin, Heidelberg, 1974.
23. W. Mikenda, J. Mol. Struct., 147 (1986) 1.
24. S. Bratos, J. Cl. Leicknam, S. Pommeret and G. Gallot, J. Mol. Struct., 708 (2004) 197.
25. B. Cabane and S. Hénon, Liquides, Solutions, Dispersions, Émulsions, Gels, Belin, Paris, 2003.
26. J. C. Dore, J. Mol. Struct., 237 (1990) 221.
27. M. C. Bellissent-Funel and L. Bosio, J. Chem. Phys., 102 (1995) 3727.
28. J. A. Pople, Proc. R. Soc., 205A (1951) 163.
29. J. J. Gilijamse, A. J. Lock and H. J. Bakker, Proc. Natl. Acad. Sci., 102 (2005) 3202.
30. J. A. Rupley and G. Careri, in C. B. Anfinsen, J. T. Edsall and F. M. Richards (Eds.), Advances in Protein Chemistry, Vol. 41, Academic Press, San Diego, 1991.
31. Handbook of Chemistry and Physics, 75 ed., CRC Press, Boca Raton, 1994.
32. C. P. Petersen and M. S. Gordon, J. Phys. Chem. A, 103 (1999) 4162.
33. J. A. White, E. Schwegler, G. Galli and F. Gygi, J. Chem. Phys., 113 (2000) 4668.
34. A. W. Omta, M. F. Kropman, S. Woutersen and H. J. Bakker, Science, 301 (2003) 347.
35. M. F. Kropman and H. J. Bakker, J. Chem. Phys., 115 (2001) 8942.
36. L.-A. Näslund, D. C. Edwards, P. Wernet, U. Bergmann, H. Ogasawara, L. G. M. Pettersson, S. Myneni and A. Nilsson, J. Phys. Chem. A, 109 (2005) 5995.
37. D. N. Chipman, J. Chem. Phys., 118 (2003) 9937.

38. M. J. Shultz, S. Baldelli, C. Schnitzer and D. Simonelli, J. Phys. Chem. B, 106 (2002) 5313.
39. K. Koga, J. Chem. Phys., 121 (2004) 7304.
40. P. Csermely, Cell. Mol. Biol., 47 (2001) 791.
41. L. F. Scatena and G. L. Richmond, J. Phys. Chem. B, 105 (2001) 1940.
42. S. Abel, F. Sterpone, S. Bandyopadhyay and M. Marchi, J. Phys. Chem. B, 108 (2004) 19458.
43. D. Langevin, J. Meunier and A.-M. Cazabat, La Recherche, 167 (1985) 720.
44. P. Lozano, A. J. Fernández, J. J. Ruiz, L. Camacho, M. T. Martin and E. Muñoz, J. Phys. Chem. B, 106 (2002) 6507.
45. C. A. Angell, in F. Franks (Ed.), Water: A Comprehensive Treatise, Vol. 7, Ch. 1, Plenum Press, New York, 1982.
46. R. A. Fine and F. J. Millero, J. Chem. Phys., 59 (1973) 5529.
47. R. J. Speedy and C. A. Angell, J. Chem. Phys., 65 (1976) 851.
48. R. Ludwig, Angew. Chem. Int. Ed., 40 (2001) 1808.
49. F. X. Prielmeier, E. W. Lang, R. J. Speedy and H.-D. Lüdemann, Phys. Rev. Lett., 59 (1987) 1128.
50. C. P. Lawrence and J. L. Skinner, J. Chem. Phys., 117 (2002) 5827.
51. A. J. Lock and H. J. Bakker, J. Chem. Phys., 117 (2002) 1708.
52. A. Pakoulev, Z. Wang, Y. Pang and D. D. Dlott, Chem. Phys. Lett., 380 (2003) 404.
53. E. Tombari, C. Ferrari and G. Salvetti, Chem. Phys. Lett., 300 (1999) 749.
54. N. F. A. van der Vegt and W. F. van Gunsteren, J. Phys. Chem. B, 108 (2004) 1056.
55. J. L. Finney, J. Mol. Liq., 90 (2001) 303.
56. B. Guillot, J. Mol. Liq., 101 (2002) 219.
57. M. P. Bassez, J. Lee and G. W. Robinson, J. Phys. Chem., 91 (1987) 5818.
58. C. H. Cho, J. Urquidi and G. Wilse Robinson, J. Chem. Phys., 111 (1999) 10171.
59. C. H. Cho, S. Singh and G. W. Robinson, Phys. Rev. Lett., 76 (1996) 1651.
60. H. E. Stanley and J. Teixeira, J. Chem. Phys., 73 (1980) 3404.
61. G. E. Walrafen, M. S. Hokmabadi and W. H. Yang, J. Chem. Phys., 85 (1986) 6964.
62. G. D'Arrigo, G. Maisano, F. Mallamace and P. Migliardo, J. Chem. Phys., 75 (1981) 4264.
63. Y. Maréchal, J. Phys. II (France), 3 (1993) 557.
64. Z. Wang, A. Pakoulev, Y. Pang and D. D. Dlott, J. Phys. Chem. A, 108 (2004) 9054.
65. G. E. Walrafen, M. R. Fisher, M. S. Hokmabadi and W.-H. Yang, J. Chem. Phys., 85 (1986) 6970.
66. J. D. Smith, C. D. Cappa, K. R. Wilson, B. M. Messer, R. C. Cohen and R. J. Saykally, Science, 306 (2004) 851.
67. Y. Maréchal, J. Mol. Struct., 322 (1994) 105.
68. O. F. A. Larsen and S. Woutersen, J. Chem. Phys., 121 (2004) 12143.
69. M. Praprotnik, D. Janezic and J. Mavri, J. Phys. Chem. A, 108 (2004) 11056.
70. N. Huse, S. Ashihara, E. T. J. Nibbering and T. Elsaesser, Chem. Phys. Lett., 404 (2005) 389.
71. M. C. Gutzwiller, Phys. Today, 48 (1995) 15.
72. I. Ohmine and S. Saito, Acc. Chem. Res., 32 (1999) 741.
73. E. Vanardou, E. Garcia-Verdugo, S. J. Barlow, Y. E. Gorbaty and M. Poliakoff, Vib. Spectrosc., 35 (2004) 103.

– 10 –

The Water Molecule in (Bio)Macromolecules

WATER MOLECULES AND THEIR DENSE HYDROGEN BOND NETWORKS

Having examined ice and its rigid hyperdense H-bond network at the origin of a single exceptional physical property, and then liquid water and its flexible and hyperdense H-bond network at the origin of many exceptional physical and chemical properties, we now turn our attention to water in large, poorly ordered systems, most of them macromolecules. These two species, ice and liquid water, illustrate how the exceptional structure of the H_2O molecule, with its two H-bond donor sites and two H-bond acceptor sites (Figure 8.1), is at the origin of the formation of a dense H-bond network. The properties of this H-bond network are, however, different in ice and in liquid water: its rigidity in ice and its great flexibility in liquid water lead to completely different chemical properties, already pointing to the important role this H-bond network plays. It may be argued that, owing to the great difference between a rigid and a highly flexible network, this difference is not very surprising. However, modifying the H-bond network of liquid water in a less drastic way, for instance, by reducing its size, also has visible consequences. Such a reduction appears in reverse micelles, described in Ch. 9, and induces changes of shapes and positions of IR bands that continuously vary when the sizes of the water nanodroplets in these reverse micelles vary (1). Also, the dynamics of H_2O molecules is substantially slower than in liquid water, as shown by measurements of relaxation times of ν_s bands (2), using nonlinear time-resolved IR spectroscopy. This slower dynamics indicates that the H-bond network developed by H_2O molecules is more rigid in these reverse micelles than in liquid water. Confined water molecules enclosed by acetone molecules, a case where this reduction is even bigger, have been shown by femtosecond pump-probe IR spectra to have a dynamics that strongly differs from that of bulk liquid water (3). These results strongly suggest that the H-bond networks developed by H_2O molecules play an important role in the emergence of macroscopic properties of samples that, other than ice or liquid water, contain water molecules.

In this chapter we examine this situation where, without having to deal with liquid water or ice, we nevertheless are still concerned with extended H-bond networks developed by H_2O molecules. This is the case for water molecules embedded in macromolecules. We shall not stress the importance of macromolecules, but the point worth noting here is that H_2O molecules are ubiquitous inside them. This is because H_2O molecules are very small and consequently most mobile, on the one hand, and on the other hand, because their exceptional structure, as shown in Figure 8.1, always provides them with the possibility of establishing H-bonds on the hydrophilic groups that exist in more or less large numbers in such macromolecules. The embedded water molecules may strongly modify the physical properties of the host macromolecule. This is already true of synthetic polymers where they may, for instance, cause unwanted aging, degrading the interesting properties of these polymers that often are "new materials". They also may be at the origin of hydrolysis of some of the constituent molecules of this polymer, thus also ruining all the interesting properties it has originally been designed for. In biological macromolecules, an H-bond network can spontaneously appear before the arrival of H_2O molecules. Thus, secondary structures of proteins described in Ch. 2, may exist with no water molecule. However, H_2O molecules are not anecdotic molecules in biomacromolecules: as will be seen throughout this chapter, they play a fundamental role there, and it is consequently crucial to know how they behave. When embedded in any macromolecule, these small and extremely efficient H-bond-establishing molecules exert this tendency to develop an extended H-bond network. This H-bond network may be reduced to a single H_2O molecule, a very extreme case, or may tend, in large H_2O clusters that may extend over or all around the host molecule, towards that of liquid water, the other extreme case.

In a given macromolecule we may pass from one situation to the other, depending on the quantity of water molecules that are embedded in it, that is on its degree of hydration. Within these H-bond networks, which are always dense but more or less extended, these H_2O molecules may adopt various H-bond configurations that can be rapidly exchanged, suggesting a rapid dynamics comparable to, but nevertheless differentiable from that of liquid water. In other words, these H-bond networks that appear in macromolecules may take many different forms with many different consequences at the macroscopic level. In the current state of our knowledge, which is presently in a very early stage of development, we scarcely understand their role(s). This is because comprehending the role of these H-bond networks in macromolecules encounters an unexpected difficulty: contrary to what we might think, the H_2O molecule is not easy to observe, especially when it is embedded in macromolecules, and only the recent methods described in Ch. 11 seem able to precisely convey enough information on it. The consequence of this difficulty is that the H_2O molecule has for a long time been ignored (4) when it is embedded in macromolecules, wrongly suggesting that it is an inert molecule that plays no special role, despite our awareness that water molecules are absolutely necessary for life to occur. This attitude has been called a scandal by John Maddox (5) in one of his "News and Views" in *Nature*, where he wrote: "Is the scandal, that so little is known about the interactions of macromolecules and their aqueous environment, about to be removed?". Calling this a scandal does not appear too strong when one realizes that water molecules are ubiquitous, on the one hand, and are far from being inert or passive, on the other. In particular, life, as we know it on Earth, cannot occur in the absence of these molecules.

In this chapter we discuss the structure of this H-bond network, examining recent experiments on hydration mechanisms of macromolecules that have contributed original information on the structures of H-bond networks developed by H_2O molecules in these macromolecules. We then describe present experiments on bioprotection that stress the impact this H-bond network has on the structure of the macromolecule itself, before mentioning protein folding, which have undergone rapid development in recent years, but have not yet reached a point where significant information could be obtained. We then proceed to discuss the role these H-bond networks are suspected to play in the reactivity of macromolecules. It may serve as a guideline for forthcoming experiments. This chapter differs from preceding ones in the sense that, contrary to the preceding ones where known facts and present research activities are presented, this one has a highly prospective character and outlines a description of what may be of importance in forthcoming years. We have seen that even in these relatively simple species such as ice or liquid water—simple from the chemist's point of view, as only H_2O molecules are present—still many questions remain unanswered. In the case examined in this chapter, of water molecules embedded in macromolecules, the system is more complicated from this same chemist's point of view and consequently still less known. The main theme of this chapter is biological aspects of water, and it is felt that progress in this field will certainly shed some new light on biological mechanisms and on the fundamental role of the up to now often disregarded H_2O molecules. All along we must keep in mind a fundamental question: "How is it that life occurs in water and in water only". We sketch an inevitably coarse answer in the conclusion of this chapter.

ARRANGEMENTS OF WATER MOLECULES IN MACROMOLECULES

The H-bond networks of liquid water and ice are very dense and extend over a very large region. All water molecules are part of these networks and nearly all of them are tetra-coordinated. This is not so in macromolecules where many different situations can be encountered, following the density of hydrophilic groups this macromolecule exhibits and also the state of hydration of the macromolecule, that is the number of H_2O molecules that are embedded in it. Some recent experiments conveyed information on the structures of the H-bond networks in these systems. They are experiments on hydration mechanisms, lyo or cryoprotection, and folding of proteins. These relatively few mechanisms that have been recently studied in some detail are described in this section. They give what could be called a preview on this H-bond network developed by H_2O molecules in macromolecules.

Hydration mechanisms

Studies on the hydration of macromolecules consist of observing the H-bond network that develops inside a macromolecule when its passes from a dried state to a completely soaked state. This can be achieved by, for instance, increasing the water vapour tension of the surrounding atmosphere, starting from a waterless atmosphere within which the dried sample is in equilibrium and ending with an atmosphere with 100% humidity, which is an atmosphere saturated with water vapour in equilibrium with liquid water in a restrained volume. This way of studying hydration has the advantage of defining the variable state of hydration with precise values, the hygrometry also called relative humidity, which we define in the next paragraph. Hydration mechanisms have formerly been studied by various methods, the most common one being thermogravimetry (6), described in Ch. 11, the most recent and precise one being IR spectrometry. This latter method is also described in detail in Ch. 11 as, when applied to water molecules, it requires specific procedures to avoid the serious drawbacks that have long hindered using IR spectroscopy to look at water molecules, despite the power

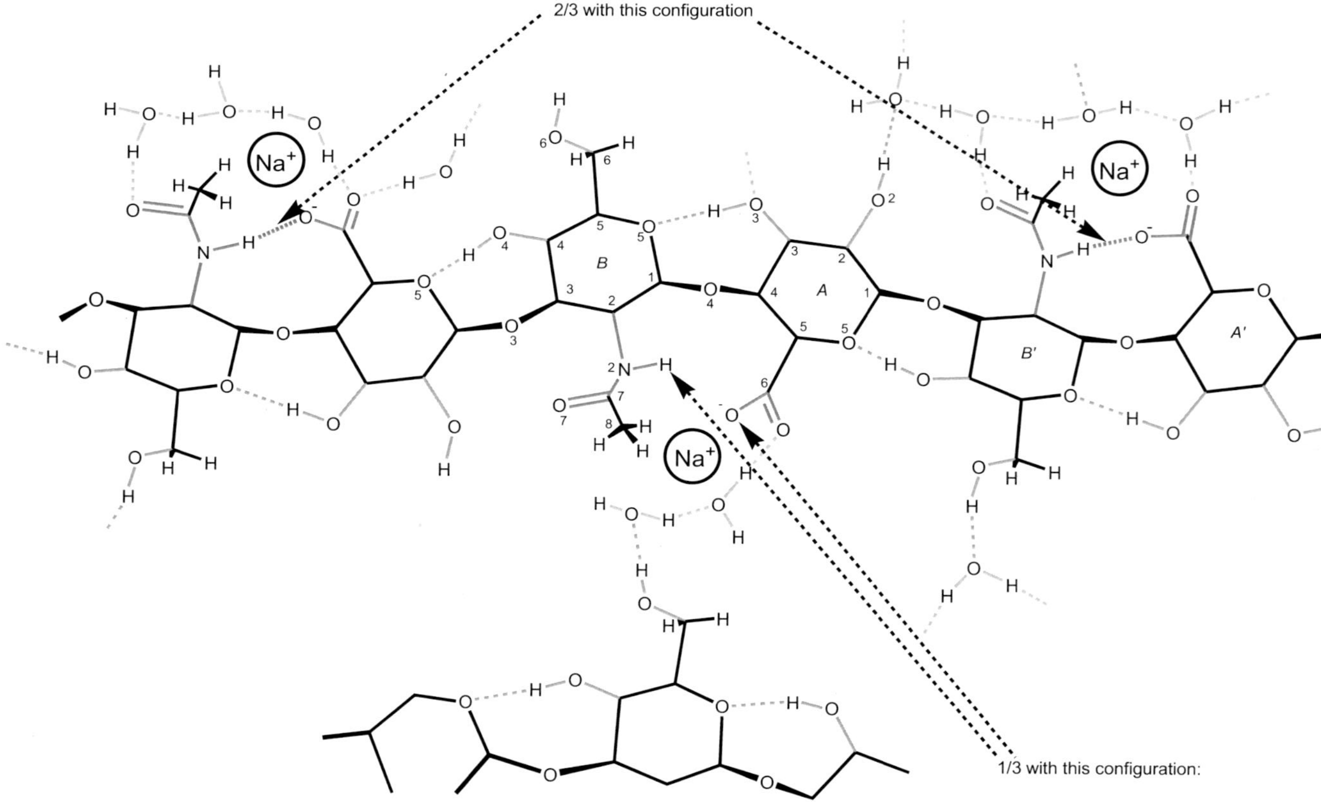

Figure 10.1 The H-bond network of dried HA, as deduced from a quantitative analysis of IR spectra (see Ch. 11). Three disaccharide repeat units A–B of a particular chain are represented. Part of another chain is shown at the bottom of the figure.

and sensitivity of this method to study H-bonded systems, as described in detail in Ch. 4. In the present chapter we analyse the results delivered by this method, often referring to equations or figures of Ch. 11, where these results are extracted from experiments.

Thermogravimetry allows drawing "sorption isotherms", an example of which appears, among other curves, in Figure 10.2, in the case of a typical macromolecule, hyaluronane (HA), a polysaccharide shown in Figure 10.1. The analysis of these former sorption isotherms has led to a classification of H_2O molecules into two types: "structure" molecules, which appear at low hygrometry of the surrounding atmosphere, and "freezable water molecules", which appear at high hygrometry and exhibit a phase transition accompanied by heat dispersion at temperatures close to but somewhat below 0 °C. This phase transition is similar to that exhibited by H_2O molecules in liquid water. Hygrometry is defined as the ratio p/p_0 of the water vapour pressure p of the surrounding atmosphere to the value p_0 of the saturating water vapour pressure at the same temperature. Thermogravimetry is a global method that is not particularly sensitive to molecules. It cannot consequently help in elucidating what these different types of water molecules it distinguishes are. As seen below, IR spectrometry can precisely define them at the molecular level, and a general scheme for hydration can then be proposed.

We examine in this chapter the hydration mechanisms of two completely different types of macromolecules: a natural biopolymer, HA and a synthetic polymer, a sulfonated polyimide that may be used in the industry. These two types of macromolecules are completely different types of macromolecules.

Hydration of a model biopolymer: hyaluronane (HA)

Sodium hyaluronate, or HA, is a biocompatible polysaccharide that is encountered in medicine and the cosmetics industry. Its macroscopic properties are highly sensitive to its hydration level, which is the reason why it has been the object of many studies. When in its dried state, or scarcely hydrated, it is stiff. In this state it may act as a skin-stiffening agent, smoothing wrinkles for instance. When highly hydrated, it acts as a lubricant that is found in synovial fluid. The elucidation of its hydration mechanism has shown us the role of the developing H-bond network in the change of its properties and, at the same time, the power and sensitivity of IR spectrometry to observe water molecules in macromolecules.

The structure of HA is that of long chains (7) of disaccharide repeat units shown in Figure 10.1. Each disaccharide is made of two β-linked six-atom sugar molecules. In Ch. 2 the β isomers of glucose are seen to allow for the appearance in polysaccharides of lateral H-bonds established by alcohol groups, a property that is fully exploited in HA, with the establishment of C3O3–H···O5 and C4O4–H···O5 bonds. Each of the sugar molecules has another one of its alcohol groups substituted by either a carboxylate COO^- group ("residue" A) or an amide group CH_3–(C=O)–NH ("residue" B). Glycoside links C–O–C in the chain are found at carbons 1 and 4 for residue A, and at carbons 1 and 3 for residue B.

Dried state of HA

It could be shown by IR spectrometry (8) that, in the dried state, about four to five H_2O molecules per repeat unit remain embedded, as shown in Figure 10.1. They are all anchored by H-bonds on C=O groups of carboxylate COO^- groups and form water wires that otherwise hydrate neighbouring hydrophilic groups, establishing H-bonds with them.

The important point for hydration at low hygrometry is that in this dried state two-third of the amide N–H groups establish H-bonds on O^- atoms of adjacent carboxylate groups, while the remaining one-third do not establish H-bonds, leaving the corresponding O^- atoms also free. The establishment of such N–H···O–C=O. H-bonds is energetically favourable. However, the dihedral angles O5–C1–O4–C4 and C1–O4–C4–C3 that define the relative arrangement of cycles A and B, together with the dihedral angles C1–C2–N–C7 and O5–C5–C6–O^- that define the relative arrangement of amido and carboxylate groups with respect to cycles B and A, should have values falling in a restricted region for such an H-bond to be established. This is an unfavourable entropic constraint and the two-third proportion of H-bonded groups N–H···O^-–C=O at room temperature is the result of the equilibrium between these two opposite forces. The presence of such H-bonds, which are compatible with the presence of O3–H···O5 H-bonds, makes the HA chain rigid. It is all the more so that 84% of these C3O3–H groups form O3–H···O5 bonds. These latter H-bonds are therefore in a greater number than the preceding N–H···O^-–C=O N–H bonds. These two numbers, 84% of C3O3–H groups establishing O3–H···O5 H-bonds and 67% of N–H groups establishing N–H···O^-–C bonds, indicate that among the 33% of N–H groups that do not establish H-bonds (cycle B in Figure 10.1), 17% (17 = 84–67) of them are due to rotations of the amido group around the C2–N2 axis or to rotations of the carboxylate group around the C5–C6 axis, or to both of them. These rotations make the establishment of N–H···O^-–C bonds impossible, the angles C1–C2–N–C7 or O5–C5–C6–O^-, or both of them, taking values significantly different from the values that allow the establishment of these H-bonds. At the same time, the angles O5–C1–O4–C4 and C1–O4–C4–C3 do not depart very much from their values in the majority of other similar sites, allowing the establishment of C3O3–H···O5 H-bonds between cycles A and B, as drawn in Figure 10.1. The absence of H-bonds for the remaining 16% of the 33% of the N–H groups that do not establish H-bonds is therefore due to improper values of either O5–C1–O4–C4 or C1–O4–C4–C3 angles that hinder both N···O^- and O3···O5 distances to be compatible with H-bonds. It could furthermore be established that all C4O4–H groups of B-type cycles establish H-bonds of the form C4O4–H···O5 that make all disaccharide repeat units rigid. The backbone of the HA chain in its dried state is consequently made of 84% (rigid) disaccharides that are also rigidly bound together with their relative orientations fixed. In other words, about five over six disaccharides establish rigid bonds between them that hinder relative rotations of the saccharide units around glycoside C–O–C links of the chain. Among these five bonds that hinder relative rotations, one is somewhat weaker, as it has its lateral N–H group free, its rigidity being only due to the presence of the lateral C3O3–H group of the A cycle, establishing an H-bond on the neighbour O5 atom of the B cycle. Only one over six bonds between disaccharides is weaker and allows relative rotations of two neighbour disaccharides. This configuration makes the chain rigid as a whole.

Successive hydration mechanisms

The water uptake of HA is represented in Figure 10.2 as a function of the hygrometry p/p_0 of the surrounding atmosphere. In this figure, the number of H_2O molecules per disaccharide that become embedded in HA following the four hydration mechanisms $H_i (i = 1–4)$, defined below, are shown. The number of H_2O molecules already present in the dried state (4–5 per disaccharide unit) is included into the total number of embedded water molecules

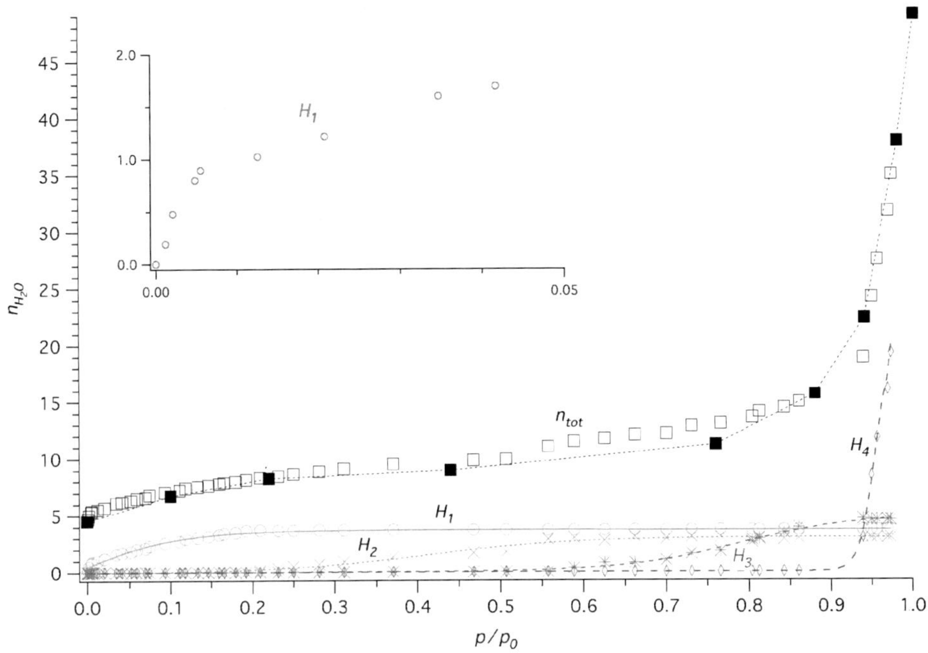

Figure 10.2 Number n_{H_2O} of water molecules per disaccharide in HA, as a function of the hygrometry p/p_0. Details at low hygrometry, where only mechanism H_1 is active, are represented in the enlarged upper left part. Numbers of molecules embedded following mechanism H_1 are represented by circles, those following mechanism H_2 by $\times$, those following mechanism H_3 by stars and those following mechanism H_4 by diamonds. n_{tot} (empty squares) is the total number of embedded H_2O molecules, obtained from IR spectra and equal to the addition of the numbers of molecules entered following all mechanisms H_1 to H_4 plus the number of residual H_2O molecules in the dried membrane (eq. (11.8)). Full squares joined by straight lines represent this total number of H_2O molecules, as measured by thermogravimetry.

represented by squares. Full squares correspond to thermogravimetry data. Empty squares are IR spectrometry data. Both types of data overlap, meaning that these two methods give comparable results. Data given by IR spectrometry are nevertheless more numerous and provide a molecular insight on the hydration mechanisms that cannot be provided by such methods as derived from calorimetry. It allows us in particular to distinguish in the hydration process four various successive steps, H_i with $i = 1$–4. They correspond to various hydration mechanisms that are seen to be active only in some definite regions of the hygrometry. Each of these various hydration mechanisms corresponds to a well-defined "elementary hydration spectrum" represented in Figure 11.4, which constitutes its spectral signature (9). Following Figure 10.2, mechanism H_1 starts at a very low hygrometry level, $p/p_0 = 0.002$ (enlarged upper drawing) and stops being active at hygrometry somewhat greater than $p/p_0 \simeq 0.25$, as within this region the number of H_2O molecules that enter the sample following mechanism H_1 increases. That of molecules hydrated by other mechanisms, H_2, H_3 or H_4, remain equal to 0, meaning that in this region of hygrometry these other mechanisms are inactive. This mechanism H_1 stops being active at $p/p_0 \simeq 0.25$, as at higher values of the hygrometry the number of water molecules that are hydrated by this mechanism no longer varies. Mechanism H_2 then becomes active, as only the number of H_2O molecules that enter the sample following mechanism H_2 increases. It stops being active at $p/p_0 \simeq 0.55$, value at which mechanism H_3 becomes active and remains so up to a hygrometry in the vicinity of 0.9. Just before it ceases being active around this value of the hygrometry, mechanism H_4 takes over, abruptly increasing the number of embedded water molecules.

First hydration mechanism H_1

The quantitative analysis of the IR spectrum H_1 displayed in Figure 11.4, which is the spectral signature of mechanism H_1, shows that at low hygrometry this hydration mechanism consists of hydration of N–H and C–O$^-$ groups that do not establish H-bonds of the type N–H···O$^-$–C. We have seen above that in the dried sample these sites represent one-third of the N–H and C–O$^-$ groups. Hydration of these groups is performed by a nanodroplet of about 25 H_2O molecules that hydrates these two sites and also the hydrophilic neighbouring sites, which encompass the water wire which remained around the carboxylate C=O groups and Na^+ ions in the dried state of HA (Figure 10.3). This hydration mechanism corresponds to a steep variation of the number of embedded H_2O molecules at very low hygrometry, as shown in the enlarged upper diagram of Figure 10.2. It is not one that we would have intuitively guessed, and which would have rather been a spreading of H_2O molecules over the many possible hydrophilic sites of the whole sample. IR spectrometry does not detect any such mechanism. It is possible that it occurs at very low hygrometry, in a region where IR is not sensitive enough to distinguish the corresponding hydrated sample from the dried one, a region with hygrometry p/p_0 less than 0.002, where hydration is already detected. At higher hygrometries, the same hydration mechanism is active as long as p/p_0 remains smaller than 0.25, and the same IR hydration spectrum appears in this whole region. Its corresponding spectrum H_1, displayed in Figure 11.4, is characterized by the absence of a ν_s(O–H) free groups and its steep decrease at high wavenumbers indicates the presence of even less O–H free groups than in liquid water. In Ch. 9 we have seen that the concentration of such O–H groups is at a maximum of some percent in liquid water. In HA it is consequently very close to 0, indicating that nearly all O–H groups are H-bonded when they enter HA at low hygrometries. This cannot be achieved by spreading H_2O molecules on various hydrophilic groups, but can easily be achieved by a nanodroplet of H_2O molecules.

Subsequent hydration mechanisms at higher hygrometries

When mechanism H_1 stops being active at $p/p_0 \simeq 0.25$, about 83% of N–H and C–O$^-$ groups are H-bonded, 66% of them by establishing N–H···O$^-$–C, as in the dried sample, and about 17% of them by establishing H-bonds on H_2O molecules that are parts of the water nanodroplets that have already entered the sample. The following mechanism, H_2, that takes over at such a hygrometry, has much in common with the preceding mechanism H_1, with, however, some clearly visible differences. Thus, C–O$^-$ groups continue being hydrated by nanodroplets of H_2O molecules, but less N–H groups are hydrated, leaving some N–H groups free. In the same way, a somewhat smaller number of C=O amide groups is hydrated. More alcohol groups C2O2–H and C3O3–H of saccharide A and C6O6–H of saccharide B are conversely hydrated. Furthermore, the number of free O–H groups established by water molecules is slightly greater than in the preceding mechanism, but the proportion of these

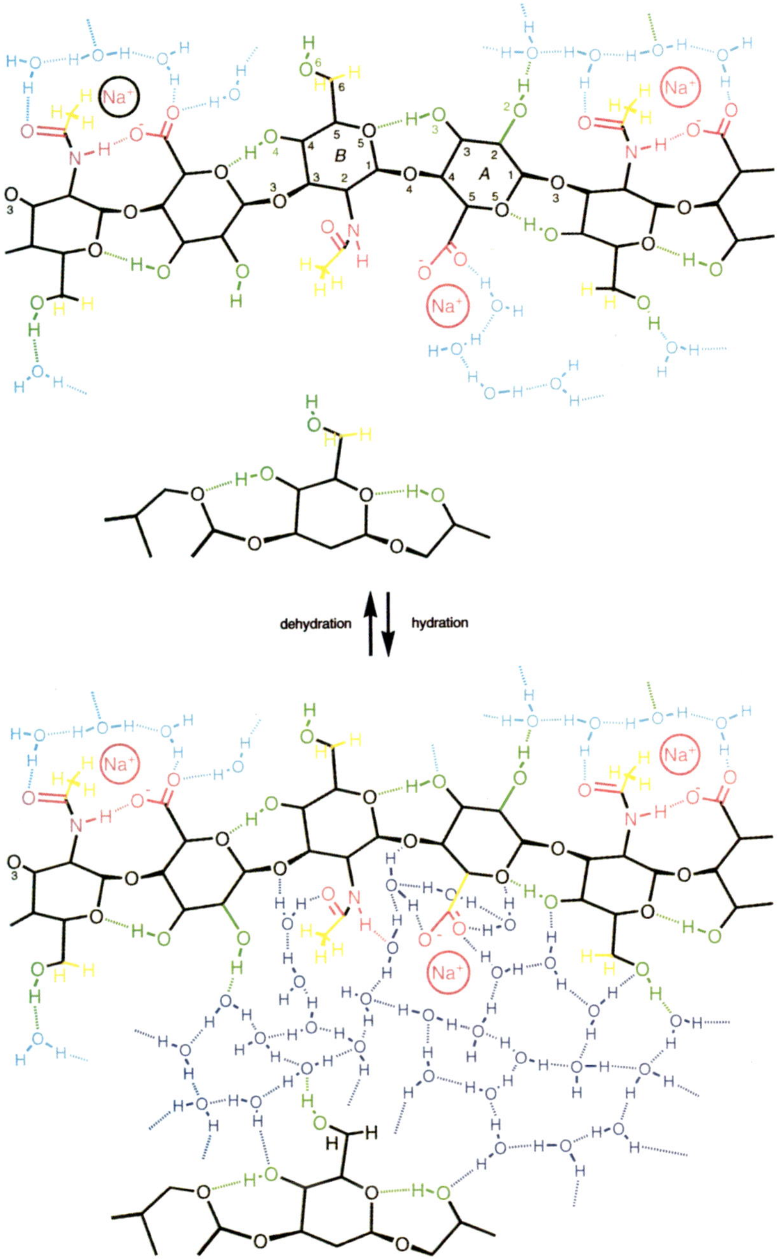

Figure 10.3 First hydration mechanism H_1 of HA: a nanodroplet of about 25 H_2O molecules binds on a site where N–H of amide group and C–O$^-$ of carboxylate group are not directly H-bonded, hydrating also neighbouring hydrophilic groups (8). Upper spectrum: before hydration. Lower spectrum: after hydration.

free O–H groups remains smaller than in liquid water. The H-bond network consequently extends from the nanodroplets that have been embedded when the previous mechanism H_1 was active by embedding new nanodroplets that hydrate other groups than N–H and C–O$^-$ that are, for most of them, already hydrated. This mechanism ceases to be active at $p/p_0 \simeq 0.55$ and mechanism H_3 then takes over. With this mechanism, the last free alcohol groups are hydrated, whereas few other hydrophilic groups that were not yet hydrated establish H-bonds on H_2O molecules. The proportion of free O–H groups has slightly increased, still remaining, however, always smaller than in liquid water. During all these successive hydration steps, the C=O groups of carboxylates COO$^-$ that were already all hydrated in the dried state increasingly accept two H-bonds. At hygrometry $p/p_0 \simeq 0.95$, these doubly hydrated C=O groups are slightly more numerous than C=O carboxylate groups that accept single H-bonds. At hygrometry greater than 0.9, beyond which all mechanisms H_1, H_2 and H_3 cease being active, all hydrophilic groups of HA are hydrated. The final step, which starts being active at this hygrometry no longer hydrates any hydrophilic group of HA. During this step the newly arriving H_2O molecules establish H-bonds on already present H_2O molecules that have developed an extended H-bond network. This arrival provokes swelling. The corresponding spectrum H_4 (Figure 11.4) much resembles that of liquid water, with the proportion of free O–H groups in the assembly of embedded water molecules being now the same as in liquid water.

Above a certain hydration level, the H-bond network is so well locally developed that it may easily fluctuate without any spectral manifestation that nevertheless induces significant structural transformation. Such a transformation is illustrated in Figure 10.4. It consists of locally limited migrations of individual H_2O molecules within the existing H-bond network. With these migrations this network is locally changed. These transformations produce barely visible effects in the IR spectra, because the number of both H_2O molecules and H-bonds remains unaltered. They may nevertheless produce a relatively important effect on the skeleton of HA and corresponding IR bands, in the form of a rotation of disaccharides of the chains, represented by rotations around axes of glycoside bonds C1–O4 and O4–C4 of saccharides B and A. These rotations are made possible by the introduction of H_2O molecules within H-bonds C3O3–H···O5 established by saccharide A on saccharide B (upper spectrum) and also within collateral N–H···O$^-$–C=O bonds. These direct C3O3–H···O5 and N–H···O$^-$–C=O bonds are replaced by a chain of H-bonds established by several H_2O molecules. The flexibility of hydrated HA comes from a mechanism that allows nearly free rotations around glycoside bonds within this H-bond network. The degree of hydration at which this transformation of the skeleton of HA chains becomes feasible has not been measured by IR spectrometry, its effects remaining small in these spectra. They might nevertheless be visible in the $\nu_{C\text{–}O\text{–}C}$ bands around 1150 cm^{-1} (7) that display some slight differences in hydration spectra H_1, H_2 and H_3 (9). These differences are nevertheless too small to be precisely analysed. In proteins, the appearance of this flexibility has been observed by electronic paramagnetic resonance (EPR) (6), at a relatively small content of H_2O molecules: around 0.2 g of H_2O for 1 g of protein.

These fluctuations of the H-bond network at constant energy strongly suggest that it is not static, but deeply dynamic in nature. It means that within this H-bond network, small and highly versatile and labile H_2O molecules are continuously moving, changing their H-bond configurations (versatility) and positions (lability). The characteristic times of the dynamics of this H-bond network are most certainly several orders of magnitude longer than those of individual H_2O molecules. The H-bond network of dried HA (Figure 10.3) is conversely

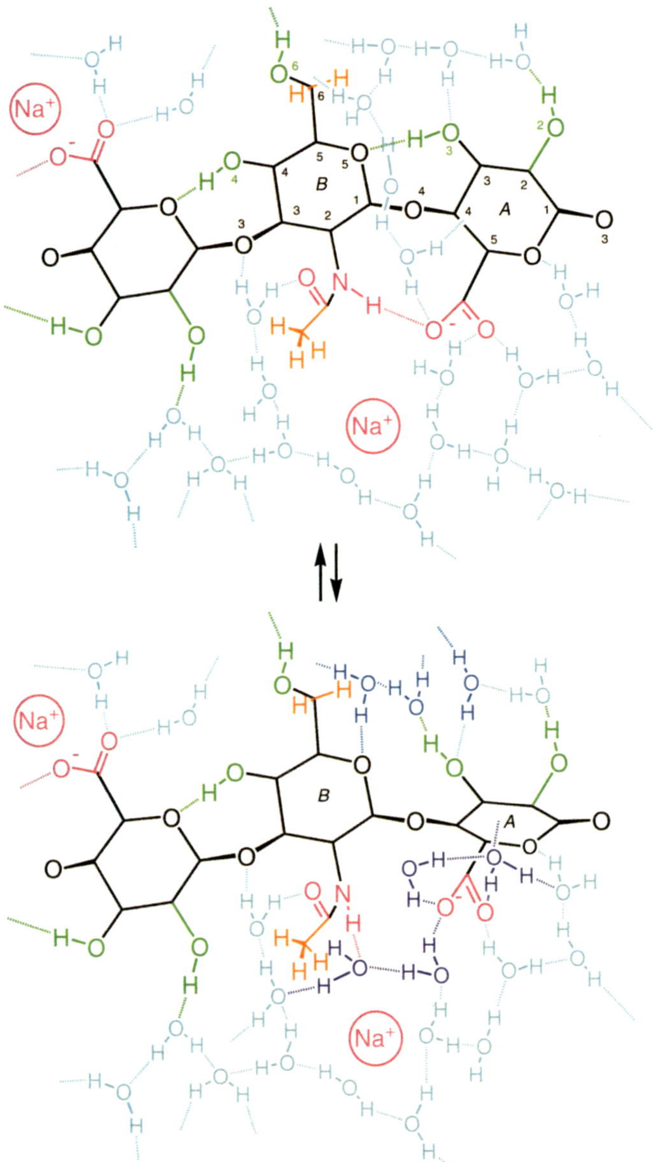

Figure 10.4 A dynamic change of the extended H-bond network around a part of an HA chain changes the structure of the chain by inducing rotations around C1–O4 and O4–C4 axes that link saccharides B and A.

rigid and does not allow much H-bond configuration and structural changes. The more this H-bond network develops and extends upon the arrival of new H_2O molecules, the more it provides structural flexibility for the host molecule, and the more its dynamic character increases.

Hydration of a synthetic polymer: a sulfonated polyimide

In the preceding subsection, hydration of a biopolymer showed how the development of the H-bond network upon the embedding of more and more H_2O molecules confers increasing flexibility on this macromolecule. In this subsection we examine hydration of another kind of a macromolecule, a synthetic polymer, before briefly sketching tentative conclusions on hydration of proteins. We will thus have examined hydration of two macromolecules of completely different kinds as deduced from precise but somewhat scarce IR experiments, described in detail in Ch. 11, which allow hydration to be interpreted in terms of chemical reactions. Many more macromolecules have been observed by global methods such as calorimetry or thermogravimetry. They provided many sorption curves that have been recorded for a lot of different macromolecules, and are for most of them similar to that shown in Figure 10.2 (6). It is then instructive to highlight the differences shown by these more precise spectroscopic methods, differences that sorption curves, which give information on the total uptake of water in these macromolecules, cannot provide. We may then hope that it becomes possible to draw general conclusions regarding the structure and role of the H-bond network in other species, on the basis of these differences and similarities that IR spectrometry shows.

In this subsection we look at hydration of sulfonated polyimide membranes, a family of "ionomers" to be possibly used in fuel cells, because they can possess both a good mechanical solidity and a good conductivity when they are hydrated. One typical example, a homopolymer, is presented in Figure 10.5. It is made of rigid and very long chains of similar repeat units composed of a tetra-aromatic rings that contains N-atoms and form its imide part, and of two benzene rings that carry two sulfonic groups that can be either neutral SO_3H or ionized form $SO_3^- \cdots$. From a quantitative analysis of their IR spectra (10), shown in Figure 11.6, it may be deduced that in the dried state, only 17% of these groups are ionized, forming $SO_3^- \cdots OH_3^+$ ion pairs, with no extra H_2O molecules (Figure 10.5, upper diagram). The 83% SO_3H groups are divided into "free groups" that do not establish H-bonds, and groups that establish H-bonds of the type SO_3–$H \cdots O{=}C$. These two types of SO_3H groups are found in a comparable proportion. In this dried state, 23% of the C=O carbonyl groups act as H-bond acceptors, the H-bond donors being both SO_3H and H_3O^+ groups.

Hydration starts with the fixation of a few H_2O molecules on these highly hydrophilic free SO_3H groups that are immediately ionized when they establish H-bonds on such H_2O molecules (intermediate diagram in Figure 10.5). These few arriving H_2O molecules establish at the same time H-bonds on C=O groups, while other H_2O molecules hydrate $SO_3^- \cdots OH_3^+$ groups, increasing the length of the water wire until it establishes an H-bond on a C=O group. These newly arriving H_2O molecules coexist with yet nonionized SO_3H groups that are most probably those that establish H-bonds on C=O groups and are therefore stabilized and certainly less hydrophilic than free SO_3H groups. When the hygrometry reaches 15%, which is the situation represented in the middle-page diagram of Figure 10.5, 80% of the sulfonate groups are ionized and consequently found in the form of $SO_3^- \cdots$ groups, whereas 52% of the C=O groups act as H-bond acceptors, most of the H-bond donors being now H_2O molecules. The total number of H_2O molecules amounts to 1.3 per sulfonate group, to which 1 H_3O^+ molecule per SO_3^- should be added. We may note that during all of the hydration process, only the O^- atom of SO_3^- groups accept H-bonds from H_2O molecules. The two other O-atoms, which are bonded to this S-atom by an S=O double bond, do not act as H-bond

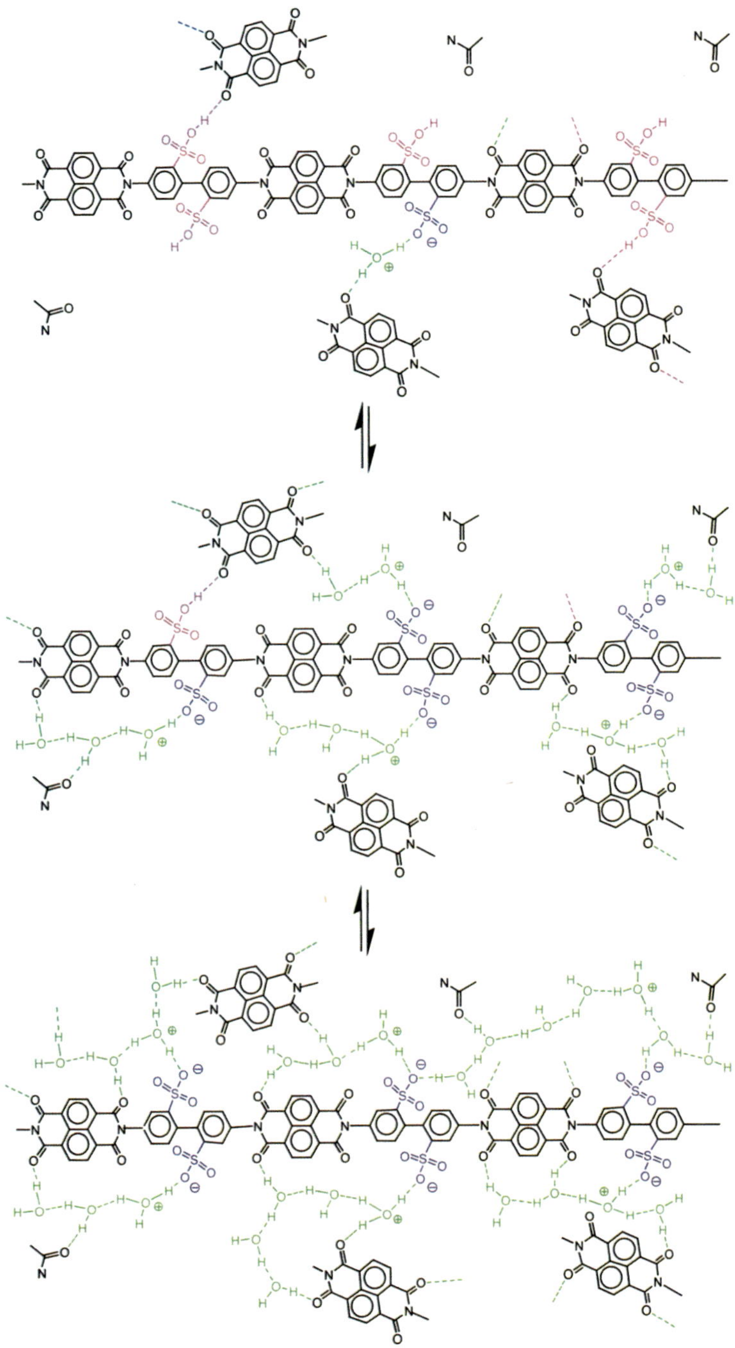

Figure 10.5 Development of the H-bond network in a sulfonated polyimide (homo)polymer made of chains of aromatic and polyaromatic substituted cycles. Three repeat units of a chain are represented. Structure of the dried membrane in the upper drawing, of the membrane in equilibrium with an atmosphere with hygrometry 15% (middle page drawing) and 65% (lower drawing).

acceptors (11). When the hygrometry increases, ionization of SO_3H groups and hydration of both $SO_3^-\cdots$ and C=O groups proceeds, till ionization of all SO_3H groups is complete at an hygrometry p/p_0 in the vicinity of 0.3–0.4, as indicated by the sorption isotherm (10) provided by the analysis of these IR spectra. At higher hygrometries, hydration consists of extending the water wires anchored on these two groups, as shown in the lower diagram of Figure 10.5, which shows the structure of the H-bond network at hygrometry $p/p_0 = 0.65$. At this value of the hygrometry, 70% of the C=O groups are hydrated and 4.3 H_2O molecules, including 1 H_3O^+ molecule per SO_3^- group, are present. At hygrometry $p/p_0 = 0.9$, these numbers are 74% for the hydrated C=O groups and six H_2O molecules, including one H_3O^+ molecule per SO_3^- group, are found.

No data are given for higher hygrometries, and the last step that has been shown to occur in HA, namely the arrival of H_2O molecules that establish H-bonds on already embedded H_2O molecules, is consequently not in evidence. It clearly appears (10) at hygrometry higher than 0.7 when instead of this "homopolymer", made of simple repeat units already mentioned, we have to do with a "block copolymer", where the repeat unit consists of x repeat units identical to those of the preceding homopolymer, which are followed by y similar units that have no sulfonate groups on their polyaromatic rings. In these block copolymers, the numbers of hydrophilic C=O groups related to one sulfonate group are thus increased. This last hydration step also appears (11) in the case of homopolymers where cations are Na^+ or triethylamine $NH^+(C_2H_5)_3$ instead of H^+, and where, consequently, no ionization step appears. It suggests, first, that the sorption isotherm most likely displays a sigmoidal shape, as illustrated in Figure 10.2 for HA, with an abrupt increase of water sorption at high values of hygrometries of the surrounding atmosphere. In the case of the homopolymer shown in Figure 10.5, this abrupt increase that goes on with swelling occurs at very high hygrometries, most likely higher than 0.95. Before this hypothetical final step, the preceding one consists of embedding a significant number of H_2O molecules that hydrate a relatively small number of C=O hydrophilic groups that have not yet been hydrated, as shown by comparing the state of hydration at hygrometries 0.65 and 0.9. The second consequence of the appearance of this final step at high hygrometries is that the conductivity of such a membrane is most likely very low at nearly all hygrometries, except those approaching 1. As may be seen in the lower diagram of Figure 10.5, diffusion of H_3O^+ ions appears most difficult at hygrometry 0.65. It requires embedding of a greater number of H_2O molecules to appear, with formation of many water wires that would connect those shown in this diagram. Recent measurements of conductivity of these membranes fully agree with this conclusion (12).

General features for hydration mechanisms?

Hydration mechanisms have up to now been scarcely looked at using precise methods that provide information at molecular scale, as described in the previous subsections. Sorption isotherms have nevertheless been measured for many macromolecules by thermogravimetry or other calorimetric methods. As mentioned above and in Ch. 11, these methods hardly give any information on the molecular structure of the sample. They have nevertheless clearly put into evidence that most of these isotherms, if not all, display a sigmoidal shape similar to that displayed by n_{tot} in Figure 10.2. This shape is characterized by a rapid increase of the number of embedded H_2O molecules at very low hygrometry, smaller than 0.01. This increase progressively slows down, becoming a moderate increase that extends over a great range of hygrometry that extends from 0.01 to 0.9–0.95. Around these latter values it displays new abrupt increases till hygrometry approaches 1. The description that calorimetric methods could provide is classification of water molecules into two classes: freezable and nonfreezing water molecules, following whether they are or not at the origin of a first-order phase transition detected by calorimetric methods. Thus, differential scanning calorimetry (DSC) shows that HA described above displays such a transition, similar to that encountered upon freezing liquid water into ice, at values for the hygrometry between 0.76 and 0.88 (13). This transition occurs at temperatures lower than that for liquid water and extends over a finite range of temperature instead of occurring at 0 °C. Figure 10.2 clearly shows that it corresponds to the H_2O molecules that are embedded following the last mechanism *D*, which are those H_2O molecules that enter the membrane and bind on other already present H_2O molecules, thus hydrating no new hydrophilic sites of the polymer. Differently said, they are H_2O molecules that become embedded in the already extended H-bond network developed by previously embedded H_2O molecules.

Calorimetric methods cannot give any information on the structure of the H-bond network that is formed by entering H_2O molecules at lower hygrometries. From the two examples given above, HA and polyimide membranes, IR spectrometry tells us that the rapid start of hydration corresponds to hydration of well-defined, especially hydrophilic groups that do not retain H_2O molecules in their dried state. It is carried out by a few H_2O molecules in the case of polyimide membranes, but by a nanodroplet of about 20–25 H_2O molecules in the case of HA. This appearance of a nanodroplet of water stands in opposition to the common thinking that this first step would be a spreading of individual H_2O molecules on the many hydrophilic groups at very low hygrometries. These well-defined and especially hydrophilic groups are SO_3H groups in sulfonated polyimide membranes, and free N–H and C–O^- carboxylate groups in HA. This hydration of particularly hydrophilic groups occurs in the form of a wire of a few H_2O molecules in polyimide, whereas it takes on the form of a nanodroplet in the case of HA. In both cases, this wire or nanodroplet also hydrates neighbouring, less hydrophilic groups. Once these particularly hydrophilic groups are all hydrated, that is at hygrometries higher than 0.2 in HA (Figure 10.2) where mechanism H_1 stops being active, hydration proceeds with the arrival of a relatively smaller number of H_2O molecules that bind on already present H_2O molecules and finish hydrating all hydrophilic groups. This or these latter step(s) correspond, in the two above described cases, to the wide hygrometric region where the slope of the total number of embedded water molecules (Figure 10.2) is weak. It finishes when the last step described above of H_2O molecules being inserted within the already developed H-bond network becomes efficient, at hygrometries above 0.9 in the case of HA. These newly embedded H_2O molecules no longer hydrate any hydrophilic group of the macromolecule, which have all been already hydrated. This last step corresponds to swelling. Is this scheme general? With the very few hydration mechanisms that have been studied in detail up to now, no firm answer can yet be given to this question, but it is likely it will appear so in the near future when more precise measurements are performed. Essential parameters that concern this mechanism can then be more precisely defined.

In addition to the two examples described above, IR spectrometric methods have been applied to a few proteins, such as bovine serum albumin (BSA) (14, 15) or lysozyme (16). They provided hydration mechanisms similar to those described above, but the relation between the determined structure of the H-bond network at a definite hygrometry and properties of macromolecules requires a deeper analysis, in view of the complexity of such macromolecules. It means that a greater number of data are necessary to establish the structure of the H-bond network in active proteins and its precise role. Clues on this role have nevertheless been given in studies on cryo- and lyoprotection of proteins described in the following subsection.

Protection of biomacromolecules against external stress (cryo and lyoprotections)

This is another field of research that has undergone appreciable development in recent years. No definitive answer concerning these problems of bioprotection has yet emerged, but interesting results shed some light on the structural role played in these mechanisms by the H-bond network that the water molecules develop in and around biomacromolecules.

Survival under freezing temperatures or hydric stress

Life occurs at temperatures in the vicinity of room temperature, and departing from this condition is often lethal. Some organisms manage to nevertheless survive in such deadly conditions where temperature is sufficiently low that other organisms cannot survive. This is the case for polar fishes that survive at temperatures down to −1.9 °C, the freezing temperature of local sea waters. The liquid water found inside these fishes is that of nearly all developed organisms. Its freezing point falls above that of the surrounding sea water, in the close vicinity of 0 °C. The same is true of some batrachians that are able to survive in waters with temperatures lower than the freezing point of their inner liquid water. Another common external stress is provided by hydric stress that manifests itself by shortage of water molecules. Some organisms may survive in these conditions, adopting a state called "*anhydrobiosis*", or life without water, within which they survive before resuming their metabolism and full activity once they are rehydrated. Among these organisms we find many bacteria, but also more elaborated organisms such as seeds, for which desiccation induces a switch of the metabolism from development to germination (6), whereas the arrival of water molecules switches from germination to development. Spores, yeast and lichens may also encounter anhydrobiotic states. Some of these organisms may stay in anhydrobiosis for decades, the greatest attainment being apparently 120 years (17). Among proteins, most of which undergo irreversible unfolding and aggregation when dried beyond some limits, some have the ability of refolding during rehydration. How do these few organisms or macromolecules survive under these conditions that are lethal for the majority of other organisms or macromolecules? A common feature of all these few surviving organisms is that they contain in their dried state a large amount of trehalose, a disaccharide shown in Figure 10.6. It offers both lyoprotection (against shortage of water molecules) and cryoprotection (against cold). Other disaccharides, such as sucrose, or maltose, also shown in Figure 10.6, offer the same type of protection with, however, less efficiency. They all have the ability to stabilize biological membranes such as the phospholipidic membranes shown in Figure 9.7, and also to stabilize the secondary structures of proteins described in Ch. 2, etc.

How do these disaccharides operate, particularly trehalose, which is the most effective one in terms of functional recovery? In order to answer such a question, many experimental methods have been used. They are all classical methods to observe water molecules and

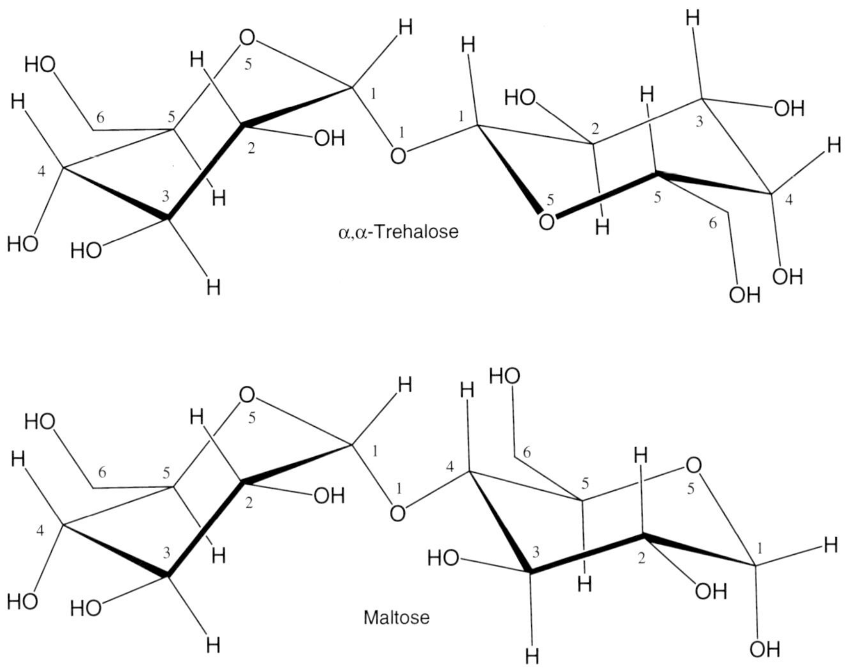

Figure 10.6 Trehalose and maltose.

are described in Ch. 11. Among them, we may mention calorimetric studies of mixtures of trehalose and water, which form glasses that exhibit transitions to vitreous states at temperatures that depend on relative concentrations of these two constituents (17). At room temperature, a mixture that has four H_2O molecules per trehalose is already in a vitreous, protecting state. Other methods include IR spectroscopy (18, 19), which allows detecting the H-bonds formed by the addition of trehalose on a membrane, particularly on the phosphate groups. Even recent nonlinear time-resolved IR spectroscopy, described in Ch. 4, has been used (20), and showed that model dipeptides immersed in trehalose glasses display a very slow dynamics, with a strongly reduced mobility of all their molecules, suggesting the possibility of hindering water molecules to escape from this highly rigid environment. Raman scattering (21) has also provided interesting clues on the destructuration of the H-bond network of liquid water by trehalose. Quasi-elastic neutron scattering (QENS) experiments (22, 23) gave information on the molecular dynamics in these glasses, which has also been investigated by molecular dynamics calculations (24–27).

Protection mechanisms

The importance for biomacromolecules of the extended H-bond network provided by liquid water clearly appears when modifications of this network are applied. Suppressing it by drying, or rigidifying it by freezing are lethal actions. The role of protectors such as trehalose is to partly avoid its collapse, by replacing it with another artificial network that maintains a minimum of its essential structural features during a sufficient time to wait for reestablishment of normal conditions for life. The main danger due to freezing for such animals as polar fishes or batrachians is to transform liquid water into crystalline ice that breaks all molecular structures with its well-known volume enhancement, a mechanism that also breaks rocks in mountains by exerting high constraints within cracks. Green and Angell (17) proposed that, in order to prevent this effect, trehalose transforms liquid water into a water–trehalose mixture they have shown to be a glass at room temperature. It avoids formation of ice crystals and is consequently as good a cryoprotector as many polyols such as glycerol. According to these authors, it also acts as a good lyoprotector thanks to the high viscosity of glasses that strongly reduces diffusion of H_2O molecules. It may thus keep them embedded for a long time, even if surrounding conditions are such that they should rapidly escape if they were in direct contact with the outside medium. In this scheme, this trehalose–water mixture acts as a protective cage. This cage-effect looks nevertheless insufficient: the biopreservative efficiencies of trehalose–glycerol mixtures do not scale with the temperatures of their vitreous transitions (25), as they should do if this was the actual mechanism. Also, dextran, a linear polysaccharide, displays no bioprotective function despite its higher glass transition temperature. Another mechanism (18) has therefore been proposed that avoids this mismatch. It puts the emphasis on the ability of trehalose or other disaccharides to establish a great number of H-bonds with a sufficient combination of flexibility and rigidity so as to preserve the structure of the macromolecule. It is thus more rigid than the H-bond network of a cluster of H_2O molecules of the same size, because the ratio of the number of H-bonds to covalent bonds is much smaller. The great number of alcohol sites it exhibits nevertheless allows it to still establish a relatively great number of H-bonds with various sites of the macromolecule, even if this number is also smaller than that of a cluster of H_2O molecules of a comparable size. This ability makes it a good lyoprotector. Furthermore, stiffness, which is limited by the possibility of free rotations around the two glycoside C1–O1 axes (Figure 10.6) that link the two glucose units, and the impossibility to establish H-bonds between the two glucose units of trehalose, keeps, for instance, phosphate groups sufficiently far apart to avoid them strongly interacting, which would lead to a destruction of phospholipid membranes by grouping all these phosphate heads. It thus roughly more or less preserves the main properties of a small water cluster, keeping some flexibility together with the essential features of the structure of the macromolecule. In the case of proteins, this mechanism nevertheless encounters greater difficulties explaining how they enter a confined region so as to keep intact their secondary structures.

From these arguments, we may conclude that trehalose is the best compromise to maintain an artificial H-bond network that temporarily replaces that developed by H_2O molecules and avoids both those catastrophic consequences, that is the formation upon freezing of crystalline ice in that part outside macromolecules that is in contact with liquid water, and the collapse of the structure of membranes or of secondary structures of proteins upon drying due to the escape of H_2O molecules. This artificial H-bond network is much less flexible than that established by H_2O molecules and considerably slows down the dynamics of H_2O molecules (23, 28). It consequently does not allow normal activity. It does not allow life to proceed in the same way as in the H_2O network of living conditions. It nevertheless avoids irreversible transformation of the structure of the macromolecule by hydric stress, thus allowing resumption of living activities by rehydration. The discussion that has appeared in the literature to decide which mechanism is the most important, the glass formation of trehalose with water or the possibilities of trehalose to establish numerous H-bonds and therefore keep the structure of macromolecules intact, does not seem fundamental, as both these effects appear to be necessary to form this temporary artificial H-bond network. Now, how is trehalose better than other similar disaccharides such as sucrose or maltose shown in Figure 10.6? This point is yet a matter of discussion, as these disaccharides display great similarities. The main difference between trehalose and maltose is the C1–O1–C1 link between the two saccharides in trehalose that is replaced by a C1–O1–C4 link in maltose. Otherwise they have the same number of alcohol groups to establish H-bonds, the same possibility of rotations around glycosidic bonds and both are unable to establish between their two glucose units H-bonds that would make them stiff. Trehalose has nevertheless been shown by QENS to be more efficient than sucrose or maltose for slowing down the dynamics of water molecules (28). It may point to the origin of its higher bioprotective effectiveness.

Despite the existence of as yet unsolved issues, these studies have clearly shown that the development of this H-bond network by H_2O molecules is at the origin of flexibility, which appears necessary to keep the structure of macromolecules intact and avoid their destruction by disruption, in the case of membranes, or denaturation, in the case of proteins. This flexibility of the H-bond network is a necessary requirement for the stability of these macromolecules. It has a strong dynamic character or, in other words, the numerous configurations that are made possible thanks to the presence of this H-bond network, continuously interchange. Characteristic times of these dynamic fluctuations are not known and may extend over several orders of magnitude when passing, for instance, from a biopolymer such as HA to proteins. Theoretical molecular dynamics methods might be able to give precision on this point. Dried proteins may remain stable. But this stability may be largely due to the remaining water molecules that might find their main role in the relaxation of many steric constraints to establish H-bonds. Activity is necessary only at very high hygrometry, when immersion in a saturating water vapour or within liquid water, which defines living conditions, are close to equivalent. How does this flexibility more precisely operate on the stability of macromolecules? Figure 10.4 gives a preliminary answer to this question. It no doubt requires more precise experiments before this point can be considered as understood.

Protein folding

The above-described fields of research have shown that H-bond networks developed by H_2O molecules play an important structural role in macromolecules. Other fields of research are promising, for example protein folding. It has been recently boosted by the appearance of neurodegenerative diseases such as Creutzfeld–Jakob's, which has shown that most neurodegenerative diseases, which are increasingly common with the current aging of populations, are due to misfoldings that trigger aggregation of proteins. Folding is the process by which a bio-inert polypeptide transforms into a living protein. As mentioned in that part of Ch. 2 that is related to the secondary structure of proteins, it is the reverse process of denaturation. Folding cannot be done by a simple test and try mechanism, which would take a time that would be close to eternity! It should be consequently driven by a chemical process, which is still unknown and occurs during syntheses of proteins. We at least know that H_2O molecules (29) play a fundamental role of "molecular recognition", the principle of which is

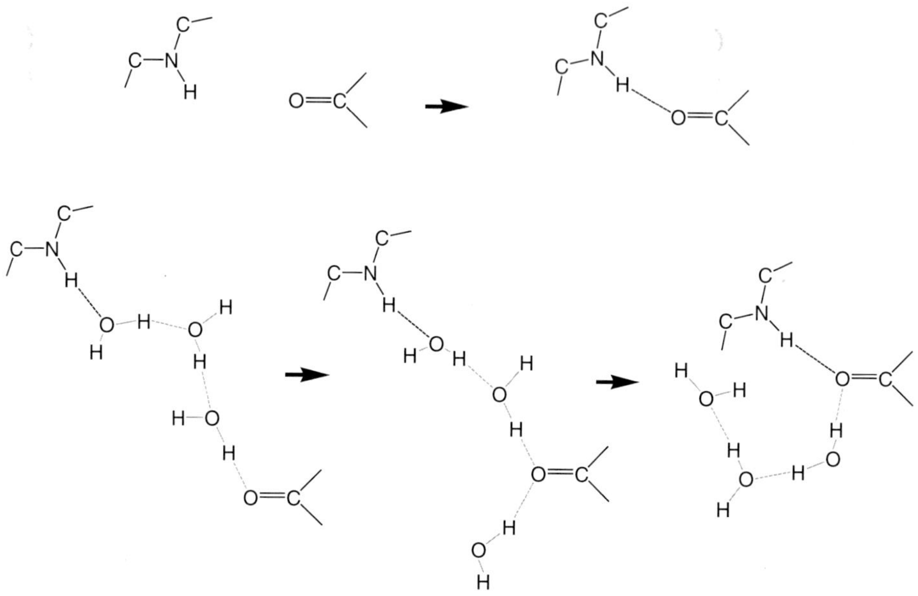

Figure 10.7 Principles of molecular recognition: in the absence of an H-bond network the two groups N–H and C=O should lie in a close vicinity to establish H-bonds (upper drawing), whereas they may initially lie much further away in its presence (lower drawing).

illustrated in Figure 10.7. In this figure the formation of such H-bonds as N–H···O=C between two groups, N–H and C=O that are far away, may already be induced when they are spatially separated by distances as great as 10 Å, a distance much greater than that of the final H-bond N–H···O=C. The number of H-bonds in the three lower diagrams is kept constant. The formation of the final N–H···O=C bond therefore occurs at the end of an isoenergetics process that does not require borrowing energy from the surroundings. This role of molecular recognition is made possible thanks to the high flexibility of this H-bond network, which we have already seen above to be at the origin of the changes of the macroscopic properties of HA during hydration. We can, however, no longer comment on the role of water in protein folding, because still too few results have been obtained in this field of research up to now and too many points related to the fundamentals of the problem are still under discussion. It nevertheless remains a highly promising field to more precisely know the general role of the H-bond network.

REACTIVITY OF WATER MOLECULES IN MACROMOLECULES

The considerations of the preceding subsections stress the importance of the structural flexibility provided by the extended and dense H-bond network developed by H_2O molecules in macromolecules to ensure the stability of some of these macromolecules, particularly biomacromolecules. The representative case is the secondary structure of proteins, which requires the presence of such a network to be stable. This H-bond network does not have a fixed structure, but a highly dynamic one (30), which induces rapid structural fluctuations within numerous isoenergetic configurations. Is this structural flexibility the only effect of this H-bond network? The answer seems to be no, there is more to it, and this H-bond network has apparently another role, which may be even more fundamental in the case of biomacromolecules. It has already been pointed in Ch. 6 that H_2O molecules are key players in the proton translocation within bacteriorhodopsin, the central mechanism of photosynthesis and vision. They are also central in the proton conduction mechanisms of membranes that are used for instance in fuel cells, where they play a key role in protonic conduction, which has many potential industrial applications.

In biomacromolecules they have a different, but certainly even more important role. In the absence of water reactivity of these macromolecules stops or is severely slowed down. This is particularly true for proteins and even more so for a particular class of them: enzymes. This is also true when proteins are transferred to other organic solvents. None of these other solvents develops a 3D H-bond network comparable to that developed by water molecules. It indicates that H_2O molecules are particularly active molecules in bioreactions. Furthermore, as described in Ch. 7, life hardly occurs when H_2O molecules are replaced by those most similar molecules, HDO or D_2O. Only some very primitive organisms, such as algae, can develop some form of life when immersed in heavy water. Their metabolism is, however, very much slowed down, the same as with ordinary water at very low (extrapolated) temperatures, well below 0 °C. If the H-bond network developed by water molecules had only the effect of conferring the necessary flexible structures to biomacromolecules, passing from H_2O to HDO or D_2O would not have such a dramatic effect. These molecules develop the same extended and flexible 3D H-bond network as H_2O, and the structural and thermodynamic properties of ordinary liquid water are very much the same as those of heavy liquid water (we do not mention a liquid made of HDO molecules, because, due to rapid H and D transfers through H-bonds these molecules only exist in the presence of H_2O and D_2O molecules). The biological properties of ordinary and heavy water are thus seen to be completely different. It shows that if H_2O molecules are an active player in bioreactions, this is by acting on the fundamental central mechanism of bioreactions: *transfers of H-atoms.*

We have seen in Ch. 6 that two kinds of transfers occurred through H-bonds: transfers of H-atoms and transfers of protons. Transfers of protons can be as dramatically affected by a H/D substitution as transfers of H-atoms. They can be nevertheless discarded in biomedia (4), because they are then most likely to release into the medium H_3O^+ and OH^- ions that would strongly modify its pH, or at least induce strong fluctuations of this pH. This pH of biomedia always remains in a close vicinity of 7, with molar concentrations of H_3O^+ and OH^- ions of about 2×10^{-9}, that is nearly zero. We may argue that we have seen H_3O^+ ions act in the case of photosynthesis or vision mechanisms, when the proton transfer is relayed by H_2O molecules of the proton wire (Figure 6.8). These are, however, not general bioreactions, but special ones where a great quantity of energy, as compared to thermal fluctuations, has been introduced into the system by absorption of a photon. In most other biological reactions no such amount of energy is present. Furthermore, in photosynthesis, these ions are not released into the medium. In order to keep the pH of the medium in the vicinity of 7, these H_3O^+ or OH^- ions, whenever they intervene in a bioreaction, can therefore only appear as a transient configuration in a proton wire, so as to avoid any change in the value of the pH by diffusion in the aqueous biomedium. All these consequences of proton transfers, which are scarcely considered when such transfers are supposed to be active, as for instance in enzyme catalysis (31), make proton transfers forming a general mechanism in a medium that has to keep its pH equal to 7 difficult, and their occurrence is consequently much scarcer than is usually thought. Said differently, the often supposed presence of H_3O^+ or OH^- ions in bioreactions is highly unlikely. This does not hinder carboxylic acids to be found at this pH in the form of anions R–COO^-, the number of which is counterbalanced by the presence of such cations as those formed from ionization of amine groups: R′–NH–R″ groups that become R′–NH_2^+–R″ groups.

H-atom transfers conversely avoid all these problems encountered with proton transfers, as they neither delocalize space charges nor create charges or release ions into the medium, with the consequent modification of its pH. These H-atom transfers occur by tautomerism, as already described in Ch. 6, where they have been shown to be responsible for H/D exchanges in H-bonded liquids, including liquid water at pH 7. It has been stressed there that they require the presence of H-bonded cyclic structures to appear. This requirement easily explains why a D_2O molecule H-bonded on the surface of ordinary (H_2O) ice does not exchange, whereas it does exchange when it is in bulk H_2O ice. It also explains why the rate of an H/D exchange of an N–H group in a protein is reduced by more than eight orders of magnitude from the rate in equivalent polypeptides (6): in the protein this N–H group is bound to C=O groups by N–H···O=C bonds at the origin of secondary structures and the C=O group acts as an acceptor only of an H-atom. No cyclic H-bonded structure is possible with this configuration, unless at least one H_2O molecule intercalates within this H-bond. Such an intercalation, similar to that shown at the bottom of Figure 10.4, is extremely difficult within these relatively stiff secondary structures of the protein. It takes consequently a very long time to appear, whereas, in the polypeptide free N–H groups, with their N-atom acting both as an H-atom donor and as an H-atom acceptor, establish H-bonds on H_2O molecules that, almost automatically, make such H-bonded cyclic structures appear. This necessary presence of H-bonded cyclic structures is therefore the most severe sterical requirement, which might make tautomerism unlikely to appear as a general mechanism, because nearly all molecules are unable to fulfil it. There is one important exception, however, which makes it a central mechanism: the small water molecule that is ubiquitous and which, in the presence of a few other water molecules, most rapidly develops an extended, dense and flexible H-bond network for which this condition is almost automatically fulfilled.

In Figure 10.8, an example is displayed of how these transfers intervene in a typical bioreaction, the hydrolysis of a peptide group R_0–CO–NH–R_1 of a protein (32). This is the most common reaction that occurs during, for instance, digestion. Such a hydrolysis requires the help of an enzyme in normal bioconditions. This enzyme is for instance chymotrypsine that is known to have an active site (31) where two special residues, serine and histidine, play a particular role. The alcohol O–H group of serine and the imidazole ring of histidine are explicitly highlighted in this figure. They establish a direct H-bond. A third residue, aspartic acid, is sometime added to this dyad. It is not represented because its position and role are a subject of controversy. The global hydrolysis reaction is written as

$$R_0\text{–CO–NH–}R_1 + H_2O \rightleftharpoons R_0\text{–COOH} + R_1\text{–NH}_2 \tag{10.1}$$

The initial R_0–CO–NH–R_1 peptide group to be hydrolysed is held fixed in a niche of the enzyme drawn as a cavity formed by the enzyme around the R_0 group in **I**. It is surrounded

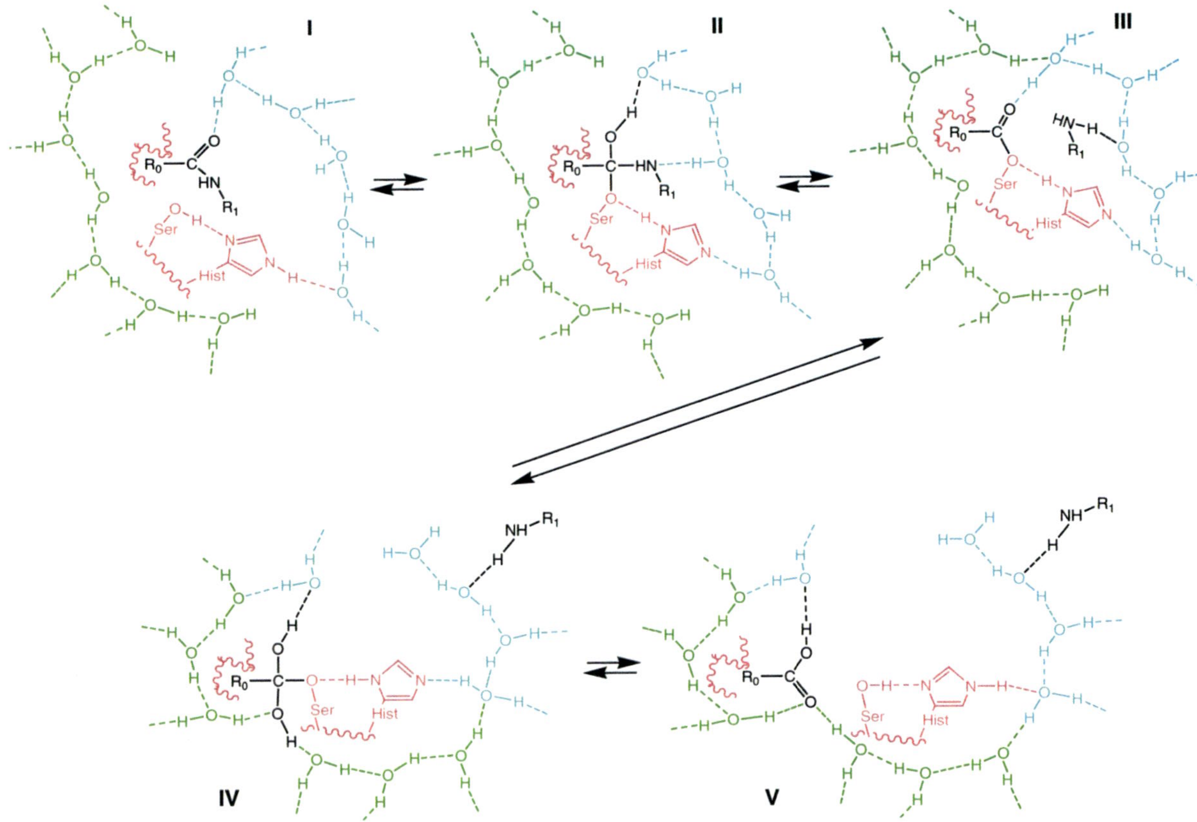

Figure 10.8 Hydrolysis of a particular peptide group R_0–CO–NH–R_1 of a protein with the help of an enzyme that has an active centre made of a serine (purple Ser–O–H) H-bonded to a histidine (purple Hist–imidazole ring) group. Parts of interest of the backbone of the enzyme are represented as purple undulated wires. Blue H_2O molecules are active during the first stage **I–III**, whereas green H_2O molecules are active during the second stage **III–V**. In the stages during which they do not participate, they may be spread further away, or may remain H-bonded one to the other one, forming a single permanent wire of blue and green water molecules. They also may be a single set, the blue ones, for instance. It would then have to slightly change its configuration between steps **III** and **IV**.

by a wire, or several wires of H_2O molecules. For simplicity we have represented them in the form of two wires, one blue and one green that are active only during steps **I–III** for the blue ones and during steps **III–V** for the green ones. These two kinds of water molecules may indeed be a single one, either blue or green. In that case, they would then have to slightly change their conformation around step **III**. In the first step (**I**), the O-atom of the alcoholic group of serine attacks the peptide at the opposite of the carbonyl C=O group, transforming the peptide group into a "tetrahedral intermediate" drawn in **II**. The formation of this intermediate is accompanied by transfers of H-atoms in the H-bonded cyclic structure made of blue H_2O molecules and of the serine–histidine dyad. This transfer appreciably lowers the energy of the tetrahedral intermediate and at the same time avoids the appearance of space charges that are energetically costly. This intermediate, at the summit of the energy barrier, may then either reverse to its initial state, or proceed to another state, **III**, where the amine part R_1–NH_2 of the former peptide is expelled. In this latter reaction, a transfer of H-atoms also occurs inside another H-bonded cyclic structure made of −NH–C–O–H and H_2O molecules drawn in blue. This state is of a relatively low energy. The serine part of the enzyme is covalently bound to the carboxyl part R_0–COO− of the former peptide. The two following steps, **IV** and **V**, consist of hydrolyzing this set so as to reconstitute the enzyme. It works in a way that is most similar to steps **I–III**, with formation of a tetrahedral intermediate that has an energy that is decreased thanks to transfers of H-atoms inside H-bonded cyclic structures: one formed by –O–C–O–H groups and H_2O molecules (most of them green on the left hand side) and another one formed of histidine, O(serine)–C–O–H and water molecules (most of them green on the bottom side).

The scheme represented in Figure 10.8 differs from that which is usually drawn by highlighting the role of H-atom transfers inside cyclic H-bonded structures that are closed with the help of a few H_2O molecules. As mentioned above, these water molecules have no difficulties in doing so, thanks to their exceptional ability to develop around them a particular dense and flexible H-bond network. These H-atom transfers avoid having to invoke the presence of supplementary amine groups in the vicinity of the serine group to stabilize the tetrahedral intermediate (31) and also avoid hypotheses that strongly limit the values of the pK_a of the various constituents, when proton transfers are considered instead of H-atom transfers. These amine groups cannot be positioned further away, with H-bond relays made of water wires. In that case, transfers of protons are likely to spread inside bulk water, thus modifying the pH of the water solvent. The price that is paid for these H-atom transfers to occur, as in Figure 10.8, is the necessary presence of several H_2O molecules that form a water wire or water nanodroplet, instead of a single one as required with the supposition of proton transfers. In a biomedium this is certainly no problem, even in those parts of a protein that are considered as hydrophobic.

It has been mentioned in Ch. 6 that these transfers of H-atoms are poorly known. They consequently need an important effort before they can be understood, so that such related quantities as typical times, numbers of participating H_2O molecules, effect of temperature, etc. become known. Such methods as IR spectrometry, both conventional and nonlinear time-resolved, will certainly be the basic experimental methods to precisely observe these transfers. Theoretical accompanying developments are likely to be also useful. On which H-bonded systems will these mechanisms be observed? No precise answer can be given, but recent hydration experiments, such as that of HA described above, may already give hints on this point. Thus, the kinetics of H/D exchange of OH and NH groups when a thin sample of HA is immersed in a vapour of heavy water, instead of ordinary water, as observed by IR spectrometry, could already be an interesting source of information on these transfers by tautomerism. The determination of the minimum number of H_2O molecules necessary to start exchange around each site could also convey interesting information on the insertion of H_2O molecules within single H-bonds established by secondary alcohols (see Figure 10.3), giving interesting conclusions on the rigidity of the HA chain, as illustrated in Figure 10.4. We may also anticipate that such fundamental reactions as hydrolyses of esters should be interesting to follow by IR spectrometry, when the solid sample of an ester is immersed in an atmosphere of variable hygrometry. It might help in determining the minimum number of H_2O molecules that are necessary to start hydrolysis. It could be a decisive test to determine which transfer, proton or H-atom, is active. Proton transfers require a single H_2O molecule to start hydrolysis (31), whereas H-atom transfers require a minimum of about five H_2O molecules to have an H-bonded cyclic structure.

CONCLUSION

At the end of this chapter it may be realized how untenable this until recently most common attitude of ignoring the H_2O molecule is, and how imperative it becomes to know much more about some of its subtle properties. This is certainly no easy task and around 20 years after Maddox (5) rang the bell to stop the attitude he called a scandal, this field of research is still in an early stage of development. We know this molecule plays an important role when it is embedded in macromolecules. This role even becomes vital when it comes to biological molecules. We are now starting to have ideas about this role. It is no doubt due to the exceptional ability this molecule has to develop around it an extended, dense and flexible 3D H-bond network by association with other H_2O molecules, even a very few of them. This exceptional ability is itself a consequence of the exceptional structure of the H_2O molecule, which, as illustrated in Figure 8.1, has two H-bond donor sites and two H-bond acceptor sites. A preliminary, unrefined answer to the question "How is it that life occurs only in presence of H_2O molecules?" is this: H_2O molecules are the only molecules, along with HDO and D_2O, that are able to provide inside macromolecules such an extended 3D H-bond network that they always develop around them.

This unique H-bond network provides a structural flexibility to many macromolecular structures that become themselves flexible and evolutive. Let us remember that this ability to evolve is a necessary requirement for biomolecules to stay alive and on a bigger scale is an essential feature of living organisms. The origin of this flexibility, illustrated in Figure 10.4, is a direct consequence of the great isoenergetic changes this H-bond network allows, which are at the origin of rapid structural fluctuations of H-bonds. These configurational fluctuations may occur on the time scale of 1 psec and have important consequences for the folding, stability and functions of proteins. In this domain of biology this structural role is important. Even more important now, however, seems to be the role this H-bond network plays in the reactivity of biomacromolecules. In the absence of H_2O molecules, no biological reactivity occurs. In their presence transfers of either protons or of H-atoms through H-bonds may happen, allowing metabolic reactions to proceed. It clearly highlights the role this extended, dense and flexible H-bond network plays in the bioreactivity in general, and more precisely in the control of these basic transfers. An immediate consequence of this role is that life is hardly possible when even a small proportion of the H-atoms of this H-bond network are replaced by D-atoms. These D-atoms do not change the structure of this H-bond network but strongly change the kinetics of these transfers, severely slowing them down. This is enough to hinder the great set of living reactions to proceed coherently, a necessary condition for life to occur—at least the kind of life we now know, which has undergone a continuous Darwinian evolution towards complexity since it started around 3–4 billion years ago.

How do these H-bond networks more precisely control these transfers? Before outlining an answer to this question, let us recall that in Ch. 6 we have distinguished two kinds of transfers: proton transfers, shown in Figures. 6.2, 6.3 and 6.5, and H-atom transfers shown in Figures. 6.2 and 6.10. Proton transfers constitute the foundations of acid–base chemistry. In biomedia with pH almost invariant at a fixed value close to 7, they are unlikely to be general mechanisms, even if they may occasionally be active in very specific reactions such as a transfer between a basic group, often an amine, and an acidic group, often a carboxylic acid. This transfer should, however, be limited along a short water wire that is isolated from the bulk liquid water, in order to avoid pH changes. The presence of a dense and extended H-bond network favours these proton transfers, certainly via a mechanism of the type shown in Figure 10.7 for molecular recognition. H-atom transfers, on the other hand, which are responsible for H/D exchange in liquid water at pH 7 or in any other H-bonded liquids, are most probably those general mechanisms at the origin of the reactivity of biomedia. They do not create space charges and do not modify the ambient pH. They consequently do not convey the above hindrances of proton transfers. In contrast, however, they require, in order to be active, the presence of H-bonded cyclic structures, such as shown in Figure 6.10. These cyclic structures are easily provided by the H-bond network developed by even a few H_2O molecules. And this might well be the crucial and unique property of H_2O molecules that make them the only molecules that offer pathways for these H-atom transfers to proceed with sufficient velocity in a biomedium.

Even if a more precise answer could be given to the above question of the importance of H_2O molecules for life, many points that concern these transfers still remain obscure. The structures of H-bond networks in many macromolecules are thus still completely unknown, and a lot of work remains consequently to be performed before we more precisely understand biomedia and the crucial, vital role of H_2O molecules. Once they are understood, we might then have an idea why all warm-blooded animals maintain their inner temperatures in the vicinity of 310 K. This temperature is a biological constant, the meaning of which is presently unknown. It might correspond to an optimal temperature for these H-atoms transfers. Which mechanisms would make it so, and more particularly what are the two antagonistic driving forces that influence these transfers and reach equilibrium at this temperature? Are they the rates of H-atom transfers within established H-bonds that are certainly favoured when temperature is raised, on one hand, and the disruption of H-bonded cyclic structures that increases with temperature, making these transfers more difficult, on the other hand? No answer can be given now but might appear in the not too distant future. We conclude that understanding the role of water molecules in macromolecules still requires much effort.

REFERENCES

1. N. V. Nucci and J. M. Vanderkooi, J. Phys. Chem. B, 109 (2005) 18301.
2. I. R. Piletic, H.-S. Tan and M. D. Fayer, J. Phys. Chem. B, 109 (2005) 21273.
3. J. J. Gilijamse, A. J. Lock and H. J. Bakker, Proc. Natl. Acad. Sci., 102 (2005) 3202.
4. Y. Maréchal, J. Mol. Struct., 700 (2004) 207.
5. J. Maddox, Nature, 332 (1988) 677.
6. J. A. Rupley and G. Careri, in C. B. Anfinsen, J. T. Edsall and F. M. Richards (Eds.), Advances in Protein Chemistry, Vol. 41, Academic Press, 1991.
7. K. Haxaire, Y. Maréchal, M. Milas and M. Rinaudo, Biopolymers (Biospectroscopy), 72 (2003) 10.
8. Y. Maréchal, M. Milas and M. Rinaudo, Biopolymers (Biospectroscopy), 72 (2003) 162.
9. K. Haxaire, Y. Maréchal, M. Milas and M. Rinaudo, Biopolymers (Biospectroscopy), 72 (2003) 149.
10. D. Jamróz and Y. Maréchal, J. Phys. Chem. B, 109 (2005) 19664.
11. D. Jamróz and Y. Maréchal, J. Mol. Struct., 693 (2004) 35.
12. X. Guo, J. Fang, T. Watari, K. Tanaka, H. Kita and K. Okamoto, Macromolecules, 35 (2002) 6707.
13. N. Jouon, M. Rinaudo, M. Milas and J. Desbrières, Carbohydr. Polym., 26 (1995) 69.
14. J. Grdadolnik and Y. Maréchal, Biopolymers (Biospectroscopy), 62 (2001) 40.
15. J. Grdadolnik and Y. Maréchal, Biopolymers (Biospectroscopy), 62 (2001) 54.
16. K. Liltorp and Y. Maréchal, Biopolymers, 79 (2005) 185.
17. J. L. Green and C. A. Angell, J. Phys. Chem., 93 (1989) 2880.
18. L. M. Crowe, Comp. Biochem. Physiol. A, 131 (2002) 505.
19. D. K. Hincha and M. Hagemann, Biochem. J., 383 (2004) 277.
20. C. H. Londergan, Y. S. Kin and R. M. Hochstrasser, Mol. Phys., 103 (2005) 1547.
21. C. Branca, S. Magazù, G. Maisano and P. Migliardo, J. Chem. Phys., 111 (1999) 281.
22. I. Köper, M. C. Bellisent-Funel and W. Petry, J. Chem. Phys., 122 (2005) 14514.
23. F. Affouard, P. Bordat, M. Descamps, A. Lerbret, S. Magazu, F. Migliardo, A. J. Ramirez-Cuesta and M. F. T. Telling, Chem. Phys., 317 (2005) 258.
24. G. Cottone, G. Cicotti and L. Cordone, J. Chem. Phys., 117 (2002) 9862.
25. A. Lerbret, P. Bordat, F. Affouard, M. Descamps and F. Migliardo, J. Phys. Chem. B, 109 (2005) 11046.
26. T. E. Dirama, G. A. Carri and A. P. Sokolov, J. Chem. Phys., 122 (2005) 114505.
27. S. B. Engelsen and S. Pérez, J. Phys. Chem. B, 104 (2000) 9301.
28. S. Magazu, F. Migliardo and M. T. F. Telling, J. Phys. Chem. B, 110 (2006) 1020.
29. P. Csermely, Cell. Mol. Biol., 47 (2001) 791.
30. C. Mattos, Trends Biochem. Sci., 27 (2002) 203.
31. C. Walsh, Enzymatic Reaction Mechanisms, Freeman, New York, 1979.
32. Y. Maréchal, J. Mol. Struct., 321 (1994) 125.

– 11 –

Observing the Water Molecule

A DIFFICULT-TO-OBSERVE MOLECULE

In this chapter, experimental and theoretical methods to observe or describe H_2O molecules are examined. These methods are also used to observe H-bonds and have already been discussed in Chs. 3 and 4. In this chapter, the specificity of the observation of H_2O molecules is emphasized. It mainly concerns IR spectrometry, which is the most precise and powerful method to study H-bonds, but has up to now scarcely been used to observe H_2O molecules for paradoxical reasons that are detailed below.

The H_2O molecule is a small and very labile molecule that easily and rapidly moves away. It is also a most versatile molecule that can rapidly change previously established H-bond configurations into new H-bond configurations. These properties are a consequence of its exceptional ability to establish a great number of H-bonds in various ways. They make it a molecule that displays a particularly rapid dynamics and is consequently not at all easy to observe using methods that are very good for observing other molecules. This is an assertion that is in opposition to the familiarity of a molecule that we manipulate everyday in so many different ways and is well known and easy to look at. This is indeed so in the case of ice, for instance, where only fixed H_2O molecules are present and their dynamics is restricted to internal vibrations, as in other crystals. This is really an exception, and even in the case of liquid water, where no other molecule is present, no experimental method has yet given any clue that would allow us to precisely understand its rapid and unique dynamics, as mentioned in Ch. 9. When embedded in macromolecules, the H_2O molecule is more difficult to observe, not only because it is no longer the only molecule that is looked at, but also because its propensity to rapidly move away or change its H-bond configurations is at the origin of disorder induced in the macromolecule itself. Powerful methods to study macromolecules, such as X-ray or neutron scattering, then encounter difficulties detecting such small molecules at the origin of random and rapidly changing H-bond networks.

As a consequence of the difficulty to observe it, this molecule has long been neglected, despite its fundamental importance, as emphasized in Ch. 10. This attitude has been called by John Maddox (1) a scandal, a qualification that implies that it can no longer be defended. It stresses the need to know this molecule better, adapting methods used to study H-bonds to its study. Macroscopic global methods are, as in Ch. 3, first examined in this chapter, followed by classical molecular methods culminating in IR spectrometry, which is described in some detail because of its relatively recent implementation.

GLOBAL METHODS

These global methods encompass classical thermodynamic methods such as thermogravimetry and calorimetry, which provide either numbers of H_2O molecules that are embedded in a macromolecule or fluxes of heat after formation or disruption of H-bonds established by these molecules that follows the change of an external parameter. They also encompass other methods that are described below. Rupley and Careri (2) have reviewed most of them in an article related to hydration of proteins. We comment only on a few of them that seem general enough to be still used. The principle of thermogravimetry consists of immersing a sample in a humid atmosphere and then waiting for equilibrium to be reached. From a precise measurement of the weight of the sample, the number of H_2O molecules that are embedded in the sample is measured. Sorption isotherms, such as those shown as dark squares in Figure 10.2 in the case of hyaluronane, can be deduced from these measurements. They represent the total number of embedded H_2O molecules when the hygrometry of this atmosphere, equal to the ratio p/p_0 of the vapour pressure p to the saturating vapour pressure p_0 at the same temperature, varies. In most thermogravimetry experiments, this hygrometry takes a limited number of values that correspond to the equilibrium of vapour in the presence of various saturated salts. In Ch. 10, it was mentioned that sorption isotherms of many macromolecules display similar sigmoidal shapes. The meaning of this master shape could be sketched on the basis of results of IR spectrometry from which such sorption isotherms could also be deduced. This method has the advantage of providing a continuous variation of this hygrometry when using a simple special hydration cell (3). It furthermore has the advantage of enabling a molecular interpretation that such a global method as thermogravimetry cannot provide. Thermogravimetry is often used in combination with calorimetry. This latter method, which is classical in chemical physics, shows that above a certain value of the numbers of H_2O molecules that are embedded in a macromolecule, a phase transition appears when temperature is varied, which is accompanied by heat dispersion, similar to that encountered upon freezing liquid water into ice. This phase transition is similar, but not exactly the same; it occurs at temperatures lower than 0 °C and extends over a finite range of temperature (4). It indicates that the corresponding clusters of H_2O molecules embedded in the macromolecule that is at the origin of this freezing have properties that are somewhat different from those of liquid water. In the case of hyaluronane HA, a polysaccharide that has been described in the preceding Ch. 10 as a typical example of a biopolymer, such clusters appear at a hygrometry above about 90%. They are made of H_2O molecules that become embedded in the macromolecule by inserting themselves into the already developed H-bond network of the macromolecule. Their corresponding IR spectrum, H_4 in Figure 11.4 below, also displays small differences from that of liquid water. With these combined thermogravimetric and calorimetric methods, the measured enthalpies of insertion of these H_2O molecules give the values of the enthalpies of formation of the H-bonds established by them. The insertion enthalpies of "freezable water molecules" are thus most often found higher than the corresponding enthalpies in liquid water (5).

Another global method is dielectric measurements at frequencies less than some GHz. It concerns collective motions in a sample. The interesting quantity is then the imaginary part $\varepsilon''(\nu)$ of the dielectric constant that, in this frequency range, corresponds to the conductivity of the sample. At low hydration levels this conductivity for various proteins is very low and abruptly starts increasing above a hydration threshold of some 0.2 for the weight ratio of water to protein (2). This increase is exponential just above this threshold when hygrometry increases. As the comparison with the conductivity measured when H-atoms have been replaced by D-atom in lysozyme (6) shows, this conductivity is due to transfers of protons inside the H-bond network, as it is highly sensitive to this isotope substitution. The interesting point is that this threshold appears at a value for the hygrometry that is close to that where enzyme reactivity also starts occurring. It has consequently been interpreted in terms of percolation of the water molecules. As the critical exponent of this conductivity strongly suggests, this percolation is 2D, meaning that it is established by a veil of H_2O molecules around the protein that continuously connects most H_2O clusters around the protein. At values of hydration lower than this critical value (around 0.2), the various H_2O clusters are not connected and therefore no conduction occurs. At higher values, protonic conduction does occur, due to connectivity.

We may conclude that these global methods are interesting for obtaining the values of various macroscopic quantities. They are not, however, sufficient to give an interpretation at the molecular level. They are consequently interesting when used in connection with some other methods that are able to provide such a molecular insight. The classification into nonfreezing and freezable water molecules they provide is not that directly evident on this molecular scale. Consequently, the interpretations of their results on the molecular level appear tentative.

CLASSICAL MOLECULAR METHODS OTHER THAN VIBRATIONAL SPECTROSCOPY

These classical "molecular methods" encompass X-ray or neutron scattering methods, NMR spectroscopy and theoretical descriptions. Their interest for the study of H-bonds in general has been the object of Ch. 3.

X-ray scattering

This is the central, widely used and most powerful method to determine the structure of ordered crystals, where fixed distances between various atoms can be determined with precision. As seen in Ch. 10, the water molecule is, however, at the origin of disorder, and except in such cases as ice or crystalline hydrates encountered in mineral chemistry or geology, where the water molecules are perfectly ordered, their positions are most often not so easy to determine. Furthermore, the scattering factor of the H-atom that has only one electron around it is one order of magnitude smaller than that of other atoms. As a consequence, H-atoms of H_2O molecules are difficult to localize. In these conditions, deciding whether a particular H_2O molecule establishes an H-bond or not is often difficult with the aid of X-ray diffraction experiments only, and even in the case of crystalline ordinary ice, ice Ih, where only O- and H-atoms are present, problems concerning its structure are not yet resolved, as mentioned in Ch. 8. As was also seen in the preceding chapters, the water molecule is found in its vapour state in the atmosphere, in solid forms in the various crystalline and amorphous ices, in a disordered form in liquid water, or as more-or-less large clusters of H_2O molecules in macromolecules. Except in crystalline ices, they are disordered. When they are the only molecules present in the studied samples, such as in liquid water or amorphous ices, X-ray diffraction can still give information on average distances, in the form of $g_{XY}(R)$ functions that can be interpreted as the probability that when sitting on an atom X one finds another atom Y at distance R (7). As illustrated in Figure 3.4, the maxima of this function give the average O···O distances of first, second, etc. neighbours of a water molecule in liquid water.

Observing the H_2O molecules that are embedded in macromolecules is also possible with X-rays. Most macromolecules, however, are poorly ordered. Some of them can be partly crystalline, in which case their structure can be determined with a fair accuracy using powerful modern tools such as X-rays delivered by synchrotron radiation and powerful computational facilities for an optimal exploitation of the numerous recorded data. Thus H_2O molecules that contribute to the proton translocation in *bR* (bacteriorhodopsin—see Ch. 6) have been positioned with fair accuracy by X-rays (8, 9). It firmly confirms the role of the proton wire that implies the involvement of several H_2O molecules in the mechanisms of photosynthesis or vision, schematically described in Figure 6.8. X-ray diffraction, which in these conditions remains a powerful tool to determine the structure of the host macromolecule, can also in favourable conditions show the existence of permanent "structural bridges" established by some of the H_2O molecules present. As illustrated in Figure 10.7, left-bottom drawing, such bridges connect two groups, N–H and C=O for instance, which lie too far away to establish an H-bond. They are at the basis of "molecular recognition", as illustrated in the two other bottom drawings of this figure. The point worth noting is that in these conditions water molecules are at the origin, not of structural disorder, but of some spatial ordering. Obtaining crystalline proteins thus requires having a minimum number of water molecules. The detection of these structured H_2O molecules remains difficult, however, because they most often represent a small part only of the numerous H_2O molecules present that are for the most part disordered and at the origin of a more intense, but also more dispersed signal.

To conclude this subsection, it may be asserted that X-ray diffraction, which is a basic and highly powerful method to determine the structure not only of perfectly ordered systems such as crystals, but also of many more-or-less disordered ones such as macromolecules (10, 11) or simple liquids, encounters severe difficulties observing these small, versatile and labile H_2O molecules that are always present within them. It may localize the O-atoms of these molecules, with a good precision in crystalline ices or mineral hydrates, in a satisfactory way in liquid water or amorphous ices, but with much less efficiency in other systems such as macromolecules. The consequence is that X-rays are often no longer predominantly used to gather information on the structure of the H-bond network in macromolecules, but are still important among the various methods used. Some special set-ups, such as by reflectivity, may occasionally give original results on the structure of H_2O molecules at interfaces (12). X-ray may thus be, in this case of surfaces or interfaces, a complementary method to SFG spectroscopy (sum-frequency generation) discussed in Ch. 4.

Neutron scattering

This method is complementary to X-ray scattering, as seen in Ch. 3. Neutron scattering offers, in some cases, a better efficiency than X-ray scattering in the observation of the H-atom or, more precisely the D-atom when the H-atoms of the sample, or more often those H-atoms that can establish H-bonds, have been exchanged with D-atoms. Contrary to X-rays, which are scattered by electrons, neutrons are scattered by nuclear spins. The neutron coherent cross-section of the D-atom is consequently of a magnitude comparable to that of other atoms. This is also true for the H-atom. But this atom has a much bigger incoherent cross-section that makes incoherent scattering much more intense than the coherent one and masks it for a great part. Let us recall that only those waves that are scattered coherently by different nuclei are at the origin of interferences from which the relative positions of these nuclei can be determined (see eq. (3.10)). D-atoms, which do not exhibit such a great incoherent scattering factor, can consequently be localized with some precision and D-bonds can therefore be put into evidence. This possibility can be fully exploited in crystals, as in the case of X-ray scattering, but in less ordered systems, such as macromolecules, the cost of having to deuterate the H-bonds of the sample, together with the intrinsic requirement to have an acceptable signal-to-noise ratio, which requires us to have large enough samples and to operate within the structures of a few specialized centres that run specialized equipment, is often too high in terms of expenses and time to make it interesting.

Despite the great incoherent cross-section of the H-atom, elastic coherent neutron scattering experiments may be used at a lower resolution to detect the presence of clusters of H_2O molecules. The method used is that of small-angle neutron scattering, often labeled by its acronym "SANS". In such an experiment, the scattered intensity of a sample is recorded as a function of the angle θ between the wave vectors of the scattered and initial waves or, equivalently, as a function of the amplitude Q of the wave vector difference $\vec{Q} = \vec{k}_1 - \vec{k}_2$ of the scattered beam defined by $\vec{k}_2$ and incident beam defined by $\vec{k}_1$ (the wave vector defines the direction of propagation of a wave, as seen in eq. (3.6)). As can be seen in Figure 11.1, these quantities are related by the equation:

$$\left|\vec{Q}\right| = 2\left|\vec{k}\right| \sin\frac{\theta}{2} = \frac{4\pi \sin(\theta/2)}{\lambda} \tag{11.1}$$

where $\vec{k}$ stands for either $\vec{k}_1$ or $\vec{k}_2$ that have, in a coherent scattering experiment, the same modulus equal to $2\pi/\lambda$, with λ the wavelength of the neutron beam (eq. (3.9)). The variation of the scattered intensity obtained in such a SANS experiment is schematically represented in Figure 11.1. The scattered intensities are recorded in all directions that make a small angle θ with the incident beam. Within these limits of small angles the measurements of distances can only be of a low-resolution type, as indicated by the mathematical properties of Fourier transform calculations, and consequently no atomic scale information can be provided. The position of the maximum of the band, at $Q_0 = 0.15$ Å^{-1}, is thus related to the average diameter d of the cluster of H_2O molecules at the origin of this scattering by the equation:

$$d = \frac{2\pi}{Q_0} \tag{11.2}$$

In the case represented in Figure 11.1, this equation gives: $d \approx 40$ Å. Such experiments have been performed in proteins (13), in carbohydrate lamellae (14), etc., where clusters of H_2O molecules that size have been put into evidence.

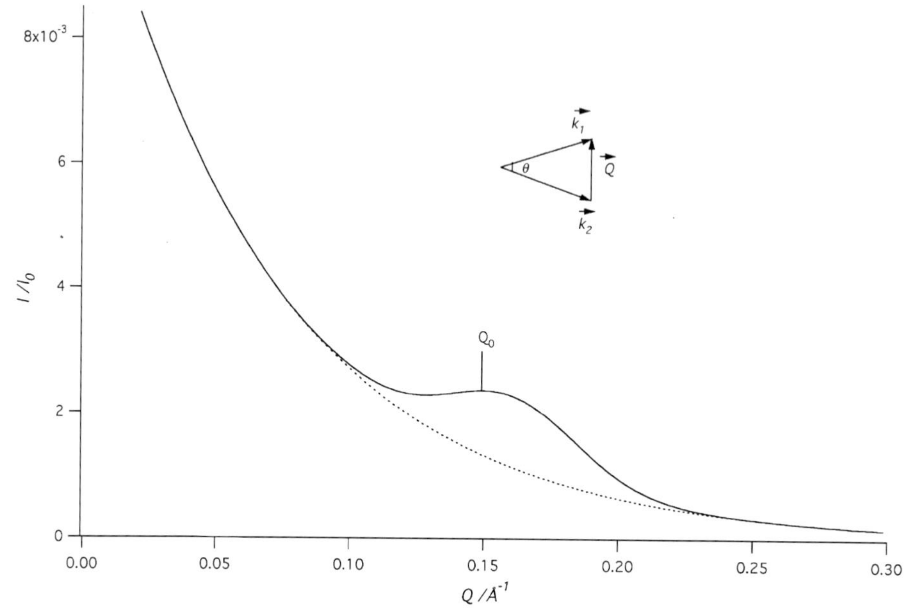

Figure 11.1 Ratio I/I_0 of the coherently scattered intensity I to the incident intensity I_0 for a neutron beam falling on a sample that contains clusters of H_2O molecules. The geometry of the wavevectors is represented in the middle-page drawing.

Furthermore, neutron scattering provides the interesting possibility to vary the contrast of H-and D-atoms by recording diffraction patterns of several samples differing only by their H/D isotopic compositions. Values of the (coherent) scattering lengths b_H and b_D of the H- and D-atoms of opposite signs: $b_H = -3.74$ fm (femtometer) and $b_D = +6.67$ fm (11, 15), are at the origin of this possibility. It implies that with appropriate proportions ρ and $1-\rho$ of the H- and D-atoms, the total scattering length $\rho b_H + (1-\rho)\, b_D$ can be made equal to zero. With these proportions, the scattering of neutrons occurs as if all H- and D-atoms were deleted from the sample. It allows a more precise determination of the positions of the other atoms. Another experiment on the same sample but with a different H/D isotopic composition then allows us to again see these H- and/or D-atoms and localize them with a better precision, already knowing the diffraction pattern of the other atoms. Positions of water molecules, which contain an especially large proportion of H- or D-atoms, inside macromolecules could thus be determined using this "isotopic contrast" variation. The few deeply buried H_2O molecules within bacteriorhodopsin (16) that are an active part of proton translocation, the central mechanism of vision and photosynthesis described in Ch. 6, could be localized using this technique.

The possibility to perform inelastic (incoherent) scattering is another interesting property of neutrons. Corresponding experiments are often called INS (inelastic neutron scattering) experiments. In these INS experiments, the high incoherent cross-section of the H-atom is used to produce a relatively intense scattered beam that is analyzed using its frequency difference with the incident beam. The spectral region that is particularly interesting is the "quasi-elastic region", or low-frequency region, where classical spectroscopic methods, IR or Raman, encounter difficulties. This is typically the region where the energy differences of scattered neutrons are of some fractions of meV, which correspond to wavenumbers of some cm^{-1} (1 meV is equivalent to 8 cm^{-1} after eq. (1.A4)). Corresponding "QENS" experiments (quasielastic neutron scattering) have for instance been used to study lyoprotective effects of some carbohydrates such as trehalose, described in Ch. 10. Such methods could be used to show that hydration increases the dynamics of proteins (13), particularly the vibrations of their α-helices. Vibrations of higher frequencies can also be observed in INS experiments. The signal-to-noise ratio is however no match with that experienced in Raman or mid-IR experiments in the higher frequency region.

NMR spectroscopy

As mentioned in Ch. 3, NMR spectroscopy, even if it is a method that is not especially sensitive to H-bonds, is able to detect H-bonds either directly by their chemical shifts, or indirectly by establishing correlations between relaxation times of various groups. The first direct method applies preferentially to simple molecules, whereas the second one applies to macromolecules. This second indirect method is useful for H_2O molecules embedded in a macromolecule, as they display a dynamics that has its own characteristic times that are different from those of the macromolecule. The H-bonds established or disrupted by water molecules adopt the rapid dynamics of these small molecules. It may consequently be separated from the slower one of H-bonds of the macromolecule itself. This particular part of NMR used for H_2O molecules in macromolecules is called relaxometry, a method that studies relaxation times as a function of frequency. It allows the separation of fast and slow dynamics. It is also known as "nuclear magnetic relaxation dispersion" (17), often labelled MRD, and has been used to study the hydration of proteins or of synthetic polymers. In the case of proteins, it has been applied to the spins of ^{17}O, an isotope of O-atoms, rather than to the usual spins of H-atoms. These latter atoms are much more numerous than the ^{17}O atoms whose natural abundance is of some 4×10^{-4} of the total amount of O-atoms. The ^{17}O resonance frequencies are different from those of the usual H-atoms and the signal due to ^{17}O atoms is consequently not masked by that due to these much more abundant H-atoms. NMR is nevertheless sufficiently sensitive method by which relaxation of ^{17}O can be studied with no problem, even at such low concentrations. The reason for studying ^{17}O relaxation dispersion rather than that of the most abundant H-atom is that relaxations of spins of these latter atoms suffer from complications that are partly due to their great numbers. Thus cross relaxations between H-atoms of proteins and those of H_2O molecules are most important. Relaxation of H-atoms also contain significant contributions from labile protein H-atoms that exchange with water, and disentangling all these interactions is often too difficult a task to give interesting results.

Relaxations of spins of ^{17}O atoms are free of these complications. A comparison of the ^{17}O relaxations in two different proteins (18), one with no buried water molecule and another with buried water molecules, showed a great difference at low frequencies. This difference was interpreted as due to these few buried molecules, offering NMR spectroscopy a way to observe them through their specific dynamics. The water molecules at the surface of proteins are highly mobile, with average orientational correlation times of some 20 psec, while buried water molecules that stabilize the protein secondary and tertiary conformations, defined in Ch. 2, have much longer correlation times. These two types of molecules furthermore are in a constant dynamic interchange on timescales from 10^{-8} to 10^{-6} sec, presumably via partial unfolding of the protein. These few buried water molecules are thought to be those that are strongly H-bonded to the protein itself, each establishing three to four H-bonds on hydrophilic sites of the protein (19) and are therefore considered as belonging to the protein.

This MRD method has also been applied to the mechanisms of photosynthesis or vision (20) that occurs in bacteriorhodopsin (bR) described in some detail in Ch. 6. In that case, it was not limited to relaxation of ^{17}O only but also incorporated relaxations of H- and D-atoms. It could thus be shown that within the bR proton conduction pathway at least seven H_2O molecules are deeply buried, which means they have long residence times, of the order of the microsecond (μs) at 277 K, a time that, even if much longer than the average exchange time of bulk H_2O molecules, remains short compared to the rate-limiting step of the photocycle. MRD has also been applied to the problem of the role of H_2O molecules in protein folding, which has been briefly covered in Ch. 10. It could thus be shown (21) that the hydration of the molten globule is much more similar to that of the native protein than was originally thought, at least as viewed from NMR spectroscopy. The molten globule, or compact denaturated protein, is an intermediate state in the unfolding process between the native protein and the denaturated polypeptide.

NMR spectroscopy is thus a potentially interesting method to observe H_2O molecules in macromolecules. It can distinguish various types of such molecules that have different relaxation times. It, however, provides direct information neither on their location nor on the H-bonds they establish. Indications can be indirectly deduced from NMR spectra. They are not however precise enough for us to consider NMR as a general method to look at the H-bond network established by H_2O molecules.

Molecular dynamics (MD)

The above considerations strongly suggest that the role of water molecules is important because of the exceptional H-bond network they establish not only when they are the only molecules present, as in liquid water, but also when they are embedded in bigger systems such as macromolecules. Theoretical methods have to then reproduce the properties of these H-bond networks. Due to their complexities this is no easy task. Furthermore, we have seen these H-bond networks are not static, but undergo rapid dynamic isoenergetic transformations that are at the origin of most of their interesting properties. The main theoretical method that can offer a description of such a dynamic set of water molecules is then molecular dynamics (MD) simulations. As seen in Ch. 3, the principle of MD computations is to calculate the evolution of a system submitted to an interaction potential that governs the dynamics of its constituent molecules following classical laws of mechanics. One calculates the state of this system when time is increased by small steps of some 10^{-15} sec, starting from a reasonable configuration at time 0. The point is that, in order to be efficient in having not too great a number of operations to perform, this interaction potential has to be simple. Applied to water molecules, this simplification consists of reducing H_2O molecules to a small number of point charges that interact with point charges of other molecules through Coulomb interactions.

These charges are so positioned in each H_2O molecule that they reproduce its dipole and quadrupole moments. However, this necessary simplification has a drawback—MD has an inherent inability to take into account the directionality of the H-bonds these water molecules establish, thus discarding one of the central properties of H-bonds. It is at the origin of the failure of MD approaches to reproduce the exceptional properties of liquid water described in Ch. 9. The same difficulty is encountered in describing the properties of the H-bond network developed by H_2O molecules that are embedded in macromolecules.

Despite these restrictions, MD simulations nevertheless offer a complementary view of the H-bonded system with many water molecules that is under study and is highly appreciated in a great number of experiments. It allows supporting hypotheses that are proposed for the interpretation of these experiments. They also give useful orders of magnitude for crucial quantities that may be otherwise hard to measure. In this sense they are highly useful. However, MD simulations, which are highly successful in the case of ordinary liquids, in the case of liquid water or of H_2O molecules in macromolecules encounter great difficulties by themselves providing original knowledge on the extended H-bond network these molecules develop. In this sense, they can be put on the same footing as most molecular experimental methods described above, which are powerful methods to observe such big systems as macromolecules but not really efficient to observe these small, labile and highly versatile H_2O molecules that come embedded in these big systems.

VIBRATIONAL SPECTROSCOPY

Vibrational spectroscopy is a method that directly gives molecular information and is consequently widespread in chemical physics. Both Chs. 4 and 5 deal with IR spectroscopy, which is part of vibrational spectroscopy and appears to be a hypersensitive and precise method to observe H-bonds. It has up to now been scarcely used to look at H_2O molecules. The reason for this surprising situation is quite paradoxical; IR is so sensitive to H-bonds and the number of H-bonds that form around H_2O molecules is so great that the absorption due to only a few H_2O molecules is already very great. In the usual absorption set-up shown in Figure 5.A1, the ratio $I(\tilde{\nu})/I_0(\tilde{\nu})$ of the transmitted intensity to the incident one becomes rapidly immeasurable in the most interesting spectral regions where it tends towards 0 in the presence of relatively few H_2O molecules. Newly arriving H_2O molecules consequently do not modify the spectrum in that region. In other words, no distinction can be made, in this region, between the spectra of two samples that only differ in their thickness l, whereas eq. (5.A8) clearly indicates that this ratio effectively depends on l. This is a typical "saturation effect". When it appears, as with the presence of a few H_2O molecules, it ruins all the qualities of IR spectra and makes IR a useless tool. Until recently, H_2O molecules were therefore considered a poison in IR spectroscopy and were avoided as much as possible. In the best cases, samples that contained H_2O molecules were studied in the relatively narrow spectral regions where these molecules exhibit bands of a low intensity only that do not provoke saturation. This was for instance the case in the 1200–2800 cm^{-1} region where H-bonded H_2O molecules do not exhibit bands of a strong intensity, as can be seen when looking at the spectrum of liquid water displayed in Figure 9.2. This spectrum is that of H_2O molecules only and in this region only the bending band $\delta_{H\text{-}O\text{-}H}$ of H_2O appears. It is the least intense of the three main bands due to

H_2O and can be used in such cases as crystalline hydrates (22) to study these H_2O molecules that are then in a relatively small number. In the case of water molecules embedded in macromolecules, this region has been used to study the interaction of water molecules with this macromolecule, as vibrations such as C=O stretching or N–H bending that are sensitive to the establishment of H-bonds appear. In the case of proteins or amide/peptide-containing samples, the corresponding normal modes are called amide I and II, respectively. They have often been studied after an H/D exchange has been performed, typically by substituting heavy water for ordinary water as a solvent. This exchange provokes the disappearance of the $\delta_{H\text{-}O\text{-}H}$ band that may saturate and its replacement by a $\delta_{D\text{-}O\text{-}D}$ band that appears at $1210\,cm^{-1}$ and does not consequently overlap or even mask these amide I and II bands. This method is also used in medicine by injection of some millilitres of heavy water in the blood of a person. From the intensities of the bands due to HDO in the somewhat later recorded IR spectra of a blood sample, together with the known quantity of injected heavy water, the total amount of body water of the patient can be measured (23, 24).

This H/D substitution/dilution method in a restricted spectral region is also often used in studies of the dynamics of water molecules in liquid water by time-resolved nonlinear IR spectroscopy described in Ch. 4 and also described below when applied to the study of the dynamics of H_2O. The $\nu_s(O\text{–}H\cdots)$ band of HDO molecules diluted within D_2O molecules may be analyzed, as they do not saturate if dilution is strong enough. This can also be done on the $\nu_s(O\text{–}D\cdots)$ band of the same HDO molecules diluted within H_2O molecules. The quantitative study of the structure of water molecules around ions (25) in an aqueous solution, or the solvation mechanism of simple organic molecules such as tetrahydrofuran (26), becomes possible with this type of spectroscopy. This method, however, rapidly reaches its limits. The H/D exchange may be difficult to perform in the case of water in macromolecules, the presence of $\delta_{D\text{-}O\text{-}D}$, $\delta_{H\text{-}O\text{-}D}$ as well as remaining $\delta_{H\text{-}O\text{-}H}$ bands complicates the analysis of the spectrum, all the more so because N–H as well as N–D groups, for instance, may be simultaneously present. Furthermore, the spectral region that is restricted by the presence of saturated bands due to O–D$\cdots$ vibrations also severely restricts the information.

Despite these problems of saturation of vibrational bands IR spectroscopy, described in the next subsection, has been recently shown to nevertheless remain an especially powerful method to observe H_2O molecules. Special recently proposed set-ups can avoid saturation in the whole conventional IR region, thus taking full advantage of the power of IR to study H-bond networks. They are first described, before the contribution of recent time-resolved nonlinear IR spectroscopy is examined. Other methods such as NIR or Raman spectroscopy, which are intrinsically free of this saturation effects can also be used to study the H_2O molecule. They are often limited to some specific problems, as they do not display the power of ordinary IR spectroscopy for the study of H-bonds or of H_2O molecules and cannot consequently be considered as general methods. They are described in the last subsection of this section on vibrational spectroscopy.

IR spectroscopy to observe H_2O molecules

In order to avoid the saturation of IR bands when H_2O molecules are present and consequently fully recover the power of IR absorption spectroscopy to observe H-bonds, two main set-ups with well-defined corresponding procedures can be used. They are described in the next two subsections. With these set-ups, IR spectroscopy becomes a highly powerful method to observe H_2O molecules. Before describing them, let us mention that hydration of relatively small molecules can occasionally be studied in gases using, among other methods, general IR set-ups as described in Ch. 4 for H-bonds. Thus hydration of 2-amino-1-phenylethanol in gases, an analogue molecule of noradrenaline, a neurotransmitter, has been studied, seeding this molecule in an expanding supersonic jet (27). The interest of this highly specialized technique is to observe gaseous molecules at very low temperatures, with consequently relatively simple spectra that can be best interpreted. In that case, the H-bond configurations of this molecule with a single, a pair, or three H_2O molecules could be accurately determined. It nevertheless remains an exceptional method to observe hydration of isolated, small molecules.

Attenuated total reflection (ATR)

An attenuated total reflection (ATR) set-up is shown in Figure 5.A1. With this set-up the IR beam does not penetrate the sample, as in an ordinary absorption set-up, but travels through a crystal of high refractive index where it performs total reflections. In the case of a liquid, the preferential symmetry of this crystal is that of a cylinder, which is the symmetry of the IR beam at the output of the interferometer of a Fourier transform apparatus. The principle of ATR is "dilution" of the IR absorption, as this absorption occurs only on the evanescent wave that travels out of the crystal whereas, in an ordinary absorption set-up, also shown in Figure 5.A1, the IR beam travels through the absorbing sample itself, causing absorption to occur on the whole propagating IR beam. In the ATR set-up, the evanescent wave, which propagates within the sample parallel to the interface with the crystal and remains localized in the near vicinity of this interface, represents but a small proportion of the whole propagating wave. The ratio of the transmitted intensity to the incident one is thus reduced to 0.1 ($\log(I_0/I) = 1$) after traveling a number of centimeters within a crystal immersed in liquid water in an ATR set-up, whereas the same ratio is obtained after traveling a few microns only in an absorption set-up (3). This absorption length of some centimeters in an ATR set-up is much easier to handle than the few microns in an absorption set-up. ATR spectra consequently display much enhanced signal-to-noise ratios as compared to absorption spectra.

The spectral range in an ATR set-up is, however, limited by the absorption region of the crystal inside which total reflections are performed. The exploitable region is that of the total transparency of the crystal, above 650–700 cm^{-1} for many crystals used with this technique. As shown in Figure 5.A2, ATR spectra resemble usual absorption spectra, but are not the same. Absorption spectra are consequently most often computed from ATR spectra and can then be compared with spectra of that type, which represent more than 95% of the spectra usually recorded in IR spectroscopy. This computation often implies calculation of the optical constants n (refractive index) and k (absorption coefficient) of the sample using Fresnel equations for the calculation of the amplitudes of the reflected and evanescent waves, and Kramers–Kronig relations defined in eq. (5.A9) between n and k (28, 29) or, more recently, between reflectivity and phase angle (30). From these n and k spectra, absorption spectra can themselves be calculated (31). These computations represent a complication that makes ATR set-ups scarcely used, especially when compared to spectra obtained with absorption set-ups that require calculations of the decimal logarithm only of the ratios of intensities (eq. (5.A8)), an easily performed operation.

In the case of liquid water, a highly absorbing medium, absorption spectra recorded with absorption set-ups most rapidly display saturation effects. An ATR set-up that "dilutes" this absorption is consequently most interesting. Absorption spectra can then be computed from experimental ATR spectra. Spectra of liquid water displayed in Figures 9.1 and 9.2 were calculated using this method. Because of the lack of a total transparency of the ATR crystal at wavenumbers smaller than 650 cm^{-1}, absorption spectra in the FIR region ($\tilde{\nu} < 700$ cm^{-1}) were added to ATR spectra so as to have the spectrum in the whole 50–7500 cm^{-1} range. This was made possible because absorption bands that have intensities proportional to wavenumbers (eq. (5.A8)) are weaker in the FIR region where saturation effects are consequently more easily avoided. Simple absorption set-ups can consequently be used in this region (32). The qualitative discussion of these IR spectra of liquid water performed in Ch. 9 led to the important conclusion that the H-bond network of liquid water is as dense as that of ice and that there are a very small number of disrupted H-bonds in liquid water. Because of rotations of individual H_2O molecules of a great amplitude in liquid water, this H-bond network is, however, much more flexible. These absorption spectra of liquid water obtained from experimental ATR spectra were precise enough to also show that the bending band $\delta_{\text{H-O-H}}$ of H_2O takes on a Lorentzian form when temperature increases (33), a result that could also have been reported while recording spectra with an usual absorption set-up. With this latter set-up the recorded spectrum is, however, not the spectrum due to only the absorption by the liquid water sample: it also includes reflections at the interface between windows and sample. The effects of these reflections usually appear in all absorption spectra under the form of a shift in the value of the absorbance. This shift is most often nearly a constant for all wavenumbers, as reflection depends on the refractive index difference $n_w - n_s$ between window (n_w) and sample (n_s), which is a quantity that displays only small variations with wavenumbers $\tilde{\nu}$. In the case of liquid water with very strong absorption bands, this is no longer true. The consequence is that the absorption spectrum of liquid water does not resemble most absorption spectra, but resemble a "distorted" absorption spectrum where the baseline is not a constant. Presenting evidence of the Lorentzian shape of the $\delta_{\text{H-O-H}}$ band was consequently more difficult than for a less absorbing liquid. Calculating the absorption only due to liquid water from experimental absorption spectra of liquid water, or equivalently the optical constants n and k of liquid water, in that case requires calculations that are very similar to those performed to transform ATR spectra into absorption spectra (32), which have been described above.

The spectra of liquid water containing organic molecules or ions can be well studied using such an ATR set-up. The number of H_2O molecules that are bound to ethylene carbonate in a solution of this species in water could thus be determined using this technique (34). Ethylene carbonate is an important component in rechargeable lithium batteries and consequently it is necessary to know how it hydrates in order to know the structure of water at charged metal interfaces more precisely. Also ionization processes of many acids, bases or salts, have been extensively studied by Max and Chapados (35–37) using ATR spectra. They showed that they could separate the bands due to "hydrating water molecules" that bind to ions from largely predominant bands due to "bulk water molecules", i.e. H_2O molecules that do not resent the presence of these ions. The concentrations of various ions could thus be measured directly using ATR spectra, without having to calculate optical constants or absorption spectra. The precision with which they measured equilibrium constants is a good illustration of the power of IR spectrometry to observe water molecules. Spectrometry, as its name indicates, consists of not only determining the wavenumbers at which bands appear, but also measuring their relative intensities so that concentrations of various components can be determined with precision. Also hydration mechanisms of many small organic molecules could be established from these ATR spectra.

ATR spectroscopy or spectrometry is well adapted to strongly absorbing liquid or nearly liquid samples, typically liquid water or aqueous solutions. In these cases, no technical problems concerning contact between sample and ATR crystal are encountered, as this contact is very good in the case of liquids. Because the interpretation of the spectra presumes this contact is perfect, having it not so may induce important errors, as is often the case with solids, for which ATR spectra should be interpreted with great care. For the study of H_2O molecules in macromolecules, the technique of thin samples described in the next subsection is therefore preferred when the macromolecule is not itself a liquid. Good contacts between crystal and solid macromolecule are very difficult to ascertain. This often hinders the performance of quantitative measurements. Applying pressure to improve this macromolecular sample contact could be a solution. It allows, for instance, the measurement of diffusion of water in polymeric common plastic membranes such as PVC (38) (polyvinyl chloride) or PET (polyethylene terephtalate) (39), using an interesting dedicated ATR set-up somewhat different from that shown in Figure 5.A1. In this set-up a thin film of the studied polymer is inserted between an ATR crystal and water. The ATR crystal is prismatic, as is usual for solid samples, instead of having a cylindrical shape that is especially designed for liquids. Such a set-up has also been applied to the study of diffusion of water in polyelectrolyte membranes (40) where the knowledge of this diffusion is important for the comprehension of its conducting properties. It has also been used in the case of epoxy-resins (41).

IR spectrometry on thin samples: hydration of macromolecules

The second way to avoid saturation of bands due to the presence of highly absorbing H_2O molecules is to use very thin samples wherein the number of H_2O molecules remains small, thus enabling the implementation of a simple absorption set-up. Thin samples are typically 1 μm thick, which is still too thick for samples of liquid water or liquid aqueous media, but is well suited for samples of macromolecules within which the thickness of the total amount of water is less than 1 μm. The structure of the H-bond network developed by H_2O molecules is then studied, varying the degree of hydration of the sample. These hydration experiments that are monitored by IR absorption are presented in detail in this section, as they illustrate quite well the power of IR spectroscopy, or more exactly IR spectrometry, to observe H_2O molecules. Furthermore results of the analysis of these spectra have already been commented upon in the preceding Ch. 10. Before presenting these experiments, let us note that absorption spectra of nanosamples of ice have been performed by J. P. Devlin and coworkers. These are also experiments on thin samples that contain a reduced amount of water molecules. The interest of nanosamples is that varying the sizes of these samples induces a variation of the S/V surface-to-volume ratio that allows separation of entangled bands due to H_2O molecules that are positioned either on the surface of ice or in the bulk (42). These studies were the origin of a wealth of information, not only on ice itself (43, 44), but also on the chemistry of the ice surface (45–47), a fundamental catalyst in the atmosphere, particularly the ice of polar stratospheric clouds that catalyzes depletion of ozone in the stratosphere, as described in Ch. 8. They also provided original data on the ionization of HF in contact with ice, which showed HF to be a strong acid at 40 K (48), a behaviour that greatly contrasts with what it displays at room temperature where its dissociation (ionization) constant, when put into contact with liquid water, is 10–13 orders of magnitude weaker than that of other hydrogen halides.

Hydration spectra of hyaluronane

The principle of hydration experiments is to record the spectra of a macromolecule that is embedded in an atmosphere with a definite water pressure once it has reached equilibrium with it. The interest of such an experiment as compared to immersion of this same macromolecule into liquid water is: first, one disposes of a varying parameter, the hygrometry of this surrounding atmosphere that takes on values between 0 (dried atmosphere) and 1 (saturating atmosphere); second, the quantity of H_2O molecules is reduced, thus avoiding the presence of bands due to bulk liquid water molecules, which are not the object of the study, as are found in ATR spectra. They consequently do not have to be eliminated to make relevant features visible, an operation that somewhat degrades the properties of the spectrum. The hygrometry, also called relative humidity of the surrounding atmosphere, is equal to the ratio p/p_0 of the water pressure p of this atmosphere to the saturating water pressure p_0 at the same temperature. In these experiments, a thin sample of the studied macromolecule is positioned within a special cell where the atmosphere and the temperature are controlled (3) and can be varied. This cell is itself positioned in the IR beam following a usual absorption set-up.

As an example of how IR spectrometry works, we take hydration of a polysaccharide, hyaluronane (HA). The structure of this polysaccharide in its dried state is shown in Figure 10.1. Its IR spectra, obtained using the set-up briefly described above, is reproduced in Figure 11.2 for two hydration stages, the dried HA recorded in equilibrium in a dry atmosphere with hygrometry equal to 0 (spectrum D), and a hydrated state of HA in equilibrium with an atmosphere of hygrometry equal to 0.51 (spectrum Hy). In this latter spectrum, a great number of hair-like narrow bands appear around 3800 and 1600 cm^{-1}. They constitute the "rotational structure" of the stretching band ν_s and bending band δ_{H-O-H} of isolated H_2O molecules of the water vapour. They also appear in spectrum W, the spectrum of the cell with no sample but filled with water vapour at a definite pressure p_c, and also in more detail in Figure 8.2. The pressure p of the water vapour is measured from the intensities of these hair-like bands that are themselves calibrated against that of spectrum W. In the case of Figure 11.2, by subtracting spectrum W multiplied by 0.73 to spectrum Hy, these bands have their intensities equal to zero. We deduce that the pressure p of the water vapour around the sample when spectrum Hy has been recorded is $p = 0.73p_c$. Reading the temperature of the cell, one finds the value of p_0 at this temperature in tables of physical constants, hence the value of the hygrometry p/p_0 for this spectrum Hy.

The bands due to water vapour are no longer useful in the analysis of the spectra. They are consequently eliminated by this subtraction procedure and only the spectrum of the dried sample D is kept, together with the *hydration spectrum Hy–D–0.73 W*. Subtraction of the water vapour features causes no degradation of the signal-to-noise ratio, as, in opposition to ATR spectra where features due to bulk H_2O molecules are predominant, they have small intensities (they are nearly all free H_2O molecules that do not establish H-bonds). The intensities of the bands of this hydration spectrum are proportional to their sensitivity

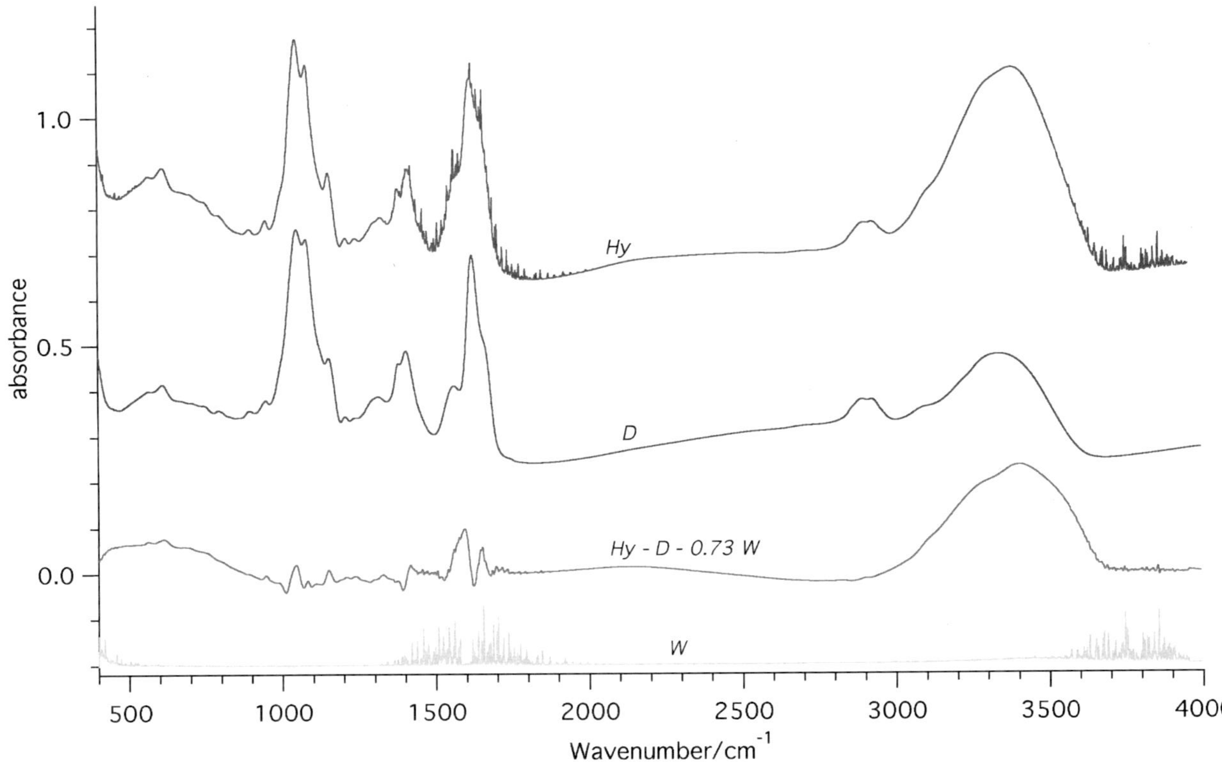

Figure 11.2 IR spectra of HA, in its dried state (*D*) when in equilibrium with the surrounding atmosphere with no water vapour ($p/p_0 = 0$), and partially hydrated when in equilibrium with the surrounding atmosphere having an hygrometry $p/p_0 = 0.51$ (*Hy*). The "hydration spectrum" labelled *Hy–D–0.73 W* is equal to the subtraction to spectrum *Hy* of spectrum *D* and spectrum *W* multiplied by a coefficient equal to 0.73, which ensures the disappearance of the narrow bands due to water vapour. These bands are shown in spectrum *W*, which is the spectrum of the cell without sample but with an atmosphere with calibrated water vapour pressure p_c.

to hydration, the mechanism we wish to observe. All bands, such as stretching and bending bands of C–C–H groups for instance, which are almost insensitive to hydration, show weak features only in these hydration spectra. In a typical experiment, 50 spectra are recorded and these span all possible hygrometry values between 0 and 1. It gives a set of (roughly) the same number of hydration spectra, in addition to the spectrum of the dried sample *D*. A selection of some of them is shown in Figure 11.3, together with spectrum *D* of the dried sample, already displayed in Figure 11.2, and drawn at the same scale as the hydration spectra.

Analysis of hydration spectra

Several steps are required to perform the analysis of 50 hydration spectra. The first one consists of reducing the number of spectra to be analyzed by decomposing them on the minimum number of "elementary hydration spectra" $H_1(\tilde{\nu})$, $H_2(\tilde{\nu})$, $H_3(\tilde{\nu})$, etc. All "experimental hydration spectra" recorded at various values of the hygrometry p/p_0, are then equal to

$$H(\tilde{\nu}, p/p_0) = a_1(p/p_0)H_1(\tilde{\nu}) + a_2(p/p_0)H_2(\tilde{\nu}) + a_3(p/p_0)H_3(\tilde{\nu}) + \cdots \quad (11.3)$$

Elementary hydration spectra $H_i(\tilde{\nu})$ ($i = 1, 2, 3$, etc.) are defined from experimental hydration spectra $H(\tilde{\nu}, p/p_0)$ and are linear combinations of them. In practice, this definition requires successive steps. Thus the first experimental hydration spectra at low hygrometries are often proportional to one another. This allows us to define a first elementary spectrum $H_1(\tilde{\nu})$. Note that $H_1(\tilde{\nu})$ could be replaced by any spectrum $\lambda H_1(\tilde{\nu})$ proportional to it, with λ a coefficient that may take any algebraic value). When hygrometry increases and is

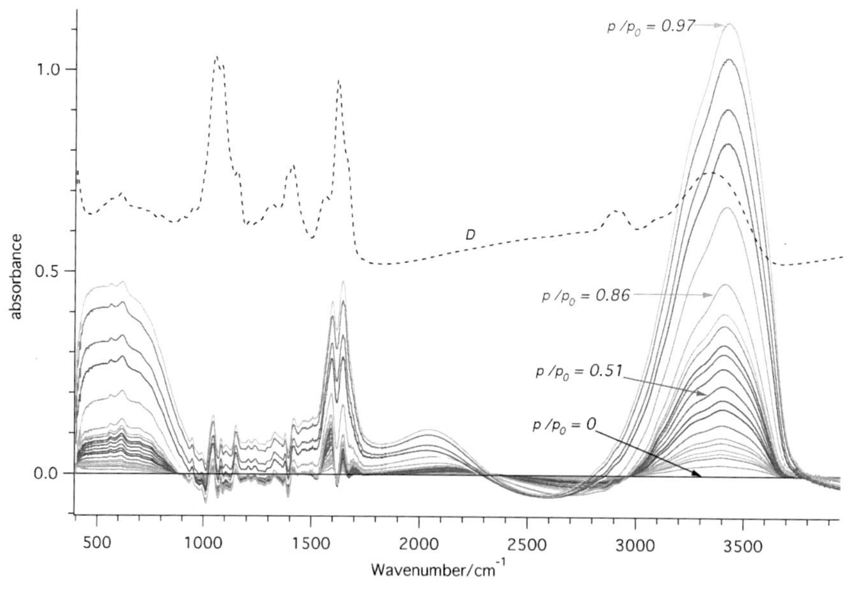

Figure 11.3 A selection of hydration spectra of HA at various values p/p_0 of the hygrometry. In dashed line and offset for clarity, spectrum *D* of the dried sample is drawn at the same scale as hydration spectra.

greater than some value p'/p'_0, this becomes no longer true, but we may nevertheless find that all spectra $H(\tilde{\nu}, p/p_0) - a_1(p'/p'_0)H_1(\tilde{\nu})$, for hygrometry p/p_0 taking on values slightly greater than p'/p'_0, are still proportional to a second elementary spectrum $H_2(\tilde{\nu})$. The process is repeated for $H_3(\tilde{\nu})$, etc. When all elementary spectra are defined, the residual spectra equal to the differences of the two members of eq. (11.3) for all experimental hydration spectra are very small. Their features are much smaller than all bands of any of the $H(\tilde{\nu}, p/p_0)$, as shown in Figure 11.4 where, together with the four "elementary hydration spectra" $H_i(\tilde{\nu})$ ($i = 1$–4) of HA, on which all $H(\tilde{\nu}, p/p_0)$ can be decomposed, the "worst" residual spectrum, i.e. that with features significantly bigger than in all other residual spectra, is displayed. The number of elementary spectra for HA, four, is typical of the up-to-now studied species. It is equal to three in BSA (bovine serum albumin) (49), lysozyme (50) or polyimide sulfonated membranes (51). Once the decomposition of all experimental hydration spectra has been performed, we may recalculate elementary spectra from all experimental hydration spectra, thus statistically improving their definition by reducing their signal-to-noise ratios. It also reduces the errors in the decomposition.

These four elementary spectra $H_i(\tilde{\nu})$ are not yet the elementary hydration spectra we are looking for, which correspond to four independent hydration mechanisms. The reason is that the definition of these four elementary spectra only rests on mathematical criteria, and mathematics tells us that instead of taking these four spectra, we may as well have taken any four independent linear combinations of them. We would then have obtained an equally good decomposition of all experimental hydration spectra, with the same statistical errors.

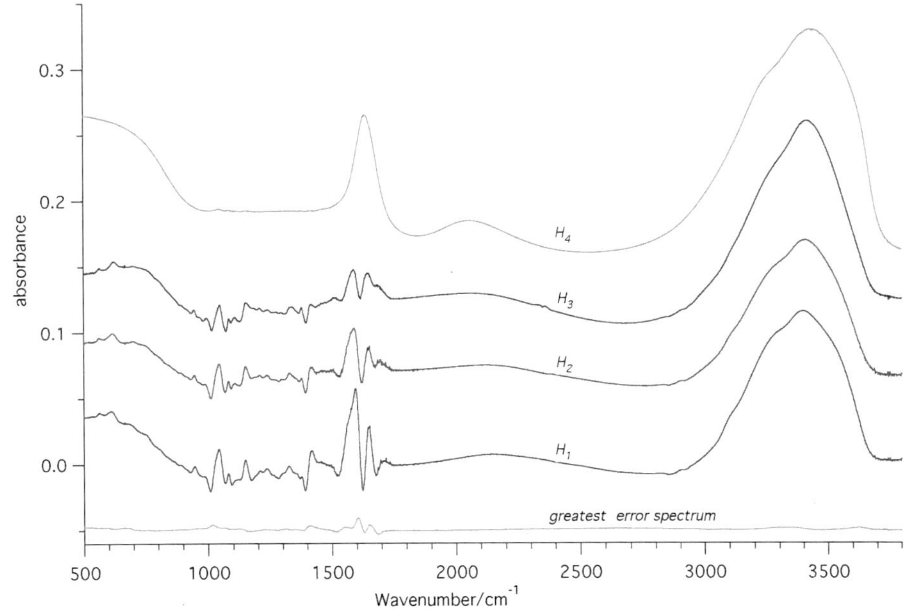

Figure 11.4 The four elementary hydration spectra $H_i(\tilde{\nu})$ ($i = 1$–4) of HA, together with the residual spectrum with greatest features, labelled "*greatest error spectrum*". Spectra are offset for clarity.

The second step therefore consists of determining at least one proper linear combination of these four elementary spectra that fulfils chemical criteria whereby they correspond to a well-defined hydration mechanism. The main such criterion is that when hygrometry p/p_0 increases, the coefficients $a_i(p/p_0)$ ($i = 1$–4 in the case of HA) display a behaviour compatible with chemical criteria. Thus the total water uptake of HA, displayed in Figure 10.2, increases, as expected, in a monotonous way when hygrometry increases. This water uptake is independent of the choice of elementary spectra: any linear combination of them gives the same n_{tot} at a given hygrometry p/p_0. The coefficients $a_i(p/p_0)$ change when performing such a linear transformation of the elementary spectra. The behaviour they display for all i in Figure 10.2, which starts from zero, and increases when hygrometry increases, reaching a plateau and remaining there for higher hygrometries, is compatible with chemical criteria. They consequently correspond to hydration mechanisms and we can keep them as elementary hydration spectra. Elementary spectra that would, for instance, have their corresponding coefficients $a_i(p/p_0)$ displaying an oscillating behaviour with p/p_0, or negative values, would not correspond to hydration mechanisms. This step of determining proper linear combinations that define elementary spectra as elementary *hydration* spectra has no mathematical significance. The spectra displayed in Figure 11.4 are successive hydration spectra, as all reach a plateau at the same value of the hygrometry where a new one starts being active. This is a result obtained for HA but does not constitute a chemical requirement. In other cases, BSA (52), polyimide block copolymers (51) or lysozyme, (50) several elementary hydration spectra are simultaneously active within some range of the hygrometry. They correspond to simpler hydration mechanisms. It would have been possible to take a linear combination of them to make them successive spectra. It was deliberately chosen, in the case of BSA and polyimide membranes (see Figure 11.6), for instance, to preferably present spectra that correspond to the smallest number of hydrophilic groups being hydrated, at the price of having several elementary hydration spectra simultaneously active in several regions of the hygrometry.

Bands due to H_2O molecules: measurement of the water uptake

The translation of these elementary hydration spectra into chemical mechanisms constitutes the third step of the analysis. This step rests on the measurements of the relative intensities of the various bands of all these spectra. In all hydration spectra, whether they are experimental spectra, as in Figure 11.3, or elementary spectra, as in Figure 11.4, two types of bands are present. The first type is that of bands due to H_2O molecules that become embedded in the macromolecule when the hygrometry increases. They are characterized by their three bands: ν_s (stretch) in the region 3000–3650 cm^{-1}, $\delta_{H\text{–}O\text{–}H}$ (bending) in the region around 1650 cm^{-1} and ρ_{H_2O} (libration) in the region 400–900 cm^{-1}. All these bands are clearly visible in the spectrum of liquid water shown in Figures 9.1 and 9.2 where they have been defined. The intense and broad ν_s and ρ_{H_2O} bands are predominant in the spectra of Figures 11.3 and 11.4, whereas the less intense $\delta_{H\text{–}O\text{–}H}$ band is masked by intense bands of the second type in most displayed spectra. This second type of band is of the macromolecule. These are difference bands, each one equal to the band after hydration minus the same band before hydration. They define the reaction of the macromolecule to hydration. In the spectra of Figure 11.4 the broad ν_s and ρ_{H_2O} bands of H_2O molecules are predominant over the difference bands of HA. This is a consequence of the high hydrophilic nature of HA that causes it to absorb a great number of H_2O molecules at any hygrometry. It is not a general feature: in BSA, for instance, difference-type bands due to the reaction of the protein to hydration are predominant over bands of H_2O in some elementary hydration spectra that are active at low hygrometry (52). It corresponds to absorption of a relatively small number of H_2O molecules at this stage of the hydration process where the corresponding mechanism is active.

The numbers of H_2O molecules at the origin of the three bands due to H_2O in each of these elementary hydration spectra can be deduced from the measurements of the intensities of one of these three bands. The most intense and noticeable band of H_2O, ν_s, can hardly be used, because, as described in Ch. 4, it is so sensitive to H-bonds that even a small modification of the strengths of these H-bonds produces appreciable changes of its integrated intensity, even when the number of H_2O molecules remains the same. It can only be used in well-defined cases where it may be ascertained that no change of the H-bond strengths occurs (51, 53). The bending band $\delta_{\text{H–O–H}}$ is in these conditions the best candidate to measure this number of H_2O molecules provided, however, that it is not masked by other bands as in Figure 11.4, for H_1, H_2 and H_3. The intensity of this band is largely independent of H-bonding (54) and consequently proportional to only the number of water molecules. It does not mean this band is insensitive to H-bonds, as its position and width, in opposition to its integrated intensity, may vary with the kind of H-bond the water molecule establishes. When this $\delta_{\text{H–O–H}}$ band is masked by another band, the libration band ρ_{H_2O} is used. Its integrated intensity is also apparently independent of the strength of the H-bonds established by the H_2O molecules, even if this point is less well documented. It can consequently be used as a calibration band. It has, however, a more important inconvenience: it extends at wavenumbers lower than 400 cm^{-1}, and a part of it is consequently absent in most recorded spectra that display their lower limit at 400 cm^{-1}. It implies that only the recorded part of this band is used, for instance by discarding half of it, the part at lower wavenumbers. This is what has been done for BSA (49) and HA and, despite this inconvenience, gave results in full agreement with other independent methods, such as thermogravimetry, as Figure 10.2 shows. The integrated intensity of the H_2O band, either $\delta_{\text{H–O–H}}$ or ρ_{H_2O} (part of it, in this case), is then calibrated against the equivalent one of a sample 1 μm thick of (liquid water displayed in Figures 9.1 and 9.2. It is carried out by subtracting the spectrum of liquid water multiplied by a coefficient λ_i from the considered elementary hydration spectrum $H_i(\tilde{\nu})$ with λ_i so determined that the integrated intensity of the difference (sample minus liquid water) $\delta_{\text{H–O–H}}$ or ρ_{H_2O} band is equal to 0. For each elementary hydration spectrum H_i this coefficient λ_i is equal to the ratio of the thickness $e^i_{H_2O}$ of an hypothetical sample of liquid water of cross-section σ that would have the same number of H_2O molecules as that measured in the sample, to the thickness $e_w = 1$ μm of a sample of liquid water with the same cross-section σ:

$$\lambda_i = \frac{e^i_{H_2O}}{e_w} \tag{11.4}$$

Furthermore, the relative density d_s of the macromolecule with respect to liquid water, equal to 1.6 in the case of dried HA, verifies the equation:

$$d_s = \frac{M_s N_s}{e_s} \frac{e^i_{H_2O}}{18 N^i_{H_2O}} \tag{11.5}$$

with M_s, the mass of the disaccharide repeat unit of HA, equal to 401 g mol^{-1}, N_s, the number of disaccharide units in the HA sample of thickness e_s, measured most often using simple but nevertheless precise interferometric methods (3), $N^i_{H_2O}$ the number of H_2O molecules that appears in spectrum $H_1(\tilde{\nu})$ and 18 g the mass of 1 mol of H_2O. From eqs (11.4) and (11.5), we deduce the number of water molecules per disaccharide unit that appear in the elementary hydration spectrum $H_i(\tilde{\nu})$:

$$\frac{N^i_{H_2O}}{N_s} = \frac{M_s}{18}\frac{e_w}{e_s}\frac{\lambda_i}{d_s} \tag{11.6}$$

Eq. (11.3) tells us that the number $n^i_{H_2O}$ of H_2O molecules per disaccharide repeat unit that have been embedded in HA at hygrometry p/p_0 by hydration mechanism i is equal to

$$n^i_{H_2O} = \frac{N^i_{H_2O}}{N_s} a_i(p/p_0) \tag{11.7}$$

These are the quantities that are represented in Figure 10.2 for the four $H_i(\tilde{\nu})$. The total number n_{tot} of embedded H_2O molecules per disaccharide is then equal to

$$n_{tot} = \sum_{i=1}^{4} n^i_{H_2O} + n_d \tag{11.8}$$

where n_d is the number of H_2O molecules per disaccharide repeat unit in the dried state. It has been found equal to 4.5 for HA, as mentioned in Ch. 10. The quantity $n_{tot} - n_d$ might also have been directly calculated, using the same method, on all the experimental hydration spectrum recorded at that value of the hygrometry, instead of on elementary hydration spectra, the above-described method. Using these latter elementary hydration spectra highly simplifies this calculation for all recorded spectra. It furthermore gives a clear insight into the mechanisms of hydration.

Intensities of the difference bands due to the macromolecule
The problem for these difference bands is to evaluate the magnitude of the changes of these various bands that appear in the elementary hydration spectra $H_i(\tilde{\nu})$ of Figure 11.4 with respect to the magnitudes of the corresponding bands in the spectrum D of the dried sample, shown in Figures 11.2 and 11.3. The magnitude of this change is equal to the percentage of each hydrophilic group that has been hydrated by the embedding of $N^i_{H_2O}$ water molecules measured above (eq. (11.6)). For simplicity we suppose that each hydrophilic group of the macromolecule, at the origin of well-defined bands, $\nu_{C=O}$ or $\delta_{C\text{-}N\text{-}H}$ for instance, takes on two states: a free state, such as C=O or CN–H, and an H-bonded state, such as C=O··· or CN–H···. The case where more than two states can be found for this group is easily generalized (50, 52). This is the case for C=O groups that can be free, H-bonded by a single H-bond or by two H-bonds, or of N–H groups of protein backbones that can be free, H-bonded to C=O groups of the same peptide chain, as in α-helices or β-sheets shown in Figures 2.5–2.7, or H-bonded to H_2O molecules. The band $s_g(\tilde{\nu})$ due to

the considered hydrophilic group is therefore, in the spectrum of the dried sample, the sum of two bands $f_g(\tilde{\nu})$ with different weights:

$$s_g(\tilde{\nu}) = N_s[(1-\sigma_g)f_g(\tilde{\nu})+\sigma_g h_g(\tilde{\nu})] \tag{11.9}$$

where $f_g(\tilde{\nu})$ is the band for one free group whereas $h_g(\tilde{\nu})$ is the band for one H-bonded group, $1-\sigma_g$ is the (unknown) proportion of free groups and σ_g the proportion of H-bonded groups. The hydration band $\delta s_g(\tilde{\nu})$ for the same hydrophilic group in any of the elementary hydration spectrum *i* is (we omit index *i* for simplicity)

$$\delta s_g(\tilde{\nu}) = N_s \delta\sigma_g [h_g(\tilde{\nu}) - f_g(\tilde{\nu})] \tag{11.10}$$

where the coefficient $\delta\sigma_g$ of this difference band is also a quantity we wish to measure for all elementary hydration spectra. These two bands, $s_g(\tilde{\nu})$ and $\delta s_g(\tilde{\nu})$, are shown in Figure 11.5 as solid lines. The measurement of σ_g and $\delta\sigma_g$ first requires determining the two components $f_g(\tilde{\nu})$ and $h_g(\tilde{\nu})$. This is done supposing that at low wavenumbers the features of both

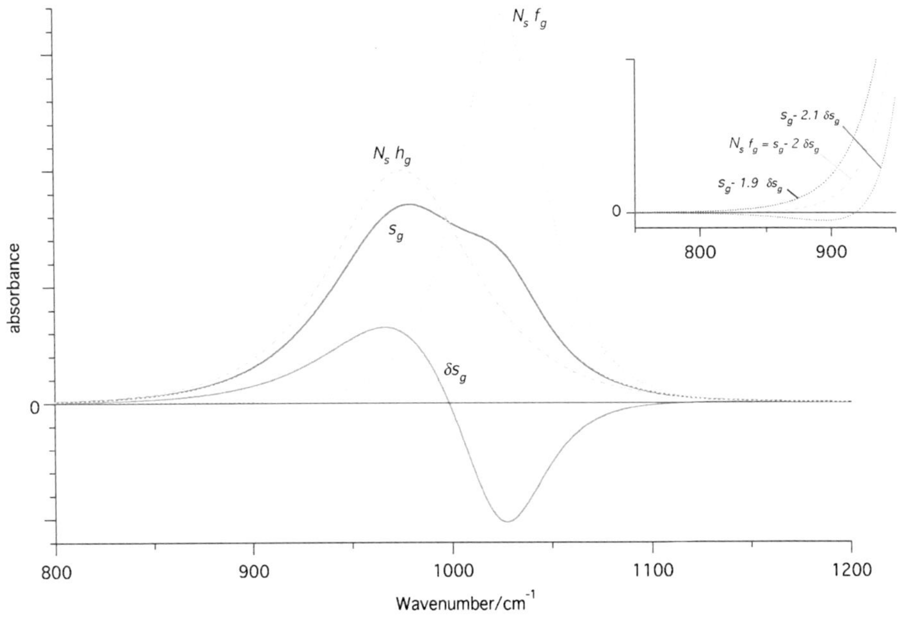

Figure 11.5 Experimental IR bands due to a particular hydrophilic group *g* of a macromolecule: $s_g(\tilde{\nu})$ is the band in the spectrum of the dried sample (spectrum *D* of Figure 11.3, for instance), $\delta s_g(\tilde{\nu})$ is the corresponding band in one of the elementary hydration spectra (shown in Figure 11.4, for instance). The enlarged (in absorbance) part in the upper right part of the figure illustrates the principle of the determination of the component $f_g(\tilde{\nu})$ at high wavenumbers: it is that which extends the less towards low wavenumbers without displaying negative values. Reproduced from the Journal of Molecular Structure, 648, Y. Maréchal, "Observing the water molecule in macromolecules and aqueous media using infrared spectrometry", 41, Copyright (2006), with permission from Elsevier.

bands $s_g(\tilde{\nu})$ and $\delta s_g(\tilde{\nu})$ are due to only the component $h_g(\tilde{\nu})$, which appears at lower wavenumbers ($\delta s_g(\tilde{\nu})$ is positive at low wavenumbers and negative at high wavenumbers) or, in other words, $f_g(\tilde{\nu})$, which appears at higher wavenumbers than $h_g(\tilde{\nu})$, does not contribute to the low wavenumber parts of $s_g(\tilde{\nu})$ and $\delta s_g(\tilde{\nu})$. In the enlarged (absorbance) diagram of the upper right part of Figure 11.5, it is clear that the band $s_g(\tilde{\nu})-2\delta s_g(\tilde{\nu})$ has the smaller contribution to the lower part of the spectrum, in the region 800–900 cm^{-1}, without, however, displaying unrealistic negative values. We therefore define $N_s f_g(\tilde{\nu})=s_g(\tilde{\nu})-2\delta s_g(\tilde{\nu})$. Repeating this procedure for the low wavenumber region we find $N_s h_g(\tilde{\nu})=s_g(\tilde{\nu})+0.5\delta s_g(\tilde{\nu})$. From these two equations, we deduce $s_g(\tilde{\nu})=N_s[0.2f_g(\tilde{\nu})+0.8h_g(\tilde{\nu})]$ and $\delta s_g(\tilde{\nu}) = 0.4N_s[h_g(\tilde{\nu})-f_g(\tilde{\nu})]$, hence the values for σ_g and $\delta\sigma_g$ of eqs. (11.9) and (11.10) are $\sigma_g = 0.8$ and $\delta\sigma_g = 0.4$.

The value $\sigma_g = 0.8$ of this example indicates that in the dried sample 20% of this group does not establish H-bonds and 80% establishes such H-bonds. The value $\delta\sigma_g = 0.4$ indicates that in the considered hydration spectrum $H_i(\tilde{\nu})$, 40% of these hydrophilic groups pass from a free configuration with no H-bond to an H-bonded configuration. This is performed through the introduction of $N^i_{H_2O}$ H_2O molecules we have seen to be at the origin of the bands due to H_2O in this elementary hydration spectrum $H_i(\tilde{\nu})$. They establish one H-bond on $N_s\,\delta\sigma_g$ hydrophilic groups. This result is the basis for the interpretation of all $H_1(\tilde{\nu})$ in terms of chemical mechanisms, as the quantity $N^i_{H_2O}/N_s$ has already been measured (eq. (11.6)) and $\delta\sigma_g$ has just been measured. With these quantitative measurements, we are then in a position to precisely interpret the four elementary hydration spectra displayed in Figure 11.4 for HA. The hydration mechanism that is at the origin of $H_1(\tilde{\nu})$, for instance, is shown in Figure 10.3. The quantitative analysis just described above, as applied to this spectrum $H_1(\tilde{\nu})$, tells us that 25 H_2O molecules hydrate one set of an hydrophilic group formed by one C–O$^-$ carboxylate group and one adjacent N–H amide group that are not linked by an H-bond and that this nanodroplet of 25 H_2O molecules at the same time hydrates (55) 0.49 alcohol I, establishes a second H-bond with 0.63 carboxyl C=O groups that are already H-bonded by a single H-bond and with 0.73 free carbonyl C=O groups of a neighbouring amide group, etc. It allows us to describe the hydration mechanism H_1 corresponding to spectrum $H_1(\tilde{\nu})$ in terms of a chemical reaction, as shown in Figure 10.3.

In Figure 11.6, the two elementary hydration spectra I^h and H^h of a sulfonated polyimide membrane, together with the spectrum D^h of the dried membrane, are displayed. The same analysis as described above for HA provides sufficient information to establish the hydration mechanism of this membrane, detailed in Ch. 10 and shown in Figure 10.5. It may be noted that these hydration spectra do not correspond to successive hydration mechanisms, as for HA, but correspond to the ionization of SO_3H groups into $SO_3^-\cdots H_3O^+$ by addition of a single H_2O molecule (I^h), and to hydration of these $SO_3^-\cdots H_3O^+$ complexes by H_2O molecules (H^h).

The above described results, obtained after a quantitative analysis of IR spectra, are quite precise. Up to now they have been obtained for a small number only of macromolecules. Let us nevertheless mention that in the case of a much simpler system, glycine, hydration mechanisms could also be determined from IR spectra in solid matrices (56) and, as can be seen in Figure 6.3, resulted in the formation of a zwitterion. These precise results may be extended to many other macromolecules besides HA or the sulfonated polyimide membranes discussed in Ch. 10. The quantitative analysis performed above on these spectra conveys a lot of information on the structures of the H-bond networks that H_2O

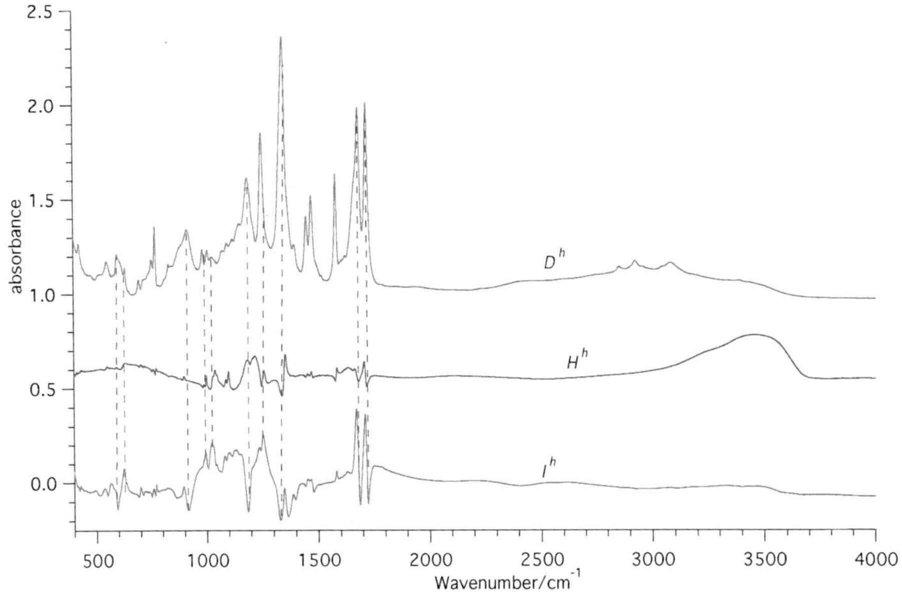

Figure 11.6 IR spectrum D^h of a dried sulfonated polyimide membrane (homopolymer) 10 μm thick and the two corresponding elementary hydration spectra I^h and H^h. Dashed vertical lines show corresponding bands.

molecules build inside them and illustrates the great variety of H-bond networks of different macromolecules. We expect to find an even greater variety of the structures of these H-bond networks when the hydrations of a greater number of macromolecules have been looked at using IR spectrometry. Furthermore, this method can be extended to other kinds of spectra. One immediate extension is to follow the H/D exchange that occurs inside these macromolecules when one hydrates them with heavy water vapour instead of normal water vapour. It is a promising but not yet completely implemented method to gather information on the H-atom transfers described in Ch. 6 by following the dynamics of H/D exchange mechanism(s) on various sites of the macromolecule (57).

Time-resolved nonlinear IR spectroscopy

Time-resolved nonlinear IR spectroscopy is a modern version of ordinary IR spectroscopy examined above. It has been referred to all along the preceding chapters. In Ch. 4 it has been shown to convey information on the dynamics of the surrounding of the studied vibration, in addition to information on this vibration itself, which is the main type of information conveyed by ordinary IR spectroscopy, also named 1D IR. This supplementary information appears through the measurements of the lifetime or relaxation time of the studied vibration. Most results obtained up to now with such methods concern liquid water. They have been described in Ch. 4, because they may be applied to any H-bonded system. They have given values of relaxation times of ν_s bands of water molecules in liquid water and have thus shown a marked isotopic dependence. Thus the relaxation time of the first excited state ν_s (O–H···) of HDO molecules diluted in heavy water is of the order of 1 psec (58, 59), whereas that of ν_s (O–D···) of HDO molecules diluted in ordinary water is 1.45 psec (60). The relaxation time of intermolecular vibrations, which are relative vibrations of two H_2O molecules in liquid water is some 0.7 psec, whereas that of rotations of individual HDO molecules is longer, around 2.5 psec. The lifetime of a D-bond is also somewhat longer, some 2 psec (61). It could also be shown that H_2O molecules that are bound to halogen cations X^- in solution in liquid water have relaxation times one order of magnitude longer than bulk H_2O molecules (62). Furthermore, the existence of two components in the ν_s band of H_2O molecules in liquid water, as shown by Raman spectroscopy, is strongly supported by femtosecond pump-probe experiments that strongly suggest that, with respect to orientational dynamics, two distinct molecular species exist in liquid water (63).

For water in macromolecules, few experiments of this type have been described in the literature. However, as for NMR spectroscopy some 30–40 years ago, time-resolved nonlinear IR spectroscopy is often limited to such chemically well-defined systems as liquid water. It is still a method for specialists, both for implementing experiments themselves and for interpreting them. However, it can be expected to have developed within a number of years to fully exploit its powerful properties so that it can be applied to observing H-bonds in much more complicated macromolecules and with dedicated apparatus that would make them more routine methods. NMR followed such a path before becoming one of the basic tools for the study of proteins. Time-resolved nonlinear IR spectroscopy has furthermore the advantage of being especially sensitive to H-bonds, a property that NMR does not have. An illustration of this tendency has been given in Ch. 10 with the use of three pulses photon echo method to study the dynamics of dipeptides in trehalose glasses (64).

NIR and Raman spectroscopies

IR spectrometry as described above is a really powerful tool to determine the structure of the H-bond network established by H_2O molecules that are embedded in macromolecules. It is most certainly due to become a basic tool, as it gives unique information on this network. It is nevertheless not the only method that will be used in the future. Two other kinds of spectroscopic methods will certainly also be of interest. They are simpler methods: NIR (near infrared) and Raman spectroscopies. These are routine methods that are complementary to IR spectrometry.

Overtones of most bands that are seen in the mid-IR region (the usual IR region) appear in the NIR region. They correspond to $0 \rightarrow n$ transitions between the ground state level 0 of a normal mode and its nth excited state. Most often these overtone bands have little interest, because in these excited states anharmonicities become important and difficult to handle as they entangle different bands. Furthermore, the intensities of these NIR bands are orders of magnitude weaker than the corresponding $0 \rightarrow 1$ transitions of the mid-IR region. In the case of an aqueous medium, which we have seen to have for long been "a poison for IR spectroscopy", because the $0 \rightarrow 1$ transitions are extremely intense and lead to catastrophic saturation effects when recorded in a routine manner with no special care, NIR bands may nevertheless be of interest in some cases. Saturation of bands is no longer a problem in the NIR region. NIR spectroscopy has thus been used to determine the difference of behaviours of two simple models of "amphiphile molecules" of similar structures, tertiobutanol $(CH_3)_3$–OH and trimethylamine-N-oxide $(CH_3)_3$–N=O that are both biologically relevant molecules. Amphiphile molecules have been looked at in detail in Ch. 9. They possess two parts, a hydrophobic one and a hydrophilic one. NIR spectroscopy clearly showed that tertiobutanol in aqueous solutions self-associates and consequently less perturbs the structures of the surrounding H_2O molecules than trimethylamine-N-oxide that does not show any self-aggregation (65). The relatively weak absorption of liquid water in this NIR region is also at the origin of the use of NIR in medicine, allowing a noninvasive measurement of intraocular pressure (66), for instance. In the food industry, the measurement of the fat content in meats (67) can be performed using this tool. NIR thus appears to be an easy-to-implement method that can be used in simple cases. As seen in the two above-mentioned examples, it is not adapted to observe H_2O molecules themselves, but macromolecules in an aqueous (bio)medium full of these molecules. In these examples, an ATR set-up in the mid-IR region, as described above, could have been used. It would have given more precise results, but would not have been as easily implemented as NIR.

Raman spectroscopy displays the same property as NIR regarding the intensities of vibrational bands. Raman bands are much less sensitive to H-bonds than IR bands, as seen in Ch. 4. For a long time, Raman spectra were therefore the most frequently used method to study liquid water, before the advent of ATR set-ups in IR spectroscopy. It has thus been possible to show that the ν_s band of liquid water is composed of two bands (68). Their significance has also for long been an object of discussion, but as indicated in Ch. 9, they are now seen as corresponding to two different states of rotations or librations of individual H_2O molecules.

CONCLUSION

In this chapter, several methods that are used to observe H_2O molecules have been described. IR spectrometry takes the lion's share among these methods. This should really be no surprise, because, contrary to common thinking, the very familiar water molecule is not easy to observe, especially when embedded in macromolecules, where it may play a fundamental but not yet completely understood role. Most classical methods of physical chemistry can give only partial views on the structure and dynamics of the extended H-bond network this small molecule always develops around itself, due to its exceptional structure that allows it to donate two H-bonds and accept two other H-bonds at the same time. The method that is by far the most precise and the most powerful to observe H-bonds, namely IR spectroscopy, is consequently the one that allows us to observe H_2O molecules with the greatest accuracy. The paradox is that it has for long been hindered as a method of observation of H_2O molecules, because of its hypersensitivity to H-bonds. This paradox was so developed that H_2O molecules were considered a poison for IR spectroscopy. This hindrance has only recently been recognized. After its origin had been identified, methods could then be devised that make IR spectrometry, or quantitative spectroscopy, a very precise tool to observe this H-bond network. It is however too recent to be a routine method. There is no doubt that its great efficiency will make it so in the near future. Relative intensities of bands can thus be accurately measured, due to its very good signal-to-noise ratio. The exploitation of these measurements allows us to translate hydration spectra, which have been considered in detail in this chapter but are not the only type of spectra that can be so analyzed, into chemical mechanisms. It is likely to be developed in the near future, and certainly combined with other methods, which are presently less efficient, but nevertheless remain interesting, as they are complementary to IR spectrometry. It is likely that one of the fields that will most benefit from the use of this tool will be the reactivity of aqueous media, particularly biomedia.

REFERENCES

1. J. Maddox, Nature, 332 (1988) 677.
2. J. A. Rupley and G. Careri, in C. B. Anfinsen, J. T. Edsall and F. M. Richards (Eds.), Advances in Protein Chemistry, Vol. 41, Academic Press, 1991.
3. Y. Maréchal, J. Mol. Struct., 648 (2003) 27.
4. N. Jouon, M. Rinaudo, M. Milas and J. Desbrières, Carbohydr. Polym., 26 (1995) 69.
5. C. Fringant, J. Desbrières, M. Milas, M. Rinaudo, C. Joly and M. Escoubes, Int. J. Biol. Macromol., 18 (1996) 281.
6. G. Careri, M. Geraci, A. Giansanti and J. A. Rupley, Proc. Natl. Acad. Sci., 82 (1985) 5342.
7. G. W. Neilson, P. E. Mason, S. Ramos and D. Sullivan, Phil. Trans. R. Soc. Lond. A, 359 (2001) 1575.
8. H. Luecke, B. Schobert, H.-T. Richter, J.-P. Cartailler and J. K. Lanyi, J. Mol. Biol., 291 (1999) 899.
9. H. Belrhali, P. Nollert, A. Royant, C. Menzel, J. P. Rosenbusch, E. M. Landau and E. Pebay-Peroula, Structure, 7 (1999) 909.
10. G. A. Jeffrey and W. Saenger, Hydrogen Bonding in Biological Structures, Springer Verlag, Berlin, 1994.
11. G. A. Jeffrey, An Introduction to Hydrogen Bonding, Oxford University Press, Oxford, 1997.
12. T. R. Jensen, M. O. Jensen, N. Reitzl, K. Balashev, G. H. Peters, K. Kjaer and T. Bjornholm, Phys. Rev. Lett., 90 (2003) 086101.
13. J.-M. Zanotti, M.-C. Bellisent-Funel and J. Parello, Biophys. J., 76 (1999) 2390.
14. T. A. Waigh, P. J. Jenkins and A. M. Donald, Faraday Discuss., 103 (1996) 325.
15. P. Chieux and J. C. Dore, in J. C. Dore (Ed.), Hydrogen Bonded Liquids, Vol. 329, Kluwer Academic, Dordrecht, 1991.
16. T. Haues, G. Papadopoulos, S. A. W. Verclas, G. Büldt and N. A. Dencher, Physica B, 234–236 (1997) 217.
17. C. Mattos, Trends Biochem. Sci., 27 (2002) 203.
18. V. P. Denisov and B. Halle, J. Mol. Biol., 245 (1995) 682.
19. V. P. Denisov, K. Venu, J. Peters, H. D. Hörlein and B. Halle, J. Phys. Chem. B, 101 (1997) 9680.
20. M. Gottschalk, N. A. Dencher and B. Halle, J. Mol. Biol., 311 (2001) 605.
21. V. P. Denisov, B.-H. Jonsson and B. Halle, Nat. Struct. Biol., 6 (1999) 253.
22. M. Falk, Spectrochim. Acta, 40A (1984) 43.
23. J. Annyas, D. Bicanic and F. Schouten, Appl. Spectrosc., 53 (1999) 339.
24. R. Bertani, G. Sergi, L. Lupoli, A. Berton, S. Volpato, K. Zambon, A. Coin and G. Enzi, Ann. Chim., 92 (2002) 135.
25. O. Kristiansson, J. Lindgren and J. de Villepin, J. Phys. Chem., 92 (1988) 2680.
26. J. Stangret and T. Gampe, J. Mol. Struct., 734 (2005) 183.
27. N. A. Macleod, E. G. Robertson and J. P. Simons, Mol. Phys., 101 (2003) 2199.
28. A. Y. Zasetsky, A. F. Khalizov, M. E. Earle and J. J. Sloan, J. Phys. Chem. A, 109 (2005) 2760.
29. R. Wagner, S. Benz, O. Möhler, H. Saathoff, M. Schnaiter and U. Schurath, J. Phys. Chem. A, 109 (2005) 7099.
30. J. E. Bertie and Z. Lan, J. Chem. Phys., 105 (1996) 8502.

31. J. E. Bertie and H. H. Eysel, Appl. Spectrosc., 39 (1985) 392.
32. H. R. Zelsmann, J. Mol. Struct., 350 (1995) 95.
33. Y. Maréchal, J. Mol. Struct., 322 (1994) 105.
34. P. A. Brooksby and W. R. Fawcett, J. Phys. Chem. A, 104 (2000) 8307.
35. J.-J. Max, M. Trudel and C. Chapados, Appl. Spectrosc., 52 (1998) 234.
36. J.-J. Max and C. Chapados, Appl. Spectrosc., 52 (1998) 963.
37. J.-J. Max and C. Chapados, Can. J. Chem., 78 (2000) 64.
38. C. Sammon, C. Mura, J. Yarwood, N. Everall, R. Swart and D. Hodgson, J. Phys. Chem. B, 102 (1998) 3402.
39. J. Yarwood, C. Sammon, C. Mura and M. Pereira, J. Mol. Liq., 80 (1999) 93.
40. S. Hajatdoost, C. Sammon and J. Yarwood, Polymer, 43 (2002) 1821.
41. P. Wu and H. W. Siesler, Chem. Phys. Lett., 374 (2003) 74.
42. J. P. Devlin, C. Joyce and V. Buch, J. Phys. Chem. A, 104 (2000) 1974.
43. J. P. Devlin, J. Sadlej and V. Buch, J. Phys. Chem. A, 105 (2001) 974.
44. M. W. Severson, J. P. Devlin and V. Buch, J. Chem. Phys., 119 (2003) 4449.
45. N. Uras, V. Buch and J. P. Devlin, J. Phys. Chem. B, 104 (2000) 9203.
46. V. Buch, J. Sadlej, N. Aytemiz-Uras and J. P. Devlin, J. Phys. Chem. A, 106 (2002) 9374.
47. L. Delzeit, K. Powell, N. Uras and J. P. Devlin, J. Phys. Chem. B, 101 (1997) 2327.
48. P. Ayotte, M. Hébert and P. Marchand, J. Chem. Phys., 123 (2005) 184501.
49. J. Grdadolnik and Y. Maréchal, Biopolymers (Biospectroscopy), 62 (2001) 40.
50. K. Liltorp and Y. Maréchal, Biopolymers, 79 (2005) 185.
51. D. Jamróz and Y. Maréchal, J. Phys. Chem. B, 109 (2005) 19664.
52. J. Grdadolnik and Y. Maréchal, Biopolymers (Biospectroscopy), 62 (2001) 54.
53. D. Jamróz and Y. Maréchal, J. Mol. Struct., 693 (2004) 35.
54. J. E. Bertie, A. M. Khalique and H. H. Eysel, J. Phys. Chem., 93 (1989) 2210.
55. Y. Maréchal, M. Milas and M. Rinaudo, Biopolymers (Biospectroscopy), 72 (2003) 162.
56. R. Ramaekers, J. Pajak, B. Lambie and G. Maes, J. Chem. Phys., 120 (2004) 4182.
57. J. Grdadolnik and Y. Maréchal, Appl. Spectrosc., 59 (2005) 1357
58. G. Gallot, S. Bratos, S. Pommeret, N. Lascoux, J. Cl. Leicknam, M. Kozinski, W. Amir and G. M. Gale, J. Chem. Phys., 117 (2002) 11301.
59. C. P. Lawrence and J. L. Skinner, J. Chem. Phys., 117 (2002) 5827.
60. T. Steinel, J. B. Asbury, J. Zheng and M. D. Fayer, J. Phys. Chem. A, 108 (2004) 10957.
61. J. B. Asbury, T. Steinel, K. Kwak, S. A. Corcelli, C. P. Lawrence, J. L. Skinner and M. D. Fayer, J. Chem. Phys., 121 (2004) 12431.
62. M. F. Kropman and H. J. Bakker, J. Chem. Phys., 115 (2001) 8942.
63. S. Woutersen, U. Emmerichs and H. J. Bakker, Science, 278 (1997) 658.
64. C. H. Londergan, Y. S. Kin and R. M. Hochstrasser, Mol. Phys., 103 (2005) 1547.
65. A. Di Michele, M. Freda, G. Onori and A. Santucci, J. Phys. Chem. A, 108 (2004) 6145.
66. D. Weissbrodt, R. Müller, J. Backhaus and J. B. Jonas, Am. J. Ophthalmol., 140 (2005) 307.
67. M. Prevolnik, M. Candek-Potokar, D. Skorjanc, S. Velikonja-Bolta, M. Skrlep, T. Znidarsik and D. Babnik, J. Near IR Spectrosc., 13 (2005) 77.
68. G. E. Walrafen, M. S. Hokmabadi and W. H. Yang, J. Chem. Phys., 85 (1986) 6964.

Part III

GENERAL CONCLUSION

– 12 –

Conclusion: H-Bond, Water Molecule and Life

The familiar water molecule, H_2O, a very simple and the most stable molecule, has not attracted all the attention it deserves and as a result is still far from being properly understood. It displays a discrete but nevertheless exceptional property that has far-reaching consequences, one of them being that without it life would not exist; its electronic structure allows it to establish around itself a far greater number of H-bonds than any other molecule. As described in this book, H-bonds have three fundamental properties that give them an up-to-now unsuspected importance: they are first directional and, second, have enthalpies of formation of the order of nkT at room temperature, with $n < 10$ for weak H-bonds. The result is that they play a unique role in our everyday life, even if this role has hardly been emphasized and we are unaware of it most of the time. Their directionality allows them to be at the origin of well-defined molecular structures, in much the same way as covalent bonds allow atoms to assemble and build well-defined molecules. Covalent bonds have enthalpies of formation one order of magnitude greater than H-bonds. They are strong bonds that make most molecules stable but rigid at room temperature, allowing only small-amplitude vibrations of atoms around their equilibrium positions. Variations of temperature have almost no effect on them. Stable (at room temperature) assemblies of H-bonded molecules are conversely flexible and evolutive, as a modification of their structures requires energies that are within reach of thermal fluctuations. The DNA double helix is a good illustration of these properties of stability, flexibility and possibility of evolution that all biomacromolecules have to possess; and thanks to H-bonds, indeed do possess.

These two properties of H-bonds are relatively well known and documented. This is not so of the third fundamental property of H-bonds, the importance of which became known only recently, but already suggests more important consequences: under certain conditions that are not yet fully understood, transfers of protons or of H-atoms occur through H-bonds. It clearly points to H-bonds as being basic centres of reactivity, particularly in aqueous media at room temperature. Theoretical descriptions of these transfers in simple and isolated H-bonded systems show that such transfers are much more complex than simple tunnellings and that many other parameters, especially the presence of other molecules and also coordinates, other than the X–H distance, influence them. These transfers are particularly important in aqueous media where the numbers of H-bonds are especially great, and where these

H-bonds form dense and highly developed "*hydrogen bond networks*". Transfers of protons are at the origin of the whole acid–base chemistry. Transfers of H-atoms are most certainly responsible for the reactivity of biomedia. In opposition to transfers of protons, they do not create space charges and consequently do not modify the pH of the medium that, in the case of a biomedium, is to stay in the close vicinity of 7. They occur via "*tautomerism*", a process shown in Figure 6.10, which requires the presence of H-bonded cyclic structures that allow simultaneous transfers of electrons to occur, thus avoiding the appearance of space charges and consequent pH changes. The sterical conditions necessary to build such cyclic H-bonded structures are most severe. As a consequence, only a very few molecules can fulfil them. Among these molecules, however, the small H_2O molecules, with their exceptional ability to establish a great number of H-bonds at the origin of a consequent H-bond network, occupy a special place. The presence of such an H-bond network, even built from a small number of H_2O molecules, makes such H-bonded cyclic structures almost automatically appear, a property that has been revealed to be crucial in aqueous media. These transfers, and more particularly H-atom transfers, are still poorly known, even in simple H-bonded systems. They are still less known in the presence of developed H-bonded networks and will most probably constitute an important field of research in the forthcoming years.

H-bonds would not have taken such a central importance if no such molecule as H_2O existed, a molecule that owes all its importance to its exceptional ability to establish a great number of H-bonds around it. A discrete property that has its origin in the simple structure of this very small molecule that has two H-bond donor sites, its two O–H bonds, and two H-bond acceptor sites, the two nonbonding pairs of electrons of its O-atom. This unique structure makes it a molecule that can establish around it a great number of H-bonds with various configurations. It is at the origin of the ubiquity of H_2O molecules in almost any samples in ordinary conditions of temperature and of surrounding atmosphere. It is also at the origin of a still poorly known species, liquid water, where the number of H-bonds is as great as that of covalent bonds, a unique situation at the origin of an exceptionally dense and 3D H-bond network. The exceptional properties of liquid water originate from the presence of this H-bond network. Its dynamics up to now eluded any precise description with the consequence that the exceptional physical properties of liquid water are not understood at the molecular level, something of a paradox for one of the most familiar species we drink and use in various ways everyday. The flexibility this H-bond network displays at room temperature, a consequence of its particular dynamics, is certainly fundamental: when liquid water becomes ice with lowering temperature, this H-bond network becomes much stiffer, with the result that ice displays much less exceptional properties than liquid water and is no longer a life-supporting medium. The origin of this flexibility in liquid water is due to the large rotational amplitudes the very small H_2O molecules can perform without breaking this H-bond network, a possibility that classical methods to describe ordinary liquids have great difficulties taking into account.

The exceptional chemical properties of liquid water are better understood at the molecular level, even if much work still remains to be done before precisely describing them. The role of H_2O molecules is, however, not limited to liquid water. It is fundamental in biomedia, where the H-bond network developed by H_2O molecules is somewhat less extended than in liquid water, due to the presence of other molecules, most of the time macromolecules that have densities of hydrophilic groups much smaller than H_2O molecules. The capacity of even a small number of H_2O molecules to develop such a nevertheless dense H-bond network around them is that property that makes water the support of life. It may be observed here that, in opposition to a common way of thinking, life requires the presence of liquid water to proceed less than the presence of a minimum number of water molecules, even if this minimum number is often provided by liquid water, a particularly well-adapted reservoir of such molecules, the exceptional properties of which can also be profitably used for some well-defined biopurposes (see Ch. 9).

A more precise answer to the question raised in the preface "how is it that life occurs in water and only in water?", more precise than that given above because it clearly formulates the role of the exceptional H-bond network that H_2O molecules develop around them, can now be proposed as suggested at the end of Ch. 10 "The water molecule in (bio)macromolecules". The dynamic flexibility of this H-bond network, which disappears at low temperatures, stabilizes the secondary structures of proteins and of other biomacromolecules. This is certainly important but does not seem to be its fundamental role, which is rather that it provides the necessary conditions for the reactivity of these macromolecules to proceed. H_2O molecules therefore appear as the molecules that control the reactivity of biomedia and without which no reactivity occurs in these media. This result is suggested by the dramatic effect an H/D substitution has in biology, which is discussed in Chs. 7 and 10. If this H-bond network had structural effects only, replacing H-atoms by D-atoms would display no such huge differences that make liquid water and liquid heavy water two species with completely different biological properties. This more precise answer to the above-fundamental question on the necessary presence of water for life to proceed may therefore be formulated as: *a few H_2O molecules are the only molecules that are able to allow and control, through the H-bond network they always develop around them, the reactivity of biomacromolecules.* We may even go a bit further in this answer: as transfers of protons are unlikely to be a general mechanism in a medium that has its pH bound to remain in close vicinity of 7, the main role of H_2O molecules is to allow transfers of H-atoms to occur between two macromolecules such as a protein and an enzyme, for instance, by providing the necessary H-bonded cyclic structures with the help of this H-bond network. It does not seem necessary to stress that still much work remains to be done: first to transform this assertion into an established point; second to know more precisely the conditions in which this H-bond network operates.

Two points should be added to these conclusions. First, if the role of H-bonds appears so fundamental in biophysics and biochemistry that H-bonds may be declared the "bonds of life", mainly thanks to the presence of H_2O molecules in all biomedia and to the fundamental role they play there, their action is not limited to these media. The H_2O molecule being ubiquitous, thanks to its exceptional possibilities to establish H-bonds, H-bonds are also often encountered in chemistry, where such terms as H-bonded solvents, hydrophilic or hydrophobic groups or molecules are currently encountered and well taken into account, even if a more precise understanding of the role that these H_2O molecules play is often needed. They are, also often encountered in physics where H_2O molecules are also currently met. Physicists, however, are less aware of their fundamental role. We have seen that the dynamics of H_2O molecules in liquid water is yet not understood at all. It is studied by recent time-resolved nonlinear IR methods that are still the domain of physicists and also by theoretical methods of molecular dynamics (MD) that have up to now not succeeded in incorporating the directionality of H-bonds in the huge H-bond network of liquid water and consequently the fundamental role rapid rotations (librations) of these very small H_2O molecules play. In another field, the properties of the atmosphere, H-bonds take on a fundamental importance. This is so for the formation of raindrops for instance, but also for such problems as ozone depletion and the catalytic role of stratospheric ice clouds. It is not so surprising then that a great journey through a large part of the present scientific world has been proposed throughout this book, a consequence of the ubiquity of H-bonds and H_2O molecules in the physics of our surroundings. This journey visits physics and extends towards molecular biology, encompassing a great part of chemistry.

The second point is that, contrary to common thinking, these highly simple and familiar H_2O molecules are not at all easy to observe, as with their ability to establish H-bonds, they are very labile and versatile molecules that easily and rapidly change their H-bond configurations. The consequence is that despite the fundamental role they play that we have seen above, particularly in biomedia but not limited to only these media, they have been ignored and often continue to be so. This is a scandalous attitude, as John Maddox put it some 20 years ago, as quoted in the preface and in Ch. 10. The situation is nowadays changing with the advent of powerful methods to observe and localize these molecules. Among these methods, IR spectroscopy, which has for long suffered from a paradoxical hindrance due to its too great sensitivity to H-bonds, occupies an important place in this book. Its power and precision to look at H-bonds make it *the* reference method to observe H_2O molecules. It has the reputation of being difficult to handle and it is true that many interpretations of IR spectra of H-bonded systems found in the literature are often approximate. In this book, qualitative descriptions of this method are given that are accompanied by more quantitative treatments in separate paragraphs. They allow us to more precisely interpret spectra and define the limits of the extracted information. IR spectroscopy should consequently not to be considered a difficult method, even if the spectroscopic set-ups are not routine and the interpretations of spectra of H-bonded systems are not classical and must be performed with care, mainly because of the strong anharmonic couplings that the main characteristic band $\nu_s(X{-}H\cdots)$ establishes with other vibrations. The adaptation of IR spectroscopy to the study of H_2O molecules, as described in Ch. 11 and used in Chs. 9 and 10 to determine the structure of the H-bond network of hydrated macromolecules and of liquid water, shows it is not so difficult to use. It is, along with NMR spectroscopy, one of the most quantitative methods in chemical physics. In the study of H_2O molecules, it has the advantage over NMR of being hypersensitive to H-bonds.

Index

MIX
Papier | Fördert gute Waldnutzung
FSC® C083411

Zeitfracht Medien GmbH
Ferdinand-Jühlke-Straße 7
99095 Erfurt, Deutschland
produktsicherheit@kolibri360.de